# 4th International Workshop on UI GreenMetric World University Rankings (IWGM 2018)



E3S Web of Conferences Volume 48 (2018)

Semarang, Indonesia
8 – 10 April 2018

**Editors:**

**R. Fitri Sari**
**H. Hadiyanto**
**N. Suwartha**

ISBN: 978-1-5108-6867-0

**Printed from e-media with permission by:**

Curran Associates, Inc.
57 Morehouse Lane
Red Hook, NY 12571

**Some format issues inherent in the e-media version may also appear in this print version.**



Printed by Curran Associates, Inc. (2018)

For additional information, please contact EDP Sciences – Web of Conferences at the address below.

EDP Sciences – Web of Conferences
17, Avenue du Hoggar
Parc d'Activité de Courtabœuf
BP 112
F-91944 Les Ulis Cedex A
France

Phone: +33 (0) 1 69 18 75 75
Fax: +33 (0) 1 69 28 84 91

contact-edps@webofconferences.org

**Additional copies of this publication are available from:**

Curran Associates, Inc.
57 Morehouse Lane
Red Hook, NY 12571 USA
Phone: 845-758-0400
Fax: 845-758-2633
Email: curran@proceedings.com
Web: www.proceedings.com

# TABLE OF CONTENTS

# PROCEEDING OF THE 4th INTERNATIONAL WORKSHOP ON UI GREENMETRIC WORLD UNIVERSITY RANKINGS (IWGM) 2018

## Universities, Impacts and Sustainable Development Goals (SDGs)

Editors:
Riri Fitri Sari
Hady Hadiyanto
Nyoman Suwartha



For Information:
**UI GreenMetric Office**
Integrated Laboratory and Research Center (ILRC) Building 4th Fl., Universitas Indonesia, Depok 16424, Indonesia
**Telephone** +622129120936
**Email:** greenmetric@ui.ac.id
**Website:** http://greenmetric.ui.ac.id

# Preface

This proceeding contains invited papers from the International Workshop on UI GreenMetric (IWGM) 2018. This 4th International Workshop on UI GreenMetric is organized by the Universitas Diponegoro, Semarang, Indonesia. The first and second IWGM were hold at University of Indonesia in November 2013 and April 2016, respectively and the 3rd IWGM was the first event to be conducted outside Indonesia in April 2017.

The workshop is an academic forum for Rectors, Vice Rectors, and Director of Sustainability and Facilities, of UI GreenMetric participants. These universities have shown a lot of development in achieving best positions in each category at the UI GreenMetric Rankings.

In this workshop university leaders share experience and effort in improving sustainable environment in their respective campuses. This forum aimed at providing an opportunity for the top leaders of participating universities to explain their university's excellence in sustainability and Green Campus. We hope that this event will provide an opportunity for developing network and collaboration on sustainability management among global university. We also hope that this workshop will be a medium in which we can hear and accommodate some feedbacks or comments from the participants to improve our questionnaire to evaluate university performance. This year there were 40 invited speakers from 25 countries, which shares their best practices in respective universities. Moreover, this event was attended by Prof. H. Mohamad Nasir, Ph.D., Ak., Minister of Research, Technology and Higher Education, Republic of Indonesia, Dr. Agus Justianto, Head of Research, Development, and Inovation of Ministry of Environment and Forestry, Republic of Indonesia, and Prof. Jatna Supriatna, Chairman of Research Center for Climate Change (RCCC), Universitas Indonesia, Indonesia

We convey our greatest appreciation to all distinguished speakers from Universitas Diponegoro - Indonesia, Universidade do Minho – Portugal, University of Sao Paulo (USP) - Brazil, Universiti Putra Malaysia – Malaysia, North Carolina Agricultural & Technical State University – USA, Universiti Teknologi Malaysia (UTM) – Malaysia, Minin University – Russia, Chaoyang University of Technology – Taiwan, University of Bologna – Italy, University of Bahrain – Bahrain, Siam University – Thailand, Institut Teknologi Sepuluh Nopember (ITS) – Indonesia, Universidade Federal de Lavras – Brazil, National Chi Nan University – Taiwan, Bogor Agricultural University – Indonesia, Universitas Negeri Semarang – Indonesia, University of Limerick – Ireland, Universitas Padjadjaran – Indonesia, Universitas Sebelas Maret - Indonesia, Shandong Normal University - Lishan College – China, University of Kashan - Iran, Chulalongkorn University - Thailand, University of Szeged – Hungary, Universiti Malaysia Sabah – Malaysia, University of Milano-Bicocca – Italy, Ton Duc Thang University – Vietnam, Universidad Autonoma de Occidente – Colombia, Ozyegin University – Turkey, Pontificia Universidad Javeriana – Colombia, National University of Colombia – Colombia, National Pingtung University of Science and Technology – Taiwan, Tra Vinh University – Vietnam, Dublin City University – Ireland, University of Malaya – Malaysia, Peoples' Friendship University of Russia (RUDN) – Russia, Universiti Tun Hussein Onn Malaysia – Malaysia, Riga Technical University – Latvia, and University of Zanjan – Iran. We thank the conference proceeding contributor for their papers. This conference has attracted active participation from many high rank officials from many universities. In total, we have participants from 73 universities of 22 countries, in which we have Indonesia, Malaysia, Bahrain, Portugal, Hungary, Taiwan, Italy, China, USA, Vietnam, Colombia, Thailand, Iran, Turkey, Kazakh, Russia, Brazil, Latvia, Ireland, France, Pakistan and Russia on the list. This large number of participants indicated that ranking has been widely acknowledge a great tool for improving standard of university infrastructure management. We thank all participants and all stakeholders for making this International Workshop on UI GreenMetric 2018 a fruitful and memorable event.

Editorial Team

Riri Fitri Sari
Hady Hadiyanto
Nyoman Suwartha

# The 4th International Workshop on UI GreenMetric (IWGM) World University Rankings

Universitas Diponegoro, Semarang, Indonesia, 8-10 April 2018

# Rector's Welcome

**Prof. Dr. Ir. Muhammad Anis. M. Met**
**Rector of Universitas Indonesia**

Dear Rectors, Vice President, University Leaders, Campus Sustainability Officers and all of the participants. Welcome to the International Workshop on UI GreenMetric 2018. This is the 4th workshop on UI GreenMetric after a successful event in 2017.

This is the highlight of all the national workshop conducted worldwide during our world tour in 2016-2017. In 2016, the national workshops of UI GreenMetric have been held by RUDN University, (Moscow – Russia), Bulent Ecevit University, (Zonguldak – Turkey), and Jordan University of Science and Technology, (Irbid – Jordan). Last year in 2017, the national workshops of UI GreenMetric have been held by Kazakh National Agrarian University, (Almaty – Kazakhstan), El Bosque University, (Bogotá – Colombia), University of Sao Paulo (USP), (Sao Paulo – Brazil), Diponegoro University, (Semarang – Indonesia), National Cheng Kung Universit (NCKU), (Nantau – Taiwan), University of Bologna, (Bologna – Italy), Aalborg University, (Aalborg – Denmark), Chulalongkorn University, (Bangkok – Thailand), RUDN University, (Moscow – Russia). We thank to host and national coordinator that has been generously committed to lead the UI GreenMetric movement in their country.

We are honoured to welcome all distinguished Rectors, Vice-Rectors, Director of Facilities and Sustainability from 25 countries in this event. We are proud to receive many positive feedbacks from many university leaders worldwide, which show that after 8 years from its initiation UI GreenMetric has become a knowledge hub. Our workshop has also become an experience sharing event in which we can learn from each other in implementing sustainable campus infrastructure.

I am pleased to see the strength of the participation here, in which we receive many prompt and direct responses from our colleagues overseas which share the same goals and priorities. This is a proof that the issue of sustainability has gained the attention of academics worldwide. We hope that Universitas Indonesia's initiative to draw attention to the sustainability issues through ranking will bring about direct changes in society. We look forward to the presentations and discussions on the issue of campus greening. I wish you all a constructive and productive time during the conference and hope your stay is enjoyable.

Semarang, 8 April 2018
Best wishes

Prof. Dr. Ir. Muhammad Anis, M.Met
Rector of Universitas Indonesia

# The 4th International Workshop on UI GreenMetric (IWGM) World University Rankings

Universitas Diponegoro, Semarang, Indonesia, 8-10 April 2018

# Rector's Welcome

**Prof. Dr. Yos Johan Utama, S.H., M.Hum.**
**Rector of Universitas Diponegoro**

We warmly welcome you to the 4th International Workshop on UI GreenMetric World University Rankings (IWGM 2018). This event is hosted by Universitas Diponegoro, one of the prestigious universities in Indonesia and the greatest University in Central Java Province. Universitas Diponegoro is currently ranked at 6th of national universities based on Ministry of Research, Technology and Higher Education and QS WUR.

Universitas Diponegoro is very much supporting UI GreenMetric movement which draws attention to sustainability and environment in higher education institutions in the world and supposed to come up with solutions to significant worldwide problems by means of research. The aim of this symposium is to provide an opportunity for universities that get top position in UI GreenMetric to explain and to share their university's excellence in UI GreenMetric and also to provide an opportunity for cooperation in sustainability management in campuses. They will share their efforts in improving the sustainable environment on their campuses. The conference program is comprised of keynote talks, invited speakers talks, and discussion among universities.

As the Rector of Universitas Diponegoro and National Coordinator of UI GreenMetric in Indonesia, I am pleased to welcome you at the International Workshop on UI GreenMetric 2018 to be held on 8-10 April in Semarang.

Semarang, 8 April 2018
Kind regards,

Prof. Dr. Yos Johan Utama, S.H., M.Hum.
Rector of Universitas Diponegoro

# A Glimpse of UI GreenMetric Rankings & Connecting Dreams to the Future and the Past

**Prof. Dr. Ir. Riri Fitri Sari, MM., MSc,**
**Chairperson of UI GreenMetric**
**(Chairperson of the IWGM 2018 Organizing Committee)**

Dear University leaders, Welcome to the 4th International Workshop on UI GreenMetric 2018. This marks the end of the 8th year since we started our university ranking. We would like to thank our host, Universitas Diponegoro for showing the hospitality and culture of central Java to our newly emerging network of green universities. This annual event takes place in the heart of Universitas Diponegoro (UNDIP), Semarang, Indonesia. Diponego is the name of an Indonesian hero from the Java war of 1825–1830 an inspiration of many Indonesia. We hope that this year we can witness many heroes of our time; those university leaders who have work night and day to make their universities beautiful and green. The aim of this conference is to provide an opportunity for UI GreenMetric participants to show their university's excellence in UI GreenMetric and also to provide an opportunity for cooperation in sustainable campus management. In this International Workshop, participating universities leaders will share their achievement in leading sustainability and conservation programs in their campuses. We are very proud to receive many warm and encouraging responses from many university Rectors/Presidents/Vice Chancellors from more than 25 countries. The spirit of UI GreenMetric networks has echoed worldwide. We are delighted with the response of many universities to make a good use of ranking for the next level of their achievement. Last year, UI GreenMetric becomes hub that connects many universities to share their aspirations for a better world in which Sustainable Development Goals are on the making.

We thank Rectors and high rank university officers from 25 countries who are willing to share their experience and efforts in improving sustainable environment in their campuses. The conference program is comprised of keynote talks, plenary talks, and Panel talks. We hope that this event will be the discussion ground among many universities. This program book includes abstract paper contributions from the speaker of IWGM 2018 which consists of 44 articles. It covers the 6 indicators used in UI GreenMetric, i.e. Setting and Infrastructure, Energy and Climate Change, Waste Management, Water Management, Transportation and Education.

We would like to thank the International Ranking Expert Group (IREG), under President Jan Sadlak for their support, although due to the time restriction no representative from IREG could come to this workshop.We thank the Minister of Research Technology and Higher Education of Republic of Indonesia, Prof. M. Nasir and Representative of the Minister of Environment and Forestry of Republic of Indonesia, Dr. Agus Justianto for the Indonesian government support to the UI GreenMetric as one of 20 university rankings in IREG inventory. We would like to thank our host and speakers for their contribution to this workshop and the proceeding.

We also thank all workshop participants for their active participation to the theme of our event "Universities, impacts and SDGs" in the workshop. Enjoy your visit to Semarang and Universitas Diponegoro campus, as well as our amazing Borobudur temple, one of the wonders of the world. We wish you a fruitful workshop and newly refreshed motivation to create the next generation of sustainable campus.

Semarang, 8 April 2018
Kind regards,
Prof. Riri Fitri Sari
Chairperson of UI GreenMetric

# Program Committee

**Advisors**

Prof. Dr. Ir. Muhammad Anis, M.Met.
Rector of Universitas Indonesia

Prof. Dr. Yos Johan Utama, S.H., M.Hum.
Rector of Diponegoro University

Prof. Dr. Ir. Ambariyanto, MSc.
Vice Rector of Diponegoro University

Prof. Dr. Ir. Tommy Ilyas, M.Eng.
Universitas Indonesia (Expert Member of UI GreenMetric)

**Organizing Committee**
**UI GreenMetric Secretariat:**

Prof. Ir. Riri Fitri Sari MM., M.Sc., IP
Chairperson of UI GreenMetric

Dr. Nyoman Suwartha, ST., MT, MAgr.
Secretary of UI GreenMetric

Prof. Gunawan Tjahjono
Expert Member of UI GreenMetric

Junaidi S.S., M.A
Expert Member of UI GreenMetric

**Dr. Baiduri Widanarko,**
Expert Member of UI GreenMetric

Sabrina Hikmah Ramadianti
Secretariat of UI GreenMetric

Arsy Imanda N. Raswati
Secretariat of UI GreenMetric

Hastin Setiani
Secretariat of UI GreenMetric

**Dr. Ruki Harwahyu**
Programmer of UI GreenMetric

**Rinoto Cahyo Utomo, S.Tr**
Programmer of UI GreenMetric

**Universitas Diponegoro:**

**Executive Chair:**
Prof. Dr. Hadiyanto, S.T., M.Sc.

**Secretariat 1:**
Kenny Dwi Purwani, S.Hum.

**Co-Chair:**
Dr.Eng. Agus Setyawan, S.Si., M.Si.

**Secretariat 2:**
Abdul Rohman, S.Hum.

**Events & Social Event:**

**Coord:**
Dr. Ir. Ita Widowati, DEA.

**Member 1:**
Susatyo Nugroho Widyo Pramono, ST., MM

**Member 2:**
Nuryanto, S.Gz, M.Gizi

**Media Publications:**

**Coord:**
Dr. Jatmiko Endro Suseno, MSi.

**Member 1:**
Wahyu Setiabudi, S.Kom., M.Kom.

**Member 2:**
Dr. Ir. Syafrudin, CES, M.T.

**Accommodation and Transportation:**

**Coord:**
Dr. Daud Syamsudewa, S.Pt., M.Si., Ph.D.

**Member 1:**
Fahmi Arifan, ST, M.Eng.

**Finance:**

**Coord:**
Dra. Arsiani Sulaksmiwati, M.Pd.

**Member 1:**
Slamet Purwanto, S.Si.

**Member 2:**
Achmad Saefuddin, S.IP.

**Supporting:**

**Coord:**
Dr.Ir. Suryanti (Meal)

**Member 1:**
Shoimatul Fitria, S.E., M.M.

**Member 2:**
dr. Sri Winarni, MKes (Safety and Health)

# UI GreenMetric World University Rankings: Background of the Ranking

## 1 Initiation of the Ranking

The UI GreenMetric World University Ranking is an initiative of Universitas Indonesia which is being launched in 2010. Prof. Gumilar Rusliwa Soemantri Stated that it is a part of strategy of raising its international standing. The University hosted an International Conference on World University Rankings on 16 April 2009. It invited a number of experts on world university rankings such as Isidro Aguillo (Webometrics), Angela Yung-Chi Hou (HEEACT), and Alex Usher (Educational Policy Canada). It was clear from the discussions that current criteria being used to rank universities were not giving credit to those that were making efforts to reduce their carbon footprint and thus help combat global climate change.

## 2 Aim of the Ranking

The aim of this ranking is to provide the result of online survey regarding the current condition and policies related to Green Campus and Sustainability in the Universities all over the world. It is expected that by drawing the attention of university leaders and stake holders, more attention will be given to combating global climate change, energy and water conservation, waste recycling, and green transportation. We hope that the ranking will be useful to university leaders in their efforts to put in place eco-friendly policies and manage behavioral change among the academic community at their respective institutions.

## 3 Creating the ranking

Universities that wish to participate are asked to provide numeric data on a number of criteria that can give a picture of their commitment to the greening of their campus and putting in place environmentally friendly policies that support sustainability. The criteria include such baseline information as the size of the university, both spatially and in terms of population, the campus location and the amount of green space; and also information on energy use, transport, water use and recycling and waste treatment. In addition, it will ask about efforts being made by the institution towards establishing green policies and management.

## 4 Methodology Used to Create the Rankings

### 4.1 The philosophy behind the rankings

We based our instrument on a broad philosophy that encompasses the three Es: Environment, Economics and Equity.

### 4.2 The criteria

We selected criteria that are generally thought to be of importance by universities concerned with sustainability. These include the collection of a basic profile of the size of the university and its zoning profile, whether urban, suburban, rural. Beyond this we want to see the degree of green space. The next category of information concerns electricity consumption because of its link to our carbon footprint. Then we want to know about transport, water usage, waste management and so on. Beyond these indicators, we want to get a picture about how the university is responding to or dealing with the issue of sustainability through policies, actions, and communication. In the first version of the methodology, used in 2010, 23 indicators were used within the five categories to calculate the ranking scores. In 2011, 34 indicators were used. Then in 2012 we leave the indicator of "smoke free and drug free campus environment" and used 33 indicators to evaluate the green campus. In 2012, we also categorize the indicators into 6 category including education criteria. One change being considered is the formation of a new category for sustainability education and research. In 2015 the theme was carbon footprint. We add two questions related this issue in the energy and climate change section.We also improved our methodology by adding a few sub-indicators that related to water and transportation in the 2015 ranking. A major change in methodology was done in 2016 by considering new trends in sustainability issues.

## 4.3 The scoring

Scoring for each item will be numeric so that our data can be processed statistically. Scores will be simple counts of things, or responses on a scale of some sort.

## 4.4 The weighting of criteria

Each of the criteria will be categorized in a general class of information and when we process the results, the raw scores will be weighted to give a final calculation. The weighting Criteria can be found in figure 1.

## 4.5 Refining and improving the research instrument

While we have put every effort into the design and implementation of the questionnaire, we realize that this seventh year-round is bound to have shortcomings. Therefore, we will be reviewing the criteria and the weightings continuously to reflect input from participants and state of the art developments in the field. We welcome your comments and input.

## 4.6 Data collection

Data will be collected through online system between June-October of the year, from the universities we have contacted and who are willing to provide information.

## 4.7 The results announcement

The results of the metrics is usually released in December.

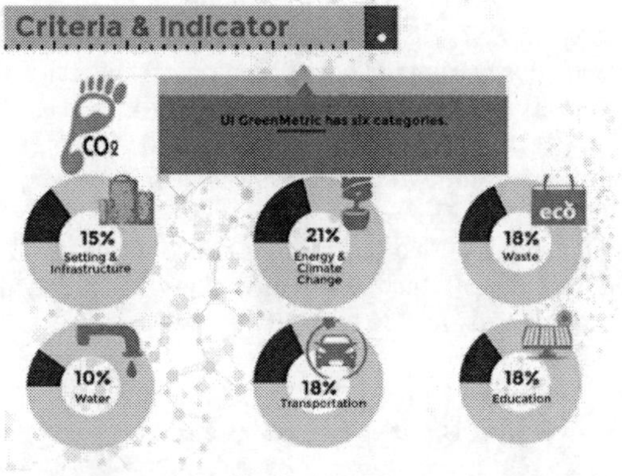


**Fig. 1** UI GreenMetric ranking criteria in 2017

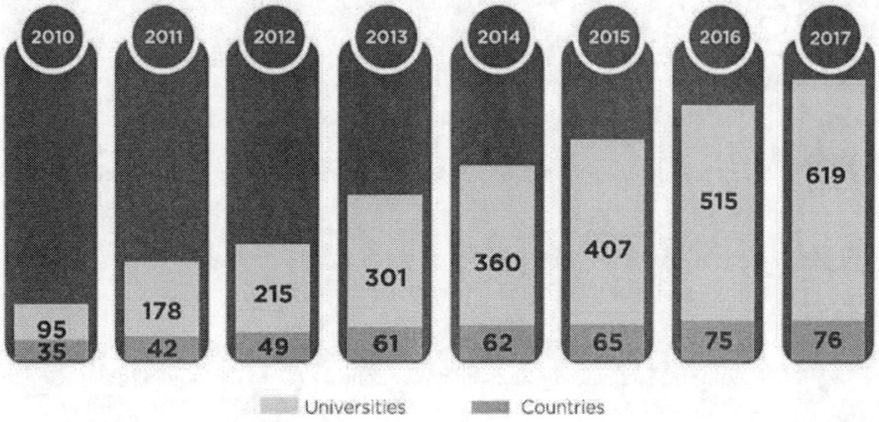


**Fig.2** Number of UI GreenMetric's participating universities 2010-201

# Speakers and Author's Profile

**Prof. H. Mohamad Nasir, Ph.D., Ak.**
**Ministry of Research, Technology and Higher Education, Republic of Indonesia**

Minister Mohammad Nasir has been entrusted to head the integration of the Ministry of Research and Technology (Ristek) and the Directorate of Higher Education (Dikti), which was considered a significant task for the national commitment in implementing a paradigm shifting in teaching and research universities. Professor Nasir has been actively involved in the academic activities at the *University of Diponegoro (UNDIP)*. He was the Vice Rector of UNDIP in 2006 – 2010 and was about to be inaugurated as the Rector of the University in December 2014, before being appointed to be the Minister. Professor Mohammad Nasir hold a bachelor's degree in Economics from *University of Diponegoro (UNDIP)*; a Master of Science (M. Si) from *Gadjah Mada University (UGM),* and a PhD in Accounting from the *Universiti Sains Malaysia (USM).*

**Prof. Dr. Ir. Muhammad Anis. M.Met.**
**Rector of Universitas Indonesia**

Muhammad Anis graduated from the Faculty of Engineering, Universitas Indonesia majoring in metallurgy in 1983. He then continued his education and earned his doctoral and master's degree in metallurgy from School of Materials, University of Sheffield, UK. Muhammad Anis has held various position within the Faculty of Engineering, among others as Vice Dean for Academic Affairs (1993-1997), Vice Dean for Cooperation (1997-2000) and Director of Extension Program for two terms (1993-2000). He was in member of the university's Senate (2000-2007). He was Universitas Indonesia's Director of Education for the terms 2003-2007 and was appointed a Vice Rector of Academic for the Term 2007-2012. In 2013, he was appointed as Interim Rector of Universitas Indonesia and was elected as a Rector of Universitas Indonesia for the term 2014-2019.

**Prof. Dr. Yos Johan Utama, S.H., M.Hum.**
**Rector, Universitas Diponegoro, Indonesia**

Prof. Yos Johan Utama is a professor of Administrative Law in Indonesia. He has been a lecturer at the University of Diponegoro for 28 years since 1987. He was graduated with a doctoral degree in 2006 at the same university. His experience in this field makes him prominent speakers in the discussion and decision-making meetings. Based on these experiences, he also gives lecture at various universities, including at the Police Academy. Prof. Yos Johan Utama's interest in the management of higher education began when he became the Vice Dean at the Faculty of Law for two periods. He then became the Dean at the same faculty, also for two terms. With the desire to bring the University of Diponegoro to be modern research and sustainable university, he became Rector since 2015.

**Prof. Dr. Ir. Riri Fitri Sari, M.M., M.Sc.**
**Chairperson of UI GreenMetric**

Riri Fitri Sari is a professor of Computer Engineering at Electrical Engineering Departement, Faculty of Engineering, Universitas Indonesia (UI). She holds PhD in Computer Networks from the School of Computing, University of Leeds, UK. Her current main teaching and research area includes Computer Network, Grid Computing, Intenet Things, and ICT for Sustainability. Since April 2010, she has been the Chairperson of UI GreenMetric Ranking of World Universities, a flagship program from the Universitas Indonesia to rank universities worldwide based on their green campus and sustainability programs. UI GreenMetric Network became an active network of more than 619 Universities from 76 Country. Since September 2015, she has been appointed a member of Special Task Force for Improving Indonesia Universities Academic Reputation for the Ministry of Research and Higher

Education. She was awarded as an honorary of professor from Kazakh National Agrarian University, Almaty – Kazakhstan on June 2017.

**Prof Adi Zakaria Afiff**
**Vice Rector for Administration, Finance and Infrastructure, University of Indonesia, Indonesia**

Prof Adi Zakaria Afiff has been Vice Rector for Administration, Finance and Infrastructure of the University of Indonesia since year 2012. He graduated from the Department of Accounting. Faculty of Economics, University of Indonesia in 1985, Earned his MBA Degree at University of Wisconsin in Madison with a double Major in Management Information System and Marketing in 1989 and his Doctorate degree in Marketing at Universitas Indonesia in 2003. In 2005. His research interest covers areas such as Social Marketing, Service Marketing, Brand Management, Marketing Communication and Consumer Behavior.

**Prof. Dr. Ir. Ambariyanto, M.Sc.**
**Vice-Rector for Research and Innovation, Universitas Diponegoro, Indonesia**

Prof. Dr. Ir Ambariyanto MSc., is Vice Rector for Research and Innovation Diponegoro University and a Professor at the Faculty of Fisheries and Marine Sciences of Diponegoro University. He received his bachelor's degree from Department of Fisheries, Faculty of Animal Husbandry of Diponegoro University in 1986. His Master's degree in Ocean Sciences gained at the University of Wales in 1990. He achieved a Ph.D. in Marine Ecology at the University of Sydney, Sydney, Australia in 1996. He completed his Postdoctoral at the University of Sydney in 1997.

**Prof. Dr. Ir. Tommy Ilyas, M.Eng.**
**Expert Member of UI GreenMetric**

Prof. Dr. Ir. Tommy Ilyas, M. Eng is a expert member in UI GreenMetric. He graduated with his bachelor's degree from Engineering of Universitas Indonesia in 1977. He received a master's degree in the field Geotechnic from Sheffield University (UK) in 1983. He completed his doctor's degree from Universitas Indonesia with Natioanl University of Singapore in 2002. He Inaugurated as Professor of Universitas Indonesia in 2006.

**Prof. Ir. Gunawan Tjahjono, M.Arch., Ph.D.**
**Expert Member of UI GreenMetric**

Prof. Gunawan Tjahjono received his BSc of Architecture from University of Indonesia, Jakarta in 1975 and MSc of Architecture from University of California at Los Angeles. He completed achieved a Ph.D. in University of California at Berkeley. Since 1989, he is Member, Advisory Board of Traditional Dwellings and Settlements Review a University of California at Berkeley based refereed journal for International Association for the Studies of Traditional Environment. He appointed as professor Professor of Architecture, Faculty of Engineering, University of Indonesia since 2002. He also is Reviewer, Directorate General of Higher Education, Department of National Education, Republic of Indonesia for competitive grant in 2002 until now. In 2011-2015, He was appointed as the Rector of the University of Pembangunan Jaya.

**Mr. Junaidi, M.A.**
**Expert Member of UI GreenMetric**

Junaidi is an expert member of UI GreenMetric World University Rankings, and Lecturer in Linguistics and Cultural Studies at English Department, Faculty of Humanities, Universitas Indonesia. He holds degrees from English Department, Universitas Indonesia (Sarjana Sastra/BA in English language and Literature); and Warwick University, UK (M.A in British Cultural Studies, with Distinction). He is currently doing PhD in Linguistics at Universitas Indonesia. He has been actively involved in the internationalization of Universitas Indonesia from 2004 to 2015 with different university positions. He is UI 2011-2015 Head of International Office and Chairperson of ASEAN University Network- ASEAN Credit Transfer System (AUN-ACTS). He is also one of the resource persons for internationalization of Indonesian universities at Ministry of Research and Higher Education, Republic of Indonesia.

**Dr. Baiduri Widanarko**
**Expert Member of UI GreenMetric**

Dr. Baiduri Widanarko completed her Bachelor of Dental Surgery (DSS) degree at the University of Indonesia in 2000. In 2001, She joined the Departement of Occupational Health and Safety at University of Indonesia as a lecturer. In 2003, She completed her Magister in Occupational Health and Safety at the University of Indonesia. Uri gained a scholarship and joined the Center for Ergonomics, Occupational Safety and Health, Massey University, New Zealand, as a PhD student in October 2008. She completed her PhD in 2013 and continued her post-doctoral program until 2015. In 2015, She was appointed as Head of Occupational Health and Safety Undergraduate Program, Faculty of Public Health, University of Indonesia until now.

**Prof. Dr. Hadiyanto, S.T., M.T.**
**Task Force Undip Ranking, Universitas Diponegoro, Indonesia**

Prof. Dr. Hadiyanto, MSc received his BSc of Chemical Engineering from Diponegoro University in 1998 and MSc of Bioprocess Engineering from Wageningen University, The Netherlands in 2003. While the degree of PhD has been obtained in 2007 from Wageningen University, Netherlands. After finishing his PhD, he had opportunity to work as scientist at NIZO Food Research BV Netherlands (2007-2009), and Research Associate at Process Intensification Group at TU DELFT Netherlands (2009-2010). Besides these works, he has also been invited as visiting research fellow at KU Leuven Belgium (2011), Kyoto University (2012) and DTU Denmark (2014). He is also actively involved in Sustainable Energy and Environmental (SEE) Forum, UNDIP green metric task force and coordinating world class university program since 2016. In 2016, he was appointed as the head of master program of environmental science, school of postgraduate studies and appointed as full professor on May 2017 in the field of bioprocess engineering.

**Prof. Jatna Supriatna, M.Sc., Ph.D.**
**Chairman of Research Center for Climate Change (RCCC), Universitas Indonesia, Indonesia**

Prof. Jatna Supriatna, M.Sc., Ph.D. is a Chairman of Research Center for Climate Change (RCCC), Universitas Indonesia, Indonesia. After finished his Master of Science (1986) and Doctorate degree (1991) from the University of New Mexico, USA, plus pre and postdoctoral at Columbia University in New York, he serves as a Senior lecturer at the Biology Dept., Director of Biodiversity and Conservation Studies, Coordinator of Graduate Program on Conservation Biology of the University of Indonesia. He also became a chief editor of Tropical Biodiversity since 1992, Editor of Asia Primate Journal (2008), Board of editor of the International Journal of Wildlife Policy and Law, board of editor of Tropical Ecology, Consulting editor of Biosphere Conservation and board of Advisor of Earthwatch Institute (2002). In 2006, elected President of South East Asia Primatologist Association, and in 2007 assigned as Chairman of IUCN-SSC PSG South East Asia.

**Prof. Riyad Y. Hamzah**
**President, University of Bahrain, Bahrain**

Prof. Riyad Y. Hamzah, is a professor of environmental biotechnology at the Arabian Gulf University and a member of the Council of Higher Education in the Kingdom of Bahrain. He earned his Ph.D. in biochemistry (enzymology) in 1984 from the University o f Houston, USA He was involved in the establishment of the Arabian Gulf University, Bahrain, since its inception in 1984 and established and directed numerous new programs in various scientific and educational fields. He servated in key academic and administrative posts including Director General of Finance and Administration, Dean of the College of Applied Sciences, and as the Vice-President of the Arabian Gulf University from 1994 to 2005.

**Prof. Tao-Ming Cheng**
**President, Chaoyang University of Technology, Taiwan**

Professor and President Tao-ming Cheng has received his Ph.D. in 1996 from the School of Civil Engineering, Purdue University. He became a faculty member of the Department of Construction Engineering, Chaoyang University of Technology (CYUT), Taiwan from 1996. He was promoted as a full professor by CYUT in 2007. He had been the Head of the Department of CE, Acting Dean of the College of Science and Engineering, Dean of Academic Affairs. and the Vice President. He is now the President of CYUT. His research interests include construction ergonomics, construction operations modeling, and higher education management. He publishes more than 100 journal or conference papers in the past decade.

**Dr. Pornchai Mongkhonvanit**
**President, Siam University, Thailand**

Dr. Pornchai Mongkhonvanit is the President of Siam University. Concurrently, he is also the chairperson of the advisory board of the Association of Universities in Asia and the Pacific as well as a member of the administrative board of the International Association of Universities. He is Honorary Professor at the Autonomous University of Guadalajara, Mexico and Guizhou University, China. He holds a Master of Business Administration in Finance Investment and Banking from the University of Wisconsin-Madison, a degree from the National Defense College of Thailand, and has studied at the Graduate School of Education, Institute of Educational Management and Harvard University.

**Prof. Ir. Joni Hermana, M.Sc.ES. Ph.D**
**Rector, Institut Teknologi Sepuluh Nopember, Indonesia**

Prof. Ir. Joni Hermana, M. Sc. ES. Ph. D is Rector of Institut Teknologi Sepuluh Nopember (ITS) since 2015, Previously, he was the Dean of Faculty of Civil Engineering and Planning, ITS, 20017-2015. He was a member of Supervisory Board of Water Supply Enterprise (PDAM), Surabaya, 2002-2006. He graduated with his bachelor's degree from Environmental Engineering at the Institute of Technology Bandung (ITB) in 1986. He received Master of Science on Environmental Sanitation from the University of Ghent, Belgium in 1991. He completed his doctoral degree from the Philosophy Department on Environmental Engineering from University of Newcastle Upon Tyne, United Kingdom in 1997.

**Prof. Yuhlong Oliver Su**
**President, National Chi Nan University, Taiwan**

Prof. Yuhlong Oliver Su is Professor of Department of Applied Chemistry and President of National Chi Nan University. He graduated with a bachelor's degree in Chemistry from National Taiwan University Taiwan. He holds a master's degree in Chemistry from National Taiwan Normal University. Prof. Yuhlong Oliver Su was a Vice President at the National Chung Hsing University in 2010-2011. He completed his PhD from Ohio State University, USA. He got Research Associate from Princeton University, USA. Since 1998 until now, he is Professor in Department of Applied Chemistry, National Chi Nan University, Taiwan. He was Secretary-General, Association of National Universities of Taiwan, 2010-2011. He expertises Electroanalytical Chemistry and Electrochemical Simulation, Porphyrin Chemistry, Organic Synthesis, Arylamine Chemistry.

**Dr. Arif Satria**
**Rector, Bogor Agricultural University, Indonesia**

Dr. Arif received his undergraduate degree majoring in Social Economics, Faculty of Agriculture (IPB), then continued his master's degree in Rural Sociology of IPB and earned his Doctorate degree at Marine Policy of Kagoshima University, Japan. He also had been a Visiting Fellow at the University of British Columbia, Canada. Dr. Arif is a lecturer of FEMA IPB since 2010 after previously serving as a lecturer in the Department of Fisheries Social Economics IPB. He also is an adviser to the Minister of Marine Affairs and Fisheries in 2012.

**Prof. Dr. Fathur Rokhman, M.Hum**
**Rector, Universitas Negeri Semarang, Indonesia**

Prof. Dr. Fathur Rokhman, M. Hum is the Rector of Universitas Negeri Semarang and a Professor of Sociolinguistics. He received his bachelor's degree from Education Program of the Indonesia Language of IKIP Bandung, 1990. He got his Master from the University of Indonesia, in 1996, and he received a Specialist Certificate in Language Curriculum and Material Development (RELC Singapore), 1998. He completed his Doctoral's degree in Linguistic at the Universitas Gadjah Mada, Jogjakarta, in 2003. He has a deep passion in environmental and cultural conservation. He embedded this passion to the university vision: to become a conservation university with international reputation.

**Prof. SeyedMohsen Najafian**
**President, University of Zanjan, Iran**

Prof. SeyedMohsen Najafian was appointed as the President of the University of Zanjan in 2017. He was graduated from Sharif Technical University in Iran and was employed as the professor of mathematics of this university in 1994. He was the vice president of Student Affairs of this university before he gets appointed as the president. He has published several papers in international and national journals and conferences, he has supervised many graduate students.

**Prof. Ravik Karsidi**
**Rector, Universitas Sebelas Maret, Indonesia**

Prof. Ravik Karsidi is expert in the field of community empowerment. He was Vice-Rector I Academic Affairs, Cooperation and Development, Universitas Sebelas Maret, 2003-2007. He graduated with his bachelor's degree in 1980 from the Faculty of Education, Department of Educational Sciences, *Universitas Sebelas Maret* (UNS). He received his magister's degree (S-2) from the Bogor Agricultural University (IPB), on Rural Sociology Program for Development Studies Bogor, in 1994. Furthermore, he completed his Doctoral program (S-3), in 1999 at the Bogor Agricultural University (IPB). He is also an alumni of the National Leadership Training Lemhamnas PPSA XVI and graduated on July 2009.

**Prof. Dr. med. Tri Hanggono Achmad, dr**
**Rector, Universitas Padjadjaran, Indonesia**

Prof. Dr. med. Tri Hanggono Achmad, dr has been Rector of Universitas Padjadjaran, Indonesia since 2015. Previously, he was Dean of Faculty of Medicine, Universitas Padjadjaran, in 2010 – 2015. He graduated Medical Doctor from Faculty of Medicine Universitas Padjadjaran in 1987. He had a took a course on Genetic enginering at the Inter University Centre – Institute of Technology Bandung, Bandung September-October 1990, course on Molecular Biology at the Inter University Centre – Institute of Technology Bandung, Bandung, October 1990 – February 1991 and short course on Animal Research Technology in Bad Onheuysen – Germany, 1992. Then, He completed his Ph.D. from Institute of Clinical Biochemistry, School of Medicine, University of Bonn, Germany in 1995.

**Dr. Des Fitzgerald**
**President, University of Limerick, Ireland**

Prior to taking up the post as President of University of Limerick in May 2017, Dr Des Fitzgerald was Professor of Molecular Medicine, UCD Vice-President for Health Affairs and Chief Academic Officer of the Ireland East Hospital Group. He is a UCD medical graduate who trained in Cardiology and Clinical Pharmacology at Vanderbilt University. He remained in the US until 1991 as Associate Professor of Medicine and Pharmacology and Director of the VA Coronary Care Unit. From 1994, he was Professor of Clinical Pharmacology and later Head of Research at RCSI. From 2004, he held various positions in UCD including Vice-President for Research. He has served on several national bodies, including the Irish Medicines Board, the Health Research Board (which he chaired) and the Irish Heart Foundation, as well as committees of the European Society of Cardiology, the UK National Institute for Health Research and the Merck Foundation International Fellowship Programme.

**Prof. Tlektes Yeszpolov**
**Rector, Kazakh National Agrarian University, Kazakhstan**

Dr. Tlektes Isabaevich Yespolov is the Rector of Kazakh National Agrarian University and the Academician of National Academy of Sciences of the Republic of Kazakhstan. Dr Yespolov obtained his post graduate degree in Technical Sciences at Scientific Research Institute of Hydraulic Engineering in Moscow in 1982. After graduation, he worked in Kazakh Institute of Agriculture as a lecturer, associate professor and the Head of Industrial Research Laboratory of Kazakh Institute of Agriculture. In 1993 he was appointed as the Director of the Almaty Industrial College, and then as the Rector of the Almaty Industrial Pedagogical Academy. In 2001 was awarded the rank of Professor.

**Prof. Paulo Jorge de Sousa Cruz**
**Pro-Rector for Quality of Life and Infrastructures, Universidade do Minho, Portugal**

Prof. Paulo Jorge de Sousa Cruz Graduated in Civil Engineering by the University of Porto in 1987. He completed his Masters in Structures by the University of Porto in 1991, and PhD in Construction Engineering by the Polytechnic University of Catalonia, Barcelona, in 1995; Habilitation in Structures by the University of Minho, in 2005. He was faculty member of the University of Minho since 1989, initially at the Civil Engineering Department, where he was Head of Department from 2003 to 2004, Dean of the School of Architecture from 2004 to 2011, Full Professor of Construction and Technology at the School of Architecture since 2008. He teaches and researches in the field of Structures, privileging the articulation between Structures and Architecture. Coordinator of the Construction and Technology Laboratory of the School of Architecture. He was Founder and President of the International Association of Structures and Architecture since 2016. Then, he also Founder of ASCP - Portuguese Association for the Safety and Maintenance of Bridges and President since 2006.

**Prof. Mirko Degli Esposti**
**Deputy Rector, Universita di Bologna, Italy**

Prof. Mirko Degli Esposti is a Deputy Rector of Universita di Bologna since November 2016. He is full Professor in Mathematical Physics. He had been Head of the Department of Mathematics and member of the Academic Senate of the University of Bologna in 2008-2015. He graduated with a bachelor's degree from Physics of University of Bologna in 1988. Subsequently, he completed his PhD in Mathematics (Caltech and PennState, USA), 1992 and his Postdoctoral fellow in Mathematical Science Research Institute (M.S.R.I), Berkely, California USA, 1993. In 1998-2000, He was visiting Professor at the Georgia Institute of Technology, USA. He is also member of the Center «L. Galvani» for integrated studies on bioinformatic, biophysics and bio-complexity. He was representing the University of Bologna as a member of the IREG Observatory, since 2016. He was a member of the QS Intelligence Unit Advisory Board (since November 2016). He was an acting as coordinator of a working group on Academic rankings (access with password) at the Conference of Italian University Rectors, CRUI (since October 2017- 34 participating universities).

**Xu Pan**
**Vice President, Shandong Normal University – Lishan College, China**

Xu Pan is mainly engaged in the research work of comprehensive utilization technology of organic waste, the technology of solar thermal utilization, low grade heat energy utilization technology. After obtaining Master of Engineering (2009) from Shandong University of Technology, he served as the general manager of Lucy Machinery Co., Ltd. In 2010, as one of the founders, he also became the general manager of Lucy New Energy Technology Co., Ltd. Ever since 2013, in Lishan College, he has served as vice president of the new energy engineering school, director of research department, and vice president, actively participating in the construction of zero-carbon campus of Shandong Normal University - Lishan College.

**Dr. Majid Monemzadeh**
**Vice-Chancellor for Research, University of Kashan, Iran**

Dr. Majid Monemzade is Vice Chancellor for Research and Technology at Kashan University. Previously, He was Head of Department of Physics, University of Kashan in 2006 - 2007. He graduated with his bachelor's degree in 1997 from Departement of Physics, Isfahan University of Technology. Subsequently, He received his magister's degree (S-2) from Isfahan University of Technology, in 2000. Furthermore, he completed his Doctoral program (S-3), in 2005 at Isfahan University of Technology. His researches are Heavy Quarks Models, Tetraquark Systems, Second-Class Constraints, BFFT embedding, Constrained systems. Until now he has write and publish many journal, some of them are indexed by Scopus.

**Associate Professor Boonchai Stitmannaithum, D.Eng.**
**Vice President for Administration, Chulalongkorn University, Thailand**

Dr. Boonchai Stitmannaithum is Associate Professor in Structural Engineering, Chulalongkorn University. He has been a Vice President of Physical Resources Management since 2012. Previouly, he was Assistant to the President (Physical Resources Management), 2008-2012. He graduated with his bachelor's degree from Civil Engineering at Chulalongkorn University, 1987. Subsequently, he received Master of Structural Engineering and Construction, AIT, Thailand, 1989, and he completed his Doctor in Civil Engineering at University of Tokyo, Japan, 1992. His research is Specialization Structural Engineering and Concrete Technology. He had got The Outstanding Technologist Award 2002, The Science and Technology Promotion Foundation and National Science and Technology Development Agency.

**Dr. Thatchavee Leelawat**
**Acting Vice President for Information Technology and Kanchanaburi Campus, Mahidol University, Thailand**

Dr. Thatchavee Leelawat is Acting Vice President for Information Technology and Kanchanaburi Campus, Mahidol University, Thailand. He graduated with his bachelor's degree from Chulalongkorn university, Thailand on Civil Engineering. He was graduated doctoral degree in 2000 University of Dundee, UK and got his master from same university in 1995. His researches are Durability of concrete, Rust reinforced concrete structure, Non-destructive, testing for infrastructure audit and evaluation., Repair of reinforced concrete structures, and Re-use of recycled concrete in civil engineering.

**Dr. Marcio Machado Ladeira**
**Representative of Rector, Universidade Federal de Lavras – UFLA, Brazil**

Bachelor at Animal Science from Universidade Federal de Viçosa (1995), Master's at Animal Science from Universidade Federal de Viçosa (1998), PhD at Animal Science from Universidade Federal de Minas Gerais (2001), and sabbatical study at Purdue University (2013). Today is Associate Professor of Universidade Federal de Lavras, acting as Advisor for Academic Development of UFLA Rector Office.

**Dr. Maria Kirrane**
**Representative of President, University College Cork, Ireland**

Dr. Maria Kirrane is a Sustainability Officer of University College Cork since May 2017 until now. In 2009, she graduated with her bachelor's degree in Ecology, her main interests being the effects of invasive species on ecosystems. This, combined with a fascination in honey bee behavior, led her to pursue research in the area of the varroa mite. She was Postdoctoral Researcher of University of Limerick in 2016 – 2017. She completed her Doctor from University College Cork, Ireland

**Dr. Mohd Fadhil Md Din**
**Representative of Vice Chancellor, Universiti Teknologi Malaysia (UTM), Malaysia**

Dr. Mohd Fadhil Md Din is Director of UTM Campus Sustainability Office. He graduated from UTM in 1999, also holding Master of Engineering in 2001 after graduated bachelor's degree in 1999 as a Tutor. Then, he completed his Master level under the supervision of Assoc. Prof. Dr. Azmi Aris (UTM, Malaysia) and Prof. Dr. Morgens Henze (Denmark Technical University, DTU). Then, further study under double-programme, between UTM (Malaysia) and TUDelft (The Netherlands). The study is comprehended with Dutch-Malaysian collaborations under the PhD scheme. Since 2006, the continuation of Research-Business was started with the aims of product commercialization. His current research Areas/Topics are Environmental Water and Wastewater Engineering, Bioenergy and Processes, Urban Heat Island

**Prof. Dr. Ahmad Zaharin Aris**
**Representative of Vice Chancellor, Universiti Putra Malaysia, Malaysia**

Professor Dr. Ahmad Zaharin Aris, CEnv, MRSC, FGS, the 2016/2017 SEARCA Regional Professorial Chair obtained his BSc (Hons) degree in Environmental Science and PhD degree in the field of Environmental Science (Hydrochemistry) from Universiti Malaysia Sabah. He is a Professor at the Department of Environmental Sciences, Faculty of Environmental Studies, Universiti Putra Malaysia (UPM) specializing in the field of Hydrochemistry, Environmental Chemistry and Analysis and Environmental Forensics. He now serves as the Dean in the Faculty of Environmental Studies, UPM. Currently, he is also the Chair for the Universiti Putra Malaysia working committee on Sustainable Green Campus.

**László Gyarmati**
**Representative of Rector, University of Szeged, Hungary**

László Gyarmati took his master's Degree at the University of Szeged Faculty of Economics and Business Administration in 2012. During his studies, he started his professional career as a PR trainee at the Management of the University of Szeged Study and Information Centre. At present, he is the Director of this Centre thus he is responsible for the environment-conscious behaviour and for the energy-efficient operation of the building. Besides, László Gyarmati has been co-ordinating and providing data on behalf of the University of Szeged for the UI GreenMetric survey for seven years. He has also been supporting the development of the Sustainability Strategy of the University; after his study tours in Europe and in the United States of America, he prepared the Sutainable Strategy Plan Suggestions for the University of Szeged with his colleagues' cooperation and has been working on realizing it ever since.

**Prof. Marcus Jopony**
**Representative of Vice-Chancellor, Universiti Malaysia Sabah, Malaysia**

Prof. Dr. Marcus Jopony is currently holding the post Assistant Vice-Chancellor as well as the Director of Centre for Strategic Management and Corporate Communication, Universiti Malaysia Sabah (UMS). He is also a member of the Board of Directors as well as Senate of UMS. He graduated with BSc. Hons in Chemistry from Universiti Kebangsaan Malaysia, MSc Soil Chemistry from University of Reading and PhD in Environmental Science from University of Nottingham. He joined UMS in 1995 and have been teaching among others, chemistry, environmental chemistry, environmental science and pollution control technology courses at the Faculty of Science and Natural Resources.

**Prof. Matteo Colleoni**
**Representative of Rector, Universita degli Studi di Milano Bicocca, Italy**

Prof. Matteo Colleoni took his dotor's degree at University of Trento in 2003 and got his 2nd Level master's Degree in Time Policies for the Quality of Life and Sustainable Mobility Politecnico of Milan at Universität Hamburg. He works as Deputy Coordinator of the Mobility Manager Network of the Italian Universities (within the Network of Italian University for Sustainability – RUS). He interests in sustainable urban mobility, environmental studies, metropolitan areas & urban change, time-space analysis and accessibility & transport exclusion.

**Prof. Godfrey A. Uzochukwu**
**Representative of Chancellor, North Carolina Agricultural & Technical State University, USA**

Godfrey Arinze Uzochukwu, PhD, Professor, is the founding director of the Interdisciplinary Waste Management Institute (WMI) of North Carolina Agricultural and Technical State University (NC A&T). Dr. Uzochukwu completed his B.S. and M.S. degrees at Oklahoma State University and his PhD degree at the University of Nebraska. He was a Post-Doctoral Scholar at Texas A&M University before joining the faculty of North Carolina Agricultural and Technical State University. He teaches geology, mineralogy, soil genesis & landuse, environmental science and has worked closely with more than 9,500 students from different disciplines at North Carolina A&T State University. Dr. Uzochukwu was a member of the Deans' Council. Dr. Uzochukwu served on the Executive Committee of the National Council of Environmental Deans and Directors (CEDD). He was a member of the North Carolina Legislative Commission on Global Climate Change.

**Dr. Ut V. Le**
**Representative of President Ton Duc Thang University, Vietnam**

At Ton Duc Thang University (TDTU), Vietnam, Dr. Ut V. Le Officially is appointed as Member, University Board, Deputy Director, Acting Director, Director of Research Affairs, Director of Foundation of Science and Technology Development in 2012 until now. He was researcher, Institute for Computation in Science and Technology, Ho Chi Minh City, Vietnam, 2012. He was Posdoctoral researcher, funded by Academy of Finland, working at University of Oulu, Finland in 2011-2012. He completed his Doctor of Philosophy in Mathematics from University of Oulu, Finland in 2011.

**Prof. Mario A. Gandini**
**Representative of Rector, Universidad Autonoma de Occidente, Colombia**

Prof. Mario A. Gandini holds a B.Sc. in Sanitary Engineering, a M.Sc in Environmental Policy and Management and a Ph.D in Environmental Engineering. He has been working in the academic sector as a lecturer and a researcher for 20 years, in the main area of pollution control, including wastewater treatment, ecological sanitation and solid waste management with a focus on leachate treatment. At the Universidad Autónoma de Occidente –UAO- (Cali, Colombia), currently he is the Head of the B.Sc. Program in Environmental Engineering, as well as, the Head of the UAO Sustainable Campus Program.

**Prof. M. Pinar Mengüç**
**Representative of Rector, Ozyegin University, Turkey**

Professor M. Pinar Mengüç received his PhD on Mechanical Engineering from Purdue University, USA in 1985. The same year he joined the University of Kentucky as an assistant professor and became a professor in 1993. He was a visiting professor at the Harvard University during 1998-99, and was awarded an Honorary Professorship at ESPOL, Ecuador in 2006. At the end of 2008, he was promoted to Engineering Alumni Association Chair Professor at Kentucky. He holds five patents and has two patent applications. He is the author of more than 130 articles in international journals, 170 conference papers and the co-author of two books. In 2009, he established the Centre for Energy, Environment and Economy

(CEEE), which he is still directing. He is a fellow of ASME and ICHMT, a Senior Member of OSA; in 2016 he was elected to Science Academy, Turkey. His research areas include radiative transfer, applied optics, nano-scale thermal transport and sustainable energy applications.

**Evgeny Zakablukovskiy, Ph.D.**
**Representative of Rector, Minin University, Russia**

Evgeny Zakablukovskiy, Ph.D. is Adviser to the Rector of Minin University since In 2015 until now. Previously, He was working in Intel Corporation (Russia) as Country Community & GR Manager Regional Public Affairs Manager in 2003 – 2014. In 2002 – 2003, He is Deputy Chief of Staff N. Novgorod Regional Government. In 2008, he got High Performance Instructor of Intel University. No. 6 in top 100 Russian PR managers in IT and Telecom. He got Doctor of Philosophical Anthropology from Minin University in 2015. He took a specialist degree in Volga Public Administration Academy, Nizhny Novgoro in 2015 and in Dobrolyubov Linguistic University, Nizhny Novgorod in 1996.

**Alexandra Leal Castellanos**
**Representative of Rector, Pontificia Universidad Javeriana, Colombia**

Currently, Alexandra Leal Castellanos is Director of University Services October since 2006 and President of the Board of JAVESALUD, since 2016. She was Director Center for Continuing Education in 1992 – 2006. She graduated from Pontificia Universidad Javeriana on Civil Engineer of Faculty of Engineering, 1983. Then, she completed her Master of Marketing Management Specialist from same university, 2002. She studied n Advanced Managment program as Senior Management Program for Business Executives in Esade Business School- Madrid, Spain, 2013, in Advanced Negotiation Program in Harvard University -Business School, 2015, in Negotiation Strategies Program in Yale University, 2017, Training in personal and executive coaching. TISOC – Spain, 2017, and in Executive leadership program, Yale University School of Management.

**Jose Herney Ramirez Franco**
**Representative of Rector, National University of Colombia, Colombia**

Jose Herney Ramirez Franco is an Associate professor at Chemical Engineering Department, Engineering Faculty, Universidad Nacional de Colombia. His research areas are in development, preparation, characterization and application of new materials based on clays, coals zeolites and silicas, and synthesis and use of these materials in the treatment of waste water. He received his bachelor's degree in Chemical Engineering from National University of Colombia, Colombia in 1998. He continued his magister in Chemical Engineering. São Paulo University, São Paulo – Brazil in 2003. He completed his Doctoral's degree in Chemical Engineering, Oporto University. Porto – Portugal in 2008.

**Dr. Jik-Chang Leong**
**Representative of President, National Pingtung University of Science and Technology, Taiwan**

Prof. J.C. Leong received his Ph.D. in Mechanical Engineering from the University of Oklahoma, USA in 2002 with his dissertation: Heat Transfer and Fluid Flow in Rotating Cylinders with a Porous Sleeve. He is a professor at the Department of Vehicle Engineering and also the Dean of the International College, National Pingtung University of Science and Technology. His research interests are: Heat and Mass Transfer, Flow Visualization Techniques, Ventilation and Fire Safety in Long Tunnels, Car Cabin Air-Conditioning and Ventilation System, Magnetic Fluid Flow.

**Juris Iljins**
**Representative of Rector, Riga Technical University, Latvia**

Mr. Juris Iljins currently is a Quality Manager and Sustainability Director at Riga Technical University. Currently he us doing his PhD in field of Change Management. Mr. Iljins is also former board member of Latvian Higher Education Council. He has been active internationally to promote sustainability with his scientific interest mainly focusing on how to effectively manage and measure change in various processes. Mr. Iljins is responsible for developing, maintaining quality management system within RTU and pursing sustainable development goals at RTU. Mr. Iljins is two-time alumni of Riga Technical University with degree in MSc. Industrial Engineering and Management, bachelor's degree and engineering qualification in Heat, Power and Thermal Engineering.

**Nguyen Thi Mai Khanh**
**Representative of Rector, Tra Vinh University, Vietnam**

Nguyen Thi Mai Khanh is Director International Collaboration and Project Promotion, Tra Vinh University, Vietnam since 2014. Previously, she was Vice – Director International Collaboration and Project Promotion of Tra Vinh University, in 2010 – 2014 and Director, Center for International Collaboration in Training and Education (CiCet) in 2009 – 2010. She graduated from bachelor's degree majoring English Teaching from Can Tho University, Vietnam. Subsequently, she continued her Master of Art in English Teaching Methodology at University of Social Science and Humanity, Ho Chi Minh City. She took her Diploma in "English as a Foreign Language: English Language and Teaching Methods for International English Teachers", Carleton University, Canada and she achieved certificate of "Theory and Practice for the Context of Teaching English as a Foreign language for Academic Purposes" from Carleton University, Canada.

**Prof. Patricia Iglecias**
**Representative of Rector, University of Sao Paulo (USP), Brazil**

Prof. Patricia Iglecias is Associate Professor, Department of Civil Law, USP Law School (2011), PhD (2007) and a master's degree (2002) by the same institution. Advisor of the master's and doctorate courses at the Faculty of Law of USP and PROCAM Environmental Science Program of USP. Currently, she is Superintendent of Environmental Management at USP and a leading researcher at the Environmental Applied Studies Group: preventive and reparative damages (USP), registered at CNPq. Main areas of activity: sustainable consumption; reverse logistic; Solid wastes and post-consumer liability; Shared responsibility; Causal link; Contaminated areas; Civil liability for damages to the environment; environmental compensation. Lecturer in Brazil and abroad, has several published works, highlighting the book; Solid waste and post-consumer civil liability, "by Editora Revista dos Tribunais. Vice President for the Southeast Region of the Institute The Right for a Green Planet. Member of the Association of Teachers of Environmental Law of Brazil APRODAB. Member of CErSOL. Multidisciplinary Center for Studies on Solid Waste at USP. Member of the European Environmental Law Association. She was the Secretary of State for the Environment of São Paulo.

**Samantha Fahy**
**Representative of President, Dublin City University, Ireland**

Samantha Fahy holds a primary Science degree from NUI Galway (1991), a Master in Optoelectronics from Queen's University Belfast (1992), a diploma in Management from Trinity College Dublin (1997) and an MBA from Dublin City University (2010). In 2004 Samantha was appointed Centre Manager for the National Centre for Plasma Science and Technology. In 2012 Samantha took on the part-time role of Manager of Sustainability at DCU, an initiative to develop and promote Sustainability across all educational functions, and moved to full-time in 2013. Samantha has over 20 year of experience in the management and administration of high technology research projects and has developed a deep understanding of research practice gained from working closely with outstanding researchers engaged in complex projects, such as SFI Strategic Research Clusters.

**Associate Professor Dr. Sumiani Yusoff**
**Representative of Vice-Chancellor, University of Malaya, Malaysia**

Associate Professor Dr, joined the Faculty of Engineering, UM in September 1990, specialising in Environmental Engineering and Management. She obtained her PhD in Environmental Engineering and Management from University of Malaya, MSc in Public Health and Environmental Control Engineering from University of Strathclyde, Glasgow, Scotland, United Kingdom and Bachelor of Civil Engineering (Hons) from Universiti Teknologi Malaysia. Currently she is Dean of Sustainability Science Research cluster in UM. Prior to that she was the Deputy Director and Chief Auditor of the Quality Management and Enhancement Centre (QMEC), UM. She teaches postgraduate and undergraduate courses in Environmental Management Systems, Solid Waste Management, Environmental Impact Assessment, and Life Cycle Management. She has published the UM Eco campus Blueprint.

**Dr. Noralfishah Sulaiman**
**Representative of Vice Chancellor, Universiti Tun Hussein Onn Malaysia, Malaysia**

Dr. Noralfishah Sulaiman is the Head of Sustainable Campus Unit (SCU) at Universiti Tun Hussein Onn Malaysia (UTHM). In February 2017. She also the head of KANZU Research: Resilient Built Environment Research Group, Universiti Tun Hussein Onn Malaysia (UTHM). Dr. Sulaiman's expertise lies in the fields of Resilient City Research (RCR), Resilience in Aging (RIA), Real Estate and Facilities Management, Information Technology in Real Estate, Internet of Things in Housing Sector, Social Care Facilities Management.

# About Universitas Indonesia

Universities, today, are uniquely positioned to address global challenges. As the result thereof, Universitas Indonesia (UI) is fully committed to seek solutions to the century's most pressing global challenges, enhance the education of future leaders and strengthen its academic endeavor. UI is innovative within our own institutions, in terms of how we are structured and governed and how we adapt to global challenge.

Having experience from year 1849 in higher education, our university is rich in history, academic excellence and its contribution to both Indonesian and international society. UI plays an active role on higher learning associations in Asia Pacific, Europe, Southeast Asia, and worldwide association such as APRU (Association of Pacific Rim Universities), ASAIHL (Association of South East Asia Institution of Higher Learning), and AUN (ASEAN University Network), in which UI is the host for AUN Credit Transfer Secretariat. Quality Culture is one of UI mottos to preserve and improve the good quality in academic and non-academic aspects, in order to provide the best services to its stakeholders. In return, UI retains its position to be amongst the top 400 universities in the world. UI is also an active member of ASEAN University Network – Quality Assurance (AUN-QA) since 2002. In 2017, UI completed the AUN University Institution Wide Assesment and received a good result. Until 2017, 28 undergraduate study programs have accomplished AUN-QA assessment, Magister Public Health Study Programme has achieved accreditation through Asia Pacific Academic Consortium for Public Health (APACPH), and Magister Management Study Programme has been accredited by The Alliance on Business Education and Scholarship for Tomorrow (ABEST 21). In 2017, UI is ranked 277 in QS World University Ranking, and ranked among 800 best universities in Times Higher Ranking.

Last but not least, the campus of UI is located in greeneries consisting of 320 hectares with 6 lakes. The university maintains the ecology conservation while developing world class academic facilities. In addition to that, as a member of IREG Observatory on Academic Ranking and Excellence (IREG observatory), UI has released UI GreenMetric University Ranking that ranks universities throughout the world according to appointed indicators of campus environmental issues such as setting and infrastructure, energy, waste management, water, transportation, and education for the past 8 years. Currently, UI have 49,000 students and approximately 6,000 academic staffs.

**Fig 1.** Landscape of Universitas Indonesia, Depok Campus

# About Diponegoro University

Diponegoro University (Undip) is a prominent university located in Semarang, the capital of Central Java province, and is considered as the best university in the province. Initially named as University of Semarang, Diponegoro University was a private university founded on January 9, 1957. On its anniversary, President Soekarno changed the name of the University of Semarang to Diponegoro University and changed its status to state university, which was then legally confirmed through government decree. The university was named after a hero of Central Java, "Prince Diponegoro", whose heroism spirit against the Dutch colonialism in the early century is tightly memorized by the Indonesian people. This spirit inspires Undip in developing research and education in science and technology as well as their application for people welfare. On its establishment, Undip had Faculty of Law as the only faculty. Soon after, several other faculties were established, i.e. Faculty of Social and Political Science in 1957, Faculty of Engineering in 1958, Faculty of Economics in 1960, Faculty of Animal Agriculture in 1964, and Faculty of Letters in1965 which was later changed into Faculty of Humanities. Due to the high demand of high school graduates to continue their study as well as the demand for developing science and technology in various fields, more faculties were established.

***In establishing its goal as research university by 2020, Diponegoro University implements its teaching and community service based on the result of its research, which is focused not only as the 'preservation of knowledge' but also as 'the advancement of knowledge'***

Undip is one of the best universities in Indonesia, as it is supported by reliable resources. It is the oldest and the biggest university in Central Java. As a state university, Undip is supported by the government of the Republic of Indonesia. At present, Undip has 7 campuses with total area of 2,444,852.48 square meters. The main campus is Tembalang campus at which all faculties, rectorate office building, university main hall, computer and language centers, research center, training center, student center, main library, and community service office are located. The second largest campus is in the heart of Semarang, i.e. in Jl. Imam Bardjo, at which offices and classrooms for most of graduate programs are located. The third biggest campus is located in Teluk Awur, Jepara, which is 100 km north east of Semarang. This campus is mainly for the Faculty of Fisheries and Marine Science. Another campus is in Jl. Dr. Soetomo, in which several classrooms and laboratories of the Faculty of Medicine are located.
In 2009, Undip started the construction of University Hospital in Tembalang Campus. The hospital is intended to facilitate more comfortable teaching and learning process for both academic staffs and students of the Faculty of Medicine. Each year Undip admits approximately 11,912 new students, which means that Undip has approximately 52,030 students.

**Fig. 1.** Landscape of Diponegoro University, Semarang

# Participant List

| TITLE | FULL NAME | NATIONALITY | INSTITUTION | POSITION | EMAIL |
|---|---|---|---|---|---|
| Dr. | **Sri Mulyani** | Indonesia | Ministry of Finance | Minister of Finance | - |
| Dr. | **Agus Justianto** | Indonesia | Ministry of Environment and Forestry, Republic of Indonesia | Head of Research, Development, and Inovation | - |
| Prof. | **Jatna Supriatna** | Indonesia | Research Center for Climate Change, UI | Chairman | jatna.supriatna@gmail.com |
| Prof. | **Muhammad Anis** | Indonesia | Universitas Indonesia | Rector | rektor@ui.ac.id |
| Prof. | **Yos Johan Utama** | Indonesia | Universitas Diponegoro | Rector | rektor@undip.ac.id |
| Prof. | **Adi Zakaria Afiff** | Indonesia | Universitas Indonesia | Vice-Rector | adi.zakaria@ui.ac.id |
| Dr. | **Mohd Fadhil Md Din** | Malaysia | Universiti Teknologi Malaysia (UTM) | Representative of Vice Chancellor | drwahid@utm.my |
| Prof. | **Riyadh Hamzah** | Bahrain | University of Bahrain | President | uobpresident@uob.edu.bh |
| Prof. | **Tao-Ming Cheng** | Taiwan | Chaoyang University of Technology | President | pres@cyut.edu.tw |
| Dr. | **Pornchai Mongkhonvanit** | Thailand | Siam University | President | pornchai@siam.edu |
| Prof. | **Joni Hermana** | Indonesia | Institut Teknologi Sepuluh Nopember | Rector | rektor@its.ac.id |
| Prof. | **Yuhlong Oliver Su** | Taiwan | National Chi Nan University | President | yosu@ncnu.edu.tw |
| Dr. | **Arif Satria** | Indonesia | Bogor Agricultural University | Rector | rektor@ipb.ac.id |
| Prof. | **Fathur Rokhman** | Indonesia | Universitas Negeri Semarang | Rector | rektor@unnes.ac.id |
| Prof. | **Mustafa Çufalı** | Turkey | Bülent Ecevit University | Rector | dincer.bicer@gmail.com |
| Prof. | **SeyedMohsen Najafian** | Iran | University of Zanjan | President | najafian@znu.ac.ir |
| Dr. | **Des Fitzgerald** | Ireland | University of Limerick | President | president@ul.ie |
| Prof. | **Tri Hanggono Achmad** | Indonesia | Universitas Padjadjaran | Rector | rektor@unpad.ac.id |
| Prof. | **Ravik Karsidi** | Indonesia | Universitas Sebelas Maret | Rector | rektor@uns.ac.id |
| Prof. | **Ayoib Che Ahmad** | Malaysia | Universiti Utara Malaysia | Deputy Vice-Chancellor | mustafa@uum.edu.my |
| Prof. | **Tlektes Yeszpolov** | Kazakhstan | Kazakh National Agrarian University | Rector | rector@kaznau.kz |
| Prof. | **Paulo Jorge de Sousa Cruz** | Portugal | Universidade do Minho | Pro-Rector | sec-pcruz@reitoria.uminho.pt |
| Prof. | **Mirko Degli Esposti** | Italy | University of Bologna | Deputy Rector | prorettore.vicario@unibo.it |
| Mr. | **Xu Pan** | China | Shandong Normal University - Lishan College | Vice President | xpan51@126.com |
| Dr. | **Majid Monemzadeh** | Iran | University of Kashan | Vice-Chancellor | monem@kashanu.ac.ir |
| Dr. | **Thatchavee Leelawat** | Thailand | Mahidol University | Acting Vice President | thatchavee.lee@mahidol.edu |
| Prof. | **Boonchai Stitmannaithum** | Thailand | Chulalongkorn University | Vice President | int.off@chula.ac.th |
| Dr. | **Maria Kirrane** | Ireland | University College Cork | Sustainability Officer | m.kirrane@ucc.ie |
| Prof. | **Ahmad Zaharin Aris** | Malaysia | Universiti Putra Malaysia | Representative of Vice Chancellor | zaharin@upm.edu.my |
| Mr. | **László Gyarmati** | Hungary | Univ of Szeged - Study and Information Centre | Representative of Rector | gyarmati.laszlo@tik.u-szeged.hu |
| Prof. | **Marcus Jopony** | Malaysia | Universiti Malaysia Sabah | Representative of Vice-Chancellor | marcj@ums.edu.my |
| Prof. | **Matteo Colleoni** | Italy | University of Milano-Bicocca | Representative of Rector | matteo.colleoni@unimib.it |
| Prof. | **Godfrey A. Uzochukwu** | USA | North Carolina Agricultural & Technical State Univ. | Representative of Chancellor | uzo@ncat.edu |
| Dr. | **Ut V. Le** | Vietnam | Ton Duc Thang University | Representative of President | levanut@tdt.edu.vn |
| Prof. | **Mario A. Gandini** | Colombia | Universidad Autonoma de Occidente | Representative of Rector | magandini@uao.edu.co |
| Dr. | **Udaya Priyantha Kankanamge Epa** | Sri Lanka | University of Kelaniya | Representative of Vice Chancellor | epa@kln.ac.lk |

| TITLE | FULL NAME | NATIONALITY | INSTITUTION | POSITION | EMAIL |
|---|---|---|---|---|---|
| **Prof.** | **Mustafa Pinar Menguc** | Turkey | Ozyegin University | Representative of Rector | pinar.menguc@ozyegin.edu.tr |
| **Dr.** | **Evgeny Zakablukovskiy** | Russia | Minin University | Representative of Rector | ezakablu@mininuniver.ru |
| **Ms.** | **Alexandra Leal Castellanos** | Colombia | Pontificia Universidad Javeriana | Representative of Rector | aleal@javeriana.edu.co |
| **Mr.** | **Jose Herney Ramirez Franco** | Colombia | National University of Colombia | Representative of Rector | jhramirezfra@unal.edu.co |
| **Dr.** | **Jik-Chang Leong** | Taiwan | National Pingtung University of Science and Technology | Representative of President | jcleong@mail.npust.edu.tw |
| **Mr.** | **Juris Iljins** | Latvia | Riga Technical University | Representative of Rector | juris.iljins@rtu.lv |
| **Ms.** | **Nguyen Thi Mai Khanh** | Vietnam | Tra Vinh University, | Representative of Rector | ntmaikhanh@tvu.edu.vn |
| **Prof.** | **Patricia Iglecias** | Brazil | University of Sao Paulo (USP) | Representative of Rector | patricia.iglecias@usp.br |
| **Ms.** | **Samantha Fahy** | Ireland | Dublin City University | Representative of President | samantha.fahy@dcu.ie |
| **Mr.** | **Habib M. Fardoun** | Saudi Arabia | King Abdulaziz University | Representative of Rector | hfardoun@kau.edu.sa |
| **Prof.** | **Sumiani Yusoff** | Malaysia | University of Malaya | Representative of Vice Chancellor | sumiani@um.edu.my |
| **Prof.** | **Margarita Redina** | Russia | RUDN University | Representative of Rector | redina_mm@rudn.university |
| **Dr.** | **Noralfishah Sulaiman** | Malaysia | Universiti Tun Hussein Onn Malaysia | Representative of Vice Chancellor | nora@uthm.edu.my |
| **Dr.** | **Esmail Karamidehkordi** | Iran | University of Zanjan | Director of Int. Scientific Coop. | e.karamidehkordi@gmail.com |
| **Dr.** | **Samar Azzi-Achkouty** | Lebanon | Holy Spirit University of Kaslik (USEK) | Representative of President | samarazzi@usek.edu.lb |
| **Dr.** | **Syed Naveed Hussain Shah** | Pakistan | Pakistan Higher Education Commission | Director | nrshah@hec.gov.pk. |
| **Prof.** | **Riri Fitri Sari** | Indonesia | UI GreenMetric | Chairperson | riri@ui.ac.id, |
| **Prof.** | **Tommy Ilyas** | Indonesia | UI GreenMetric | Expert Member | t.ilyas@eng.ui.ac.id |
| **Mr.** | **Junaidi** | Indonesia | UI GreenMetric | Expert Member | junaidi@ui.ac.id |
| **Prof.** | **Gunawan Tjahjono** | Indonesia | UI GreenMetric | Expert Member | virtueray@gmail.com |
| **Dr.** | **Baiduri Widanarko** | Indonesia | UI GreenMetric | Expert Member | baiduri@ui.ac.id |
| **Dr.** | **Ruki Harwahyu** | Indonesia | UI GreenMetric | Expert Member | ruki.harwahyu11@ui.ac.id |
| **Mr.** | **Rinoto Cahyo Utomo** | Indonesia | UI GreenMetric | Programmer | rintocahyo@gmail.com |
| **Ms.** | **Sabrina Hikmah Ramadianti** | Indonesia | UI GreenMetric | Secretariat | sabrina.hikmah@ui.ac.id |
| **Ms.** | **Arsy Imanda N. Raswati** | Indonesia | UI GreenMetric | Secretariat | arsy.imanda@ui.ac.id |
| **Ms.** | **Hastin Setiani** | Indonesia | UI GreenMetric | Secretariat | hastin.setiani@ui.ac.id |
| **Prof.** | **Ambariyanto** | Indonesia | Universitas Diponegoro | Vice Rector | ambariyanto.undip@gmail.com |
| **Prof.** | **Hadiyanto** | Indonesia | Universitas Diponegoro | Executive Chair of IWGM 2018 | hadiyanto@live.undip.ac.id |
| **Dr.** | **Agus Setyawan** | Indonesia | Universitas Diponegoro | Co-Chair of IWGM 2018 | asetyawan@gmail.com |
| **Ms.** | **Kenny Dwi Purwani** | Indonesia | Universitas Diponegoro | Organizing Com. | kenny dwi kenpurwani@gmail.com |
| **Mr.** | **Abdul Rohman** | Indonesia | Universitas Diponegoro | Organizing Com. | rohman@live.undip.ac.id |
| **Dr.** | **Ita Widowati** | Indonesia | Universitas Diponegoro | Organizing Com. | ita.widowati@undip.ac.id |
| **Mr.** | **Susatyo Nugroho Widyo Pramono** | Indonesia | Universitas Diponegoro | Organizing Com. | susatyo_nwp@live.undip.ac.id |
| **Dr.** | **Jatmiko Endro Suseno** | Indonesia | Universitas Diponegoro | Organizing Com. | jatmikoendro@gmail.com |
| **Mr.** | **Wahyu Setiabudi** | Indonesia | Universitas Diponegoro | Organizing Com. | iwgm@live.undip.ac.id |
| **Dr.** | **Syafrudin** | Indonesia | Universitas Diponegoro | Organizing Com. | iwgm@live.undip.ac.id |
| **Dr.** | **Daud Syamsudewa** | Indonesia | Universitas Diponegoro | Organizing Com. | daudreproduksi@gmail.com |
| **Mr.** | **Fahmi Arifan** | Indonesia | Universitas Diponegoro | Organizing Com. | iwgm@live.undip.ac.id |
| **Mr.** | **Nuryanto** | Indonesia | Universitas Diponegoro | Organizing Com. | iwgm@live.undip.ac.id |
| **Dra.** | **Arsiani Sulaksmiwati** | Indonesia | Universitas Diponegoro | Organizing Com. | iwgm@live.undip.ac.id |
| **Mr.** | **Slamet Purwanto** | Indonesia | Universitas Diponegoro | Organizing Com. | iwgm@live.undip.ac.id |
| **Mr.** | **Achmad Saefuddin** | Indonesia | Universitas Diponegoro | Organizing Com. | iwgm@live.undip.ac.id |
| **Dr.** | **Suryanti** | Indonesia | Universitas Diponegoro | Organizing Com. | iwgm@live.undip.ac.id |
| **Ms.** | **Shoimatul Fitria** | Indonesia | Universitas Diponegoro | Organizing Com. | iwgm@live.undip.ac.id |
| **dr.** | **Sri Winarni** | Indonesia | Universitas Diponegoro | Organizing Com. | iwgm@live.undip.ac.id |

| TITLE | FULL NAME | NATIONALITY | INSTITUTION | POSITION | EMAIL |
|---|---|---|---|---|---|
| **Mr.** | **Ting-Fang Lee** | Taiwan | Chaoyang University of Technology | Program Manager | tflee@cyut.edu.tw |
| **Mr.** | **Pungut Ibrahim** | Malaysia | Universiti Malaysia Sabah | Director | pungut@ums.edu.my |
| **Dr.** | **How Siew Eng** | Malaysia | Universiti Malaysia Sabah | Director, EcoCampus Man. | sehow@ums.edu.my |
| **Dr.** | **Lay-Gaik Teoh** | Taiwan | National Pingtung University of Science and Technology | Chairperson, Department of Mech..Eng. | n5888107@mail.npust.edu.tw |
| **Dr.** | **Pomin Li** | Taiwan | National Pingtung University of Science and Technology | Dean, Office of Int. Affairs | kokann@mail.npust.edu.tw |
| **Dr.** | **Indera Syahrul Mat Radzuan** | Malaysia | Universiti Tun Hussein Onn Malaysia | Senior Lecturer | syahrul@uthm.edu.my |
| **Mr.** | **Helmi Syaifuddin** | Indonesia | UIN Maulana Malik Ibrahim Malang | Secretary of QA | helmi.syaifuddin@uin-malang.ac.id |
| **Dr.** | **Syed Nadeem Akhtar** | India | Integral University | Director, Planning & Research | snadeem@iul.ac.in |
| **Prof.** | **Amin Retnoningsih** | Indonesia | Universitas Negeri Semarang | Head Depaof Conservation | aminretnoningsih2016@mail.unnes.ac.id |
| **Dr.** | **HM. Zainuri** | Indonesia | Muria Kudus University | Vice-Rector II | zainuri@umk.ac.id |
| **Drs.** | **RM. Hendy Hendro Hadi Sridjono** | Indonesia | Muria Kudus University | Vice-Dean I of the Fac. of Agriculture | hendy.hendro@umk.ac.id |
| **Mr.** | **Budi Utomo, S.Kom.** | Indonesia | Muria Kudus University | Head of General Adm. Bureau | budi.utomo@umk.ac.id |
| **Prof.** | **Muhammad Zamrun F.** | Indonesia | Halu Oleo University | Rector | rektor@uho.ac.id |
| **Prof.** | **Gusti R. Sadimantara** | Indonesia | Halu Oleo University | Head of Global Partnership Agency | gusti5@yahoo.com |
| **Dr.** | **La Ode Santiaji Bande** | Indonesia | Halu Oleo University | Head of the development & QA | ls.bande@yahoo.co.id |
| **Mr.** | **Dinçer Biçer** | Turkey | Bülent Ecevit University | International Affairs Coordinator | dincer.bicer@gmail.com |
| **Prof.** | **Chanita Rukspollmuang** | Thailand | Siam University | Vice President | chanita@siam.edu |
| **Mrs.** | **Erry Ika Rhofita** | Indonesia | UIN Sunan Ampel Surabaya | Lecturer | erryikarhofita@uinsby.ac.id |
| **Mr.** | **Nurwijayadi Wiratmo** | Indonesia | Telkom University | Log. & Assets Dir. | nurwijayadi@gmail.com |
| **Dr.** | **Zahra Zamani Nokabadi** | Iran | University of Isfahan | Coord. for Green Documentation | z_zamani85@yahoo.com |
| **Dr.** | **Mehran Zeynalian Dastjerdi** | Iran | University of Isfahan | Secretary of the Green Management | m.zeynalian@eng.ui.ac.ir |
| **Dr.** | **Noor Amila Wan Abdullah Zawawi** | Malaysia | Universiti Teknologi Petronas | Dir., Inst. for Sustainable Living | docamila11@gmail.com |
| **Mr.** | **Ananto Aji** | Indonesia | Universitas Negeri Semarang | Coord. of Nat. Resources and Env. | ajiananto@yahoo.com |
| **Drs.** | **Said Sunardiyo** | Indonesia | Universitas Negeri Semarang | Coord. of Tech. & Innov. | saidelektro@gmail.com |
| **Mr.** | **Teguh Husodo** | Indonesia | Universitas Padjadjaran | Head of Camp. Env | t.husodo@unpad.ac.id |
| **Mr.** | **Cipta Endyana** | Indonesia | Universitas Padjadjaran | School of Postgrad. | cipta.endyana@unpad.ac.id |
| **Ms.** | **Beta Paramita** | Indonesia | Universitas Pendidikan Indonesia | Lecturer | betaparamita@upi.edu |
| **Dr.** | **Mochamad Syamsiro** | Indonesia | Janabadra University | Vice Dean, Faculty of Eng. | syamsiro@janabadra.ac.id |
| **Ms.** | **Fen Shia Lee** | Taiwan | National Chi Nan University | Project Manager | fxli@ncnu.edu.tw |
| **Dr.** | **Masoud Salavati Niasar** | Iran | University of Kashan | Head of Inst. of Nano Science and Nano Tech. | salavati@kashanu.ac.ir |
| **Dr.** | **Naser Alnaser** | Bahrain | University of Bahrain | Assistant Professor | nalnaser@uob.edu.bh |
| **Mr.** | **Pratya Chaitientong** | Thailand | Mahidol University | Management and Adm. officer | pratya.chi@mahidol.ac.th |
| **Prof.** | **Dadan Ramdan** | Indonesia | Univ. of Medan Area | Rector | dadan@uma.ac.id |
| **Dr.** | **Mufti Sudibyo** | Indonesia | Univ. of Medan Area | Head of GreenMetric | msud16@gmail.com |
| **Mr.** | **João Múcio Amado Mendes** | Brazil | University of Sao Paulo (USP) | PhD candidate | joaomucio@hotmail.com |
| **Ms.** | **Ana Carolina Corberi Famá Ayoub e Silva** | Brazil | University of Sao Paulo (USP) | PhD candidate | fama@igleciasfama.com |
| **Dr.** | **Nuzhat Afsar** | Pakistan | University of Karachi | Assistant Professor | nafsar@uok.edu.pk |
| **Ms.** | **Anna Galushkina** | Russia | Minin University | Student Leader | galushkina@st.mininuniver.ru |

| TITLE | FULL NAME | NATIONALITY | INSTITUTION | POSITION | EMAIL |
|---|---|---|---|---|---|
| Dr. | **Ming-Ta Tsai** | Taiwan | Fo Guang University | Dean of Gen. Affairs | mttsai@mail.fgu.edu.tw |
| Dr. | **Paul Bolger** | Ireland | University College Cork | Representative of President | |
| Mr. | **Mulyono** | Indonesia | Universitas Airlangga Surabaya | Head of Health, Safety & Env. | mulyono@fkm.unair.ac.id |
| Mr. | **Karnaji** | Indonesia | Universitas Airlangga Surabaya | Director of Infrastructure and Env. Facility | karnaji@fisip.unair.ac.id |
| Mr. | **Caizergues** | France | Inseec U. | Quality & RSE Manager | pascal.caizergues@esce.fr |
| Dr. | **Muhdi** | Indonesia | Universitas PGRI Semarang | Rector | muhdi_sh@yahoo.com |
| Mr. | **Jin Delu** | China | Shandong Normal University - Lishan College | The chairman of Univ.& Lucy Group | xpan51@126.com |
| Mr. | **Jin Zebang** | China | Shandong Normal University - Lishan College | | xpan51@126.com |
| Prof. | **Okid Parama Astirin** | Indonesia | Universitas Sebelas Maret | Lecturer | parama_astirin@yahoo.com |
| Dr. | **Winny Astuti** | Indonesia | Universitas Sebelas Maret | Chairman of Greencampus | winnyast64@gmail.com |
| Prof. | **Ellen Kumaat** | Indonesia | Sam Ratulangi University | Rector | rektor@unsrat.ac.id |
| Ms. | **Tayebeh Eftekhari** | Iran | University of Guilan | Supervisor of Interantional Office | int.support@guilan.ac.ir |
| Mrs. | **Normaliza Abdul Manaf** | Malaysia | University Sains Malaysia | Senior Science Officer | normaliza@usm.my |
| Ms. | **Siti Fairuz Mohd Radzi** | Malaysia | University Sains Malaysia | Research Officer | fairuzradzi@usm.my |
| Drs. | **Moh Muzammil** | Indonesia | Universitas Terbuka | Vice Rector | wr4@ecampus.ut.ac.id |
| Mr. | **Mutdi Ismuni** | Indonesia | STIE PERTIWI Bekasi | Vice Head | mutdi.ismuni@stie-pertiwi.ac.id |
| Prof. | **Mohamad Farizal Rajemi** | Malaysia | Universiti Utara Malaysia | Representative of Vice-Chancellor | farizal@uum.edu.my |
| Prof. | **Hasanuddin** | Indonesia | Gorontalo Univ. | Rector | milamalango@yahoo.co.id |
| Dr. | **M. Zainuddin** | Indonesia | UIN Maulana Malik Ibrahim Malang | Vice Rector | helmi.syaifuddin@gmail.com |
| Dr. | **I Wayan Dasna** | Indonesia | Univ Negeri Malang | Vice Rector | wr4@um.ac.id |
| Dr. | **Senam** | Indonesia | UNY | Vice Rector | rezahalimah@gmail.com |
| Mr. | **Abzal Abdramanov** | Kazakhstan | Kazakh National Agrarian University | Head of the Dep. for Ratings | abdramanov.abzal@gmail.com |
| Ms. | **Lutfiye Kozlan** | Turkey | Yeditepe University | Director of Env. & Landscape | lkozlan@yeditepe.edu.tr |
| Dr. | **Pinar Karakas** | Turkey | Yeditepe University | Associate Professor | pinarkarakas@hotmail.com |
| Mr. | **Jatmmiko Wahyudi** | Indonesia | Reg. Plan. Develop. Agency of Pati | Researcher | jatmiko_tkuns@yahoo.com |
| Dr. | **Dede Lia Zariatin** | Indonesia | Universitas Pancasila | Lecturer | dedeliazariatin@univpancasila.ac.id |
| Prof. | **Agus Hartoko** | Indonesia | University of Bangka Belitung | Vice Rector | info@ubb.ac.id |
| Prof. | **Bustami Syam** | Indonesia | Univ. Sumatera Utara | Vice Rector IV | |
| Dr. | **Marcio Machado Ladeira** | Brazil | Universidade Federal de Lavras | Adviser of the Rector | mladeira@dzo.ufla.br |
| Dr. | **Irhamah** | Indonesia | Institut Teknologi Sepuluh Nopember | Ka UP4 ITS | up4@its.ac.id |
| Prof. | **Akbar Tahir** | Indonesia | Univ. of Hasanuddin | Sustainability Off. | akbar_tahir@mar-sci.unhas.ac.id |
| Mr. | **Bima Andika Pratama** | Indonesia | Univ. of Hasanuddin | Sustainability Off. | akbar_tahir@mar-sci.unhas.ac.id |
| Mr. | **Sasmito Djati** | Indonesia | Universitas Brawijaya | Vice Rector | msdjati@ub.ac.id |
| Ms. | **Sentagi S Utami** | Indonesia | Univ. Gadjah Mada | GreenMetri Team | sentagi@gmail.com |
| Mr. | **Randy Frans Fela** | Indonesia | Univ. Gadjah Mada | GreenMetric Team | randyrff@gmail.com |
| Mr. | **Ahmad Seng** | Indonesia | Khairun University | Head of Adm. and Finance | ahmadseng61@gmail.com |
| Ms. | **Razsera Binti Hassan Basri** | Malaysia | International Islamic University Malaysia (IIUM) | Deputy Dir., Develop. & Plan. | razsera@iium.edu.my |
| Ms. | **Hazlina Binti Mazlan** | Malaysia | International Islamic University Malaysia (IIUM) | Deputy Dir.(QS), Develop. Div. | mhazlina@iium.edu.my |

# Workshop Program

**THE 4th INTERNATIONAL WORKSHOP ON UI GREENMETRIC WORLD UNIVERSITY RANKINGS**
*"Universities, Impacts and Sustainable Development Goals (SDGs)"*
**8-10 April 2018**
Universitas Diponegoro, Semarang, Indonesia

**DAY 1: 8 APRIL 2018 – SUNDAY**
Venue: Meeting Room, Grand Candi Hotel, Semarang

| Timing | Session |
|---|---|
| **Whole day** | **Arrival of participants** |
| 01:00 PM – 02:00 PM | Networking Lunch (by invitation) |
| 02:00 PM - 05:45 PM | **UI GWURN Steering Committee Meeting (by invitation)**<br>Venue: Meeting Room, Grand Candi Hotel, Semarang<br>1. Welcome of National Coordinators, quorum established<br>2. Secretariat UI GreenMetric (Prof. Riri Fitri Sari, Mr. Junaidi, M.A., and Prof. Tommy Ilyas)<br>3. Adoption of the agenda<br>4. Remarks by **Prof. Adi Zakaria Afiff,** Vice Rector of Finance, Logistics and Facilities, Universitas Indonesia<br>5. Adoption of the minutes of the 1st Steering Committee Meeting in Istanbul, 9 April 2017 (Doc. 3)<br>6. Report of Activities of UI GWURN and national coordinators:<br>• **Prof. Patricia Iglecias,** Representative of Rector, University of Sao Paulo (USP), Brazil<br>• **Mr. Xu Pan,** Shandong Normal University - Lishan College, China<br>• **Mr. Caizergues,** Representative of Dean, Inseec U., France<br>• **Prof. Yos Johan Utama**, Rector, Universitas Diponegoro, Indonesia<br>• **Prof. Mirko Degli Esposti**, Deputy Rector, University of Bologna, Italy<br>• **Prof. SeyedMohsen Najafian**, President, University of Zanjan, Iran<br>• **Dr. Maria Kirrane,** Representative of President, University College Cork, Ireland<br>• **Mr. Abzal Abdramanov,** Representative of Rector, Kazakh National Agrarian University, Kazakhstan<br>• **Mr. Juris Iljins,** Representative of Rector, Riga Technical University, Latvia<br>• **Dr. Syed Naveed Hussain Shah,** Pakistan Higher Education Commission, Pakistan<br>• **Prof. Paulo Jorge de Sousa Cruz,** Pro-Rector for Quality of Life and Infrastructures, Universidade do Minho, Portugal<br>• **Dr. Thatchavee Leelawat,** Acting Vice President for Information Technology and Kanchanaburi Campus, Mahidol University, Thailand<br>• **Mr. Andy Nolan,** Representative of Vice-Chancellor, University of Nottingham, United Kingdom (by skype)<br>• **Dr. Mohamad Farizal Bin Rajemi,** Representative of Vice-Chancellor, Universiti Utara Malaysia, Malaysia (by skype)<br>• **Mr. Habib M. Fardoun**, Representative of Rector, King Abdul Aziz, Saudi Arabia (by skype)<br>• **Dr. Margarita Redina,** Representative of Rector, People's Friendship University Russia (RUDN), Russia (by skype)<br>7. Discussion on the future activities of UI GWURN<br>- Shaping Global Higher Education and Research in Sustainability<br>- Creating Global Sustainability Leaders<br>- Partnering on Solutions to Sustainability Challenges<br>8. Any other business<br>9. Closing |
| 07:00 PM - 09:00 PM | UI GWURN Steering Committee Dinner, hosted by Rector Universitas Diponegoro (by invitation)<br>Venue: Koena-Koeni Restaurant |

<table>
<tr><th colspan="3">DAY 2: 9 APRIL 2017 – MONDAY<br>Venue: Auditorium of Rectorate Building, Universitas Diponegoro</th></tr>
<tr><td>07:00 AM</td><td colspan="2">Bus Transfer from Hotel to Rectorate Building, Universitas Diponegoro</td></tr>
<tr><td>07:00 AM -<br>08:00 AM</td><td colspan="2">Registration for Delegates</td></tr>
<tr><td>08:00 AM -<br>08:10 AM</td><td colspan="2">• Safety Induction<br>• Indonesian National Anthem "Indonesia Raya"</td></tr>
<tr><td>08:10 AM -<br>08:35 AM</td><td colspan="2">Remarks<br>Venue: Rectorate Building, Universitas Diponegoro<br>• Prof. Yos Johan Utama, Rector of Diponegoro University<br>• HE Sri Mulyani Indrawati, S.E., M.Sc., Ph.D., Minister of Finance, Republic of Indonesia<br>• Prof. Riri Fitri Sari, Chairperson of UI GreenMetric World University Rankings</td></tr>
<tr><td>08:35 AM -<br>09:15 AM</td><td colspan="2">Keynote Speeches<br>• Dr. Harry Purwanto, Expert Staff to Minister of Research, Technology and Higher Education, Republic of Indonesia<br>• Dr. Agus Justianto, Head of Research, Development, and Inovation of Ministry of Environment and Forestry, Republic of Indonesia</td></tr>
<tr><td>09:15 AM –<br>09:30 AM</td><td colspan="2">Technical Address (Special Issues on University Environment)<br>• Biodiversity Index<br>Prof. Jatna Supriatna, Chairman of Research Center for Climate Change (RCCC), Universitas Indonesia, Indonesia</td></tr>
<tr><td>09:30 AM -<br>09:45 AM</td><td colspan="2">Signing of Declaration of Membership of UI GreenMetric Network</td></tr>
<tr><td>09:45 AM -<br>10:15 AM</td><td colspan="2">Group photo and Coffee Break</td></tr>
<tr><td>10:15 AM -<br>11:15 AM</td><td>Parallel Session 1 A<br>Issues and Innovation in Managing Setting and Infrastructure<br>Venue: Room 1, Rectorate Building, Universitas Diponegoro<br><br>Chair:<br>Dr. Esmail Karamidehkordi<br>International Scientific Cooperation Office, University of Zanjan, Iran<br>(4 presentations – 15 Minutes Each)<br>• Managing University Landscape and Infrastructure towards Green and Sustainable Campus<br>Prof. Adi Zakaria Afiff, Vice Rector of Finance, Logistics and Facilities, Universitas Indonesia<br>• The Challenges of Adopting BIM for Setting and Infrastructure Management of the University of Minho<br>Prof. Paulo Jorge de Sousa Cruz, Pro-Rector for Quality of Life and Infrastructures, Universidade do Minho, Portugal<br>• The University of São Paulo on the 2017's GreenMetric Ranking<br>Prof. Patricia Iglecias, Representative of Rector, University of Sao Paulo (USP), Brazil<br>• Green at Universiti Putra Malaysia: Cultivating the Green Campus Culture<br>Prof. Dr. Ahmad Zaharin Aris, Representative of Vice Chancellor, Universiti Putra Malaysia, Malaysia<br>Question and Answer</td><td>Parallel Session 1 B<br>Issues and Innovation in Managing Setting and Infrastructure<br>Venue: Room 2, Rectorate Building, Universitas Diponegoro<br><br>Chair:<br>Prof. Dr. Hadiyanto, S.T., M.T.<br>TaskForce Undip Ranking, Universitas Diponegoro, Indonesia<br>(3 presentations – 15 Minutes Each)<br>• Setting and Infrastructure at North Carolina Agricultural and Technical State University<br>Prof. Godfrey A. Uzochukwu, Representative of Chancellor, North Carolina Agricultural & Technical State University, USA<br>• Greening Campus Experience: Moving towards Living Laboratory Action Plan<br>Dr. Mohd Fadhil Md Din, Representative of Vice Chancellor, Universiti Teknologi Malaysia (UTM), Malaysia<br>• Making an Urban University 'Green': Uniting Administration and Students towards Synergy<br>Evgeny Zakablukovskiy, Ph.D., Representative of Rector, Minin University, Russia<br><br>Question and Answer</td></tr>
</table>

| | | |
|---|---|---|
| 11:15 AM -<br>12:15 AM | **Parallel Session 2 A**<br>**Issues and Innovation in Managing Energy**<br>Venue: Room 1, Rectorate Building, Universitas Diponegoro<br><br>Chair:<br>**Prof. Dr. Sudarto**<br>Previous Rector, Universitas Diponegoro, Indonesia<br>*(3 presentations – 15 Minutes Each)*<br>• ***Creating a Low-carbon Campus in Chaoyang University of Technology (CYUT)***<br>**Prof. Tao-Ming Cheng,** President, Chaoyang University of Technology, Taiwan<br>• ***The Unibo Energy Saving and Climate Change Approach***<br>**Prof. Mirko Degli Esposti,** Deputy Rector, Universita di Bologna, Italy<br>• ***UniTO Performance in UI GreenMetric Energy and Climate Change***<br>**Prof. Marcello Baricco,** Vice-rector for Simplification, University of Turin, Italy<br>• ***Evaluation of Electricity Consumption and Carbon Footprint of UI GreenMetric Participating Universities using Regression Analysis***<br>**Prof. Riri Fitri Sari,** Chairperson of UI GreenMetric World University Rankings<br>*Question and Answer* | **Parallel Session 2 B**<br>**Issues and Innovation in Managing Energy**<br>Venue: Room 2, Rectorate Building, Universitas Diponegoro<br><br>Chair:<br>**Prof. Patricia Iglecias**<br>Representative of Rector, University of Sao Paulo (USP), Brazil<br>*(3 presentations – 15 Minutes Each)*<br>• ***Design and Operation Experience of Zero - Carbon Campus***<br>**Xu Pan,** Vice President, Shandong Normal University - Lishan College, China<br>• ***Industrial Revolution 4.0: The Universiti Malaysia Sabah Perspective***<br>**Dr. How Siew Eng,** Representative of Vice-Chancellor, Universiti Malaysia Sabah, Malaysia<br>• ***Expansion of Renewable Energy Resources and Energy-conscious Behavior at the University of Szeged***<br>**László Gyarmati,** Representative of Rector, University of Szeged, Hungary<br><br>*Question and Answer* |
| 12:15 AM -<br>01:00 PM | Networking Lunch | |
| 01:00 PM -<br>02:00 PM | **Parallel Session 3 A**<br>**Issues and Innovation in Managing Waste**<br>Venue: Room 1, Rectorate Building, Universitas Diponegoro<br><br>Chair:<br>**Prof. Ir. Joni Hermana, M. Sc. ES., Ph.D.**<br>Rector, Institut Teknologi Sepuluh Nopember, Indonesia<br>*(4 presentations – 15 Minutes Each)*<br>• ***Challenges of Sustainability Effors of Universities Regarding the Sustainable Development Goals: A Case Study in the University of Zanjan, Iran***<br>**Prof. SeyedMohsen Najafian,** President, University of Zanjan, Iran<br>• ***Current Practices of Waste Management in Universitas Diponegoro Campus***<br>**Prof. Yos Johan Utama,** Rector, Universitas Diponegoro, Indonesia<br>• ***Methodological Proposal for Environmental Aspects Evaluation in Higher Education Institution (HEI)***<br>**Jose Herney Ramirez Franco,** Representative of Rector, National University of Colombia, Colombia | **Parallel Session 3 B**<br>**Issues and Innovation in Managing Waste**<br>Venue: Room 2, Rectorate Building, Universitas Diponegoro<br><br>Chair:<br>**Prof. Dr. Ir. Tommy Ilyas, M.Eng.**<br>Expert Member of UI GreenMetric<br>*(5 presentations – 15 Minutes Each)*<br>• ***Waste Water Treatment for Achieving Sustainable***<br>**Prof. Dr. Ravik Karsidi MS,** Rector, Universitas Sebelas Maret, Indonesia<br>• ***BASE: A Sustainable Path for Milano-Bicocca University***<br>**Prof. Matteo Colleoni,** Representative of Rector, Universita degli Studi di Milano Bicocca, Italy<br>• ***Sustainability in Motion at UAO: Integrated Waste Management***<br>**Prof. Mario A. Gandini,** Representative of Rector, Universidad Autonoma de Occidente, Colombia<br>• ***Toward Integrated and Sustainable Waste Management System in University of Malaya: UM Zero Waste Campaign*** |

| | | |
|---|---|---|
| | *Question and Answer* | **Associate Professor Dr. Sumiani Yusoff,** Representative of Vice Chancellor, University of Malaya, Malaysia<br>• ***The Sustainability Efforts of Ton Duc Thang University in the South of Vietnam*** **Dr. Ut V. Le,** Representative of President Ton Duc Thang University, Vietnam |
| 02:00 PM - 03:00 PM | **Parallel Session 4 A**<br>**Issues and Innovation in Managing Water**<br>Venue: Room 1, Rectorate Building, Universitas Diponegoro<br><br>Chair:<br>**Prof. Gunawan Tjahjono**<br>Expert Member of UI GreenMetric<br>*(3 presentations – 15 Minutes Each)*<br>• ***Water Conservation and Management Policies and Practices: Case study in Siam University*** **Dr. Pornchai Mongkhonvanit,** President, Siam University, Thailand<br>• ***Designing Catchment Area for Water Resources Management in Institut Teknologi Sepuluh Nopember Campus*** **Prof. Ir. Joni Hermana, M. Sc. ES., Ph.D.** Rector, Institut Teknologi Sepuluh Nopember, Indonesia<br>• ***Water Conservation and Water Recycling Program Implementations at University of Kashan*** **Dr. Majid Monemzadeh,** Vice-Chancellor for Research, University of Kashan, Iran<br>*Question and Answer* | **Parallel Session 4 B**<br>**Issues and Innovation in Managing Water**<br>Venue: Room 2, Rectorate Building, Universitas Diponegoro<br><br>Chair:<br>**Dr. Maria Kirrane**<br>Representative of President, University College Cork, Ireland<br>*(3 presentations – 15 Minutes Each)*<br>• ***Universitas Padjadjaran Concern for Sustainable Water Resources from West Java to National and to the World*** **Mr. Cipta Endyana,** Representative of Rector, Universitas Padjadjaran, Indonesia<br>• ***Dublin City University – Toward a Sustainable Campus*** **Samantha Fahy,** Representative of President, Dublin City University, Ireland<br>• ***Tra Vinh University and Strategies Heading to Green Campus*** **Nguyen Thi Mai Khanh,** Representative of Rector, Tra Vinh University, Vietnam<br><br>*Question and Answer* |
| 03:00 PM - 04:00 PM | **Parallel Session 5 A**<br>**Issues and Innovation in Managing Education**<br>Venue: Room 1, Rectorate Building, Universitas Diponegoro<br><br>Chair:<br>**Mr. Junaidi, M. A.**<br>Expert Member of UI GreenMetric<br>*(4 presentations – 15 Minutes Each)*<br>• ***How Universities Can Work Together with Local Communities to Create a Green, Sustainable Future*** **Prof. Yuhlong Oliver Su,** President, National Chi Nan University, Taiwan<br>• ***Accelerating the Transformation to a Green University: University of Bahrain Experience*** **Dr. Naser Alnaser,** Representative of President, University of Bahrain, Bahrain<br>• ***The Internalization of Conservation Mindset in the Disruptive Era*** **Prof. Dr. Fathur Rokhman, M. Hum,** Rector, Universitas Negeri Semarang, Indonesia | **Parallel Session 5 B**<br>**Issues and Innovation in Managing Education**<br>Venue: Room 2, Rectorate Building, Universitas Diponegoro<br><br>Chair:<br>**Dr. Baiduri Widanarko**<br>Expert Member of UI GreenMetric<br>*(4 presentations – 15 Minutes Each)*<br>• ***Mainstreaming Sustainability Science to Enhance the Impacts of "Tridharma" in Higher Education Institution: IPB Lessons Learned and Future Direction*** **Dr. Arif Satria,** Rector, Bogor Agricultural University, Indonesia<br>• ***NPUST's Practice on Achieving Sustainable Development Goals*** **Dr. Jik-Chang Leong,** Representative of President, National Pingtung University of Science & Technology, Taiwan<br>• ***Bringing a Sustainable University in Turkey through Teaching, Research and Service*** **Prof. M Pinar Mengüç,** Representative of Rector, Ozyegin University, Turkey<br>• ***RTU Approach to Sustainable Development: Governance and Measuring Impact*** |

| | | |
|---|---|---|
| | • ***How the Environmental Planning of Federal University of Lavras Impact Higher Education***<br>**Dr. Marcio Machado Ladeira,** Representative of Rector, Universidade Federal de Lavras – UFLA, Brazil<br>*Question and Answer* | **Juris Iljins,** Representative of Rector, Riga Technical University, Latvia<br>*Question and Answer* |
| 04:00 PM - 04:15 PM | Coffee Break | |
| 04:15 PM - 05:15 PM | **Panel Session 6**<br>**Issues and Innovation in Managing Transportation**<br>Venue: Room 1, Rectorate Building, Universitas Diponegoro<br><br>Chair:<br>**Prof. Mirko Degli Esposti**<br>Deputy Rector, Universita di Bologna, Italy<br>*(2 presentations – 15 Minutes Each)*<br>• ***Tackling Energy, Transportation and Sustainability on Campus***<br>**Dr. Des Fitzgerald,** President, University of Limerick, Ireland<br>• ***Green Transportation System to Promote Sustainable Lifestyle in Chulalongkorn University***<br>**Associate Professor Boonchai Stitmannaithum, D.Eng.**, Vice President for Administration, Chulalongkorn University, Thailand<br>• ***Good Mobility Practices at Poltificia Universidad Javeriana (Transportation)***<br>**Alexandra Leal Castellanos,** Representative of Rector, Pontificia Universidad Javeriana, Colombia<br>• ***Reengineering Sustainability at Universiti Tun Hussein Onn Malaysia (UTHM): Travel & Transportation***<br>**Dr. Noralfishah Sulaiman,** Representative of Vice-Chancellor, Universiti Tun Hussein Onn Malaysia, Malaysia<br>*Question and Answer* | |
| 05:15 PM - 05:20 PM | Closing Remarks<br>• **Prof. Yos Johan Utama,** Rector of Diponegoro University<br>• **Prof. Riri Fitri Sari,** Chairperson of UI GreenMetric<br>• **Report of University Colllage Cork, Ireland (Host of IWGM 2019)** | |
| 05:20 PM - 06:00 PM | Campus Tour | |
| 07:00 PM - 09:00 PM | Dinner (hosted by Governor of Central Java)<br>Venue: Wisma Perdamaian Building | |
| 09:00 PM | Transfer to Hotel | |
| **DAY 3: 10 APRIL 2017 – TUESDAY**<br>**Complimentary Tour** | | |
| 06:00 AM - 07:00 AM | Registration at Hotel Lobby | |
| 07:00 AM - 09:30 AM | Transfer to Borobudur Temple | |
| 09:30 AM – 12:00 AM | Exploring Borobudur Temple | |
| 12:00 AM – 13:30 PM | Lunch<br>Venue: Manohara Hotel | |
| 13:30 PM – 17:00 PM | Transfer to Hotel/Semarang Airport | |

edp sciences

E3S Web of Conferences

# Statement of Peer review

In submitting conference proceedings to E3S Web of Conferences, I certify to the Publisher that I adhere to the **Policy on Publishing Integrity** of the journal in order to safeguard good scientific practice in publishing.

1. All articles have been subjected to peer review administered by the proceedings editors.
2. Reviews have been conducted by expert referees, who have been requested to provide unbiased and constructive comments aimed, whenever possible, at improving the work.
3. Proceedings editors have taken all reasonable steps to ensure the quality of the materials they publish and their decision to accept or reject a paper for publication has been based only on the merits of the work and the relevance to the journal.

Title, date and place of the conference

The 4th International Workshop on UI GreenMetric World University Ranking (IWGM)
Semarang, April 8 - 10, 2018

Proceedings editor(s): Hadiyanto.

Date and editor's signature

Prof. Dr. Hadiyanto.

E3S Web of Conferences **48**, 01001 (2018) https://doi.org/10.1051/e3sconf/20184801001
*IWGM 2018*

# Biodiversity Indexes: Value and Evaluation Purposes

*Jatna* Supriatna[1]

[1]Institute for Sustainable Earth and Resources (I-SER) and Dept of Biology, Faculty of Math and Sciences, Universitas Indonesia

**Abstract.** Biodiversity is a word recently introduced by experts in the field of biology. This word became more meaningful after Edward O. Wilson of Harvard University introduced it in a book entitled Biodiversity, an extension of biological diversity, in 1989 [11]. In subsequent developments, it became very popular and used not only by environmental biologists but also by researchers, environmentalists, funders, educators, social experts, economists, policy makers, and many others, although many do not know what that means. Biodiversity includes variations within the biological community, where living species, and ecosystems, where communities are located, as well as interaction between them (Pri. The science of biodiversity has emerged rapidly since then included monitoring and evaluation systems which is measuring the value of biodiversity components, such as the number of species present, the population of species, a habitat or the sum of all such components within a given area or site. Such monitoring and evaluation may be carried out for a variety of reasons, included identification of a given area for biodiversity richness, evenness or healthy ecosystems. The richness is the number of species per sample, the more species present in a sample, the richer the sample. Evenness is a measure of the relative abundance of the different species making up the richness of an area. Two commonly used to measure biodiversity Simpson index Ds and Shannon's index H'. Simpson's index DS is similarity index (the higher the value the lower in diversity). While Shannon index is combining evenness and richness and less weighted on dominant species. Both indexes are more reflective in nature and can predict the environment health. Therefore, it may be good to have one of those biodiversity indexes to be used for UI GreenMetric to understand the environment healthiness in the campus.

## 1 Introduction

When the UI GreenMetric chairperson asked me how to do we measure biodiversity in the campus, of course, I have so many things in my mind. As a professor who has taught Conservation Biology for the last 3 decades at UI, I have so many possible explanations to measure biodiversity, either scientifically measured that has to be carried out by biologists or other model that may be friendly for anyone in the campus who can fill in the form as long as they know what kind of biodiversity locally known. Therefore, I took the liberty to propose by describing biodiversity first then what model need to be used in measuring biodiversity in the campus.

Green Campus is varied in many ways. Local university's outdoor-education program, local bike paths connect more to the city area, fishing, beach walking, disc golf, triathlon, shooting sports team, rowing, sailing, and more traditional sports are found mostly in the open spaces of the campus. Those are open spaces created by most of the university campuses all over the world. However, some city campus, of course, will be difficult to get more open spaces for those activities. Those many open spaces hosted full with biodiversity in it. But we need to understand which biodiversity we may measure, for example whether we count common species, richness or endemic biota, or proportion of those then we call it an index.

In this short paper, I want to share what should UI GreenMetric do in adding biodiversity on its formula. There is not going to be easy to use those indexes with mathematical equation unless we develop a computational tool that may be user-friendly by even small campuses. Therefore, I recommend having a new formula instead the current index in order to get more holistic view on biodiversity such as included richness, endemicity or native, representativeness, and others.

## 2 What is biodiversity and why we measure.

The term of biodiversity is not so long only a few decades ago when the person who come up with the term of biological diversity then to be shortening to become biodiversity. Although the science of flora, fauna and small creatures have been recognized as oldest one, biology, but the genuine term was proposed by Edward O Wilson of Harvard University in his paper in 1985 entitled "Biological diversity crisis" then a book called Biodiversity in 1989 [10, 11]. Soon after than book called Conservation Biology written by Michael Soule (1996) used biodiversity in many chapters [7]. It becomes very popular worlds after UN Earth Summit in 1992 at Rio de Janairo Brazil. The summit results many world's agenda to tackle the environmental problems included; Agenda 21, Forest for People, Climate Change and Convention on Biological Diversity (CBD). This CBD convention has been developing many commitments to conserve biodiversity in every 4 years meetings. Those commitments are now becoming a major driving force behind efforts to reform land management and development practices worldwide and to establish a more harmonious relationship between people and nature [8].

With working definition of biodiversity as the variety of life and its processes, it means including the variety of living organism, the genetic differences among them, the communities and ecosystems in which they occur [11]. The importance of this definition is that it draws attention to the many dimensions of biodiversity. It explicitly recognizes that every biota can be characterized by its taxonomic, ecological, and genetic diversity and that the way these dimensions of diversity vary over space and time is a key feature of biodiversity. Thus, only a multidimensional assessment of biodiversity can provide insights into the relationship between changes in biodiversity and changes in ecosystem functioning and ecosystem service. Biodiversity is the foundation of ecosystem service to which human well-being is intimately linked. No feature of Earth is more complex, dynamic, and varied than the layer of living organisms that occupy its surfaces and its seas, and no feature is experiencing more dramatic change at the hands of humans than this extraordinary, singularly unique feature of Earth. This layer of living organisms—the biosphere—through the collective metabolic activities of its innumerable plants, animals, and microbes physically and chemically unites the atmosphere, geosphere, and hydrosphere into one environmental system within which millions of species, including humans, have thrived. It follows that large-scale human influences over this biota have tremendous impacts on human well-being. It also follows that the nature of these impacts, good or bad, is within the power of humans to influence [1].

There are many measures on biodiversity; species richness (the number of species in a given area) represents a single but important metric that is valuable as the common currency of the diversity of life—but it must be integrated with other metrics to fully capture biodiversity. Because the multidimensionality of biodiversity poses formidable challenges to its measurement, a variety of surrogate or proxy measures are often used. These include the species richness of specific taxa, the number of distinct plant functional types (such as grasses, forbs, bushes, or trees), or the diversity of distinct gene sequences in a sample of microbial DNA taken from the soil. Species- or other taxon-based measures of biodiversity, however, rarely capture key attributes such as variability, function, quantity, and distribution—all of which provide insight into the roles of biodiversity [2].

Ecological indicators are scientific constructs that use quantitative data to measure aspects of biodiversity and ecological condition, services, or drivers of change, but no single ecological indicator captures all the dimensions of biodiversity. Ecological indicators form a critical component of monitoring, assessment, and decision-making and are designed to communicate information quickly and easily to policy makers [3]. In a similar manner, economic indicators such as GDP are highly influential and well understood by decision makers. Some environmental indicators, such as global mean temperature and atmospheric $CO_2$ concentrations, are becoming widely accepted as measures of anthropogenic effects on global climate. Ecological indicators are founded on much the same principles and therefore carry with them similar pros and cons [4].

## 3 Measuring biodiversity by index

Many indices have been applied on environmental issues. Those are the Living Planet Index (LPI), Ecological Footprint (EF), City Development Index (CDI), Human Development Index (HDI), Environmental Sustainability Index (ESI), Environmental Performance Index (EPI), Environmental Vulnerability Index (EVI), Index of Sustainable Economic Welfare/Genuine Progress Index (ISEW/GPI), Well-Being Index (WI), Genuine Savings Index (GS), and Environmental Adjusted Domestic Product (EDP). And Biodiversity index. The latest is not so commonly used but it is starting to get traction after Singapore made City Biodiversity index [2].

For Biodiversity index, it would be easier if biodiversity could be measured by the quantity of birds in a forest, wildflowers in a meadow or beetles in a log. Unfortunately, the simplicity is not one of the virtues of biodiversity [4]. Ecosystem are more complex then we can imagine. A common misconception is that biodiversity is equivalent to species diversity, more species in an area, it means greater its biodiversity. Biodiversity is not a number of game; the quality is more important than quantity. It is not so much in number of species but in its identity [2]. For example, fragmenting old growth forest with clear cut, would increase species richness at local scale but not contribute to species richness at a broader scale if sensitive species were lost from the landscape [4].

Diversification can all too easily become homogenization. The greatest cause of homogenization is the introduction of non-native species of plants and animals, often called exotics. Exotics are species that have invaded new areas due to accidental or deliberate transport by human. In many cities and also in the campus, those exotics have been commonly found due to deliberately planted or released it. The exotics polluted the native flora and fauna, but their contribution was nothing to biodiversity. Regions invaded by exotics lose their distinctive characters, the results is global impoverishment [4].

In spite of many tools and data sources, biodiversity remains difficult to quantify precisely. But precise answers are seldom needed to devise an effective understanding of where biodiversity is, how it is changing over space and time, the drivers responsible for such change, the consequences of such change for ecosystem service and human well-being, and the response options available. Ideally, to assess the conditions and trends of biodiversity either globally or sub-globally, it is necessary to measure the abundance of all organisms over space and time, using taxonomy (such as the number of species), functional traits (for example, the ecological type such as nitrogen-fixing plants like legumes versus non-nitrogen-fixing plants), and the interactions among species that affect their dynamics and function (predation, parasitism, competition, and facilitation such as pollination, for instance, and how strongly such interactions affect ecosystems). Even more important would be to estimate turnover of biodiversity, not just point estimates in space or time. Currently, it is not possible to do this with much accuracy because the data are lacking. Even for the taxonomic component of biodiversity, where information is the best, considerable uncertainty remains about the true extent and changes in taxonomic diversity [2].

When diversity indices are used in ecology, the types of interest are usually species, but they can also be other categories, such as genera, family, functional type or haplotypes. The entities of interest are usually individual plants or animals, and the measure of abundance can be, for example, number of individuals, biomass or coverage. In demography, the entities of interest can be people, and the types of interest various demographic groups. In information science, the entities can be characters and the types the different letters of the alphabet. The most commonly used diversity indices are simple transformations of the effective number of types (also known as 'true diversity'), but each diversity index can also be interpreted in its own right as a measure corresponding to some real phenomenon (but a different one for each diversity index) [3].

Many indices only account for categorical diversity between subjects or entities. Such indices, however, do not account for the total variation (diversity) that can be held between subjects or entities which occurs only when both categorical and qualitative diversity are calculated. True diversity, or the effective number of types, refers to the number of equally abundant types needed for the average proportional abundance of the types to equal that observed in the dataset of interest (where all types may not be equally abundant). Globally threatened species that have been assessed following the IUCN Red List criteria as having a high risk of extinction, restricted-range species with small global distributions, assemblages of species concerned to a particular broad habitat type, or biome and congregations of species that gather in large numbers at species sites during some stage in their life cycle [4].

### 3.1 Species Richness

Richness R simply quantifies how many different types the dataset of interest contains. For example, species richness (usually noted S) of a dataset is the number of different species in the corresponding species list. Richness is a simple measure, so it has been a popular diversity index in ecology, where abundance data are often not available for the datasets of interest. Because richness does not take the abundances of the types into account, it is not the same thing as diversity, which does take abundances into account. However, if true diversity is calculated with $q = 0$, the effective number of types (D) equals the actual number of types (R) [2].

### 3.2 Shannon and Simpson indexes

The Shannon index has been a popular diversity index in the ecological literature, where it is also known as Shannon's diversity index, the Shannon–Wiener index, the Shannon–Weaver index and the Shannon entropy. The measure was originally proposed by Claude Shannon to quantify the entropy (uncertainty or information content) in strings of text.[5] The idea is that the more different letters there are, and the more equal their proportional abundances in the string of interest, the more difficult it is to correctly predict which letter will be the next one in the string. The Shannon entropy quantifies the uncertainty (entropy or degree of surprise) associated with this prediction. It is most often calculated as follows [2]:

$$H= \Sigma \text{ pi ln pi or } H' = -\Sigma(n_i/N) * \log(n_i/N) \quad (1)$$

where $p_i$ is the proportion of characters belonging to the ith type of letter in the string of interest. In ecology, $p_i$ is often the proportion of individuals belonging to the ith species in the dataset of interest. Then the Shannon entropy quantifies the uncertainty in predicting the species identity of an individual that is taken at random from the dataset. Although the equation is here written with natural logarithms, the base of the logarithm used when calculating the Shannon entropy can be chosen freely. Shannon himself discussed logarithm bases 2, 10 and e, and these have since become the most popular bases in applications that use the Shannon entropy.

Since mean proportional abundance of the types increases with decreasing number of types and increasing abundance of the most abundant type, $\lambda$ obtains small values in datasets of high diversity and large values in datasets of low diversity. This is counterintuitive behavior for a diversity index, so often such transformations of $\lambda$ that increase with increasing diversity have been used instead [2].

The most popular of such indices have been the inverse Simpson index ($1/\lambda$) [6] and the Gini–Simpson index ($1 - \lambda$). Both of these have also been called the Simpson index in the ecological literature, so care is needed to avoid accidentally comparing the different indices as if they were the same. The formula for diversity (D) is as follows:

$$D = 1 - \Sigma(n/N)^2 \quad (2)$$

where n is the abundance of the i-th species in an area and N the ii, total number of said species living in the same area.

This simply equals true diversity of order 2, i.e. the effective number of types that is obtained when the weighted arithmetic mean is used to quantify average proportional abundance of types in the dataset of interest. The original Simpson index equals the probability that two entities taken at random from the dataset of interest (with replacement) represent the same type. Its transformation $1 - \lambda$ therefore equals the probability that the two entities represent different types. This measure is also known in ecology as the probability of interspecific encounter (PIE) and the Gini–Simpson index [9].

## 4 Other biodiversity indexes related to city

The biodiversity areas out of species index have been used by many cities but pioneered by Singapore Government with developing formula for the so called "The city Biodiversity index" [12]. This formula may fit with the campus model (UI GreenMetric) with specific modification in the future included not only species per se but also other criteria or indicators included:

1. Proportion of natural area in the city
2. Connectivity measure or ecological networks to counter fragmentation
3. Native biodiversity in built up area (bird species)
4. Change number of native species
5. Proportion of protected area
6. Proportion of exotics or invasive species
7. Regulation on quantity of water use
8. Climate regulation, carbon storage and cooling effects
9. Recreational and education
10. Budget allocated for biodiversity
11. Number of biodiversity projects annually
12. Policies, rules and regulations: existence of Biodiversity Strategic action Plan
13. Institutional capacity
14. Participation and partnership
15. Education and awareness

Those indicators have already developed a user manual so that proposed cities can use it in order to see whether they are can make a planned against the indicators.

The criteria for identifying key biodiversity areas need to be agreed upon internationally backbone university at UI GreenMetric and these standards must be consistently applied. However, to guarantee ownership of the indicator plans and the sustainability of efforts to implement them, the process of identifying these indicators must be led at a national or international levels.

## References

1. Groom, M.J., K.K. Meffe, R. Caroll. *Principles of Conservation Biology*. Sinauer Associates, Icn, Sundeland, M.A (2006)
2. Hill, D., M. Fashmam, P. Shaw, *Handbook of Biodiversity methods: Survey, Evaluation and Monitoring*, Cambridge University Press, Cambridge (2005)
3. Noss, R.F., *Indicators for monitoring biodiversity: A hierarchial. Conser,* Biol 4: 355-364 (1990)
4. Noss, R.F & A.Y. Cooperrider, *Saving nature's legacy: Protecting and Restoring Biodiversity,*. Island Press, Washington, D.C. (1994)
5. Primack, R., Primer of Conservation Biology. Sinauer Associates, Sunderland, M.A. (2004)
6. Simpson, E. H., "Measurement of diversity", Nature. 163: 688. doi:10.1038/163688a0 (1949)
7. Soule, M., *Conservation Biology: the Science of Scarcity and Diversity,* Sinauer Asscoaites, Sunderland, M.A. (1986)
8. Supriatna, J., Melestarikan alam Indonesia, Yayasan Obor, Jakarta (2009)
9. Tuomisto, H., *"A consistent terminology for quantifying species diversity? Yes, it does exist"*, Oecologia. 4: 853–860. doi:10.1007/s00442-010-1812-0 (2010)
10. Wilson, E.O., *The Biological Diversity Crisis,* Bioscience 35: 700-706 (1985)
11. Wilson, E.O., *The current state of biological diversity*. In: E.O. Wilson (ed): Biodiversity, pp3-18, National Academy Press, Washington, D.C. (1989)
12. www.cbd.int/.../city/.../subws-2014-01-singapore-index.

# Managing university landscape and infrastructure towards green and sustainable campus

*Muhammad* Anis[1], *Adi Zakaria* Afiff[2], *Gandjar* Kiswanto[3], *Nyoman* Suwartha[4,*] and *Riri Fitri* Sari[5]

[1]Department of Metallurgy and Material Engineering, Faculty of Engineering, Universitas Indonesia, Kampus UI Depok, Depok 16424, Indonesia
[2]Department of Accounting, Faculty of Economy and Business, Universitas Indonesia, Kampus UI Depok, Depok 16424, Indonesia
[3]Department of Mechanical Engineering, Faculty of Engineering, Universitas Indonesia, Kampus UI Depok, Depok 16424, Indonesia
[4]Department of Civil Engineering, Faculty of Engineering, Universitas Indonesia, Kampus UI Depok, Depok 16424, Indonesia
[5]Department of Electrical Engineering, Faculty of Engineering, Universitas Indonesia, Kampus UI Depok, Depok 16424, Indonesia

**Abstract.** Landscape and infrastructure are two main basic aspects which play a significant role for any universities in achieving environmental, social-behavioural, or aesthetic outcomes. This paper describes the setting and infrastructure management at the Universitas Indonesia (UI) as a continuous effort towards green and sustainable campus. Some developments and improvements have been made during the last three years, such as forest city maintenance, development of pedestrian boulevard, and installation of road marks and traffics signs. In addition, infrastructures have been improved by replacing conventional lamps with LED type ones, development of real-time energy consumption monitoring system, green chilling system, solar photo-voltaic development, solar cooling absorption chiller system, air conditioning with VRF/VRV system, and preparation for green certified building. These improvements on campus setting and landscape are reflected on the UI GreenMetric 2017 results which ranked at #23 worldwide with the score of 957 out of 1500 total score for "setting and infrastructure" category. With regards to these achievements, some challenges and strong commitment in sustainability policy need to be implemented in the forthcoming years to maintain the UI's vision to be a "world class sustainable university".

## 1 Introduction

Massive development worldwide in recent years brings significant indications that the future of the world is very much depending on the environment status. Global warming and climate change impacts at various sectors is a natural prove of the unsynchronized development and un-environment-friendly resources exploration. In order to mitigate serious impacts of climate change, several environmentalists and activists of various elements (government, NGO, society, higher education institutions, schools, etc.) have moved together to promote the save the environment movement. One of the example is the campaign by the Climate Reality Project, through a 24 Hours of Reality: Be the Voice of Reality [1]. Furthermore, as to raising international public awareness of global warming and reenergizing the environmental movement, Al Gore had presented a documentary movie ("An Inconvenient Sequel: Truth to Power") on the threat that climate change poses to the Earth - it's causes, effects and history and potential solutions to it. The recent COP 23 UNFCC conference in Bonn showcased the multiparty effort on achieving the Climate Change mitigation program.

At the higher education institutions (HEIs) level, universities play an important role as contributor to Greenhouse Gas Emission (GHG) from daily teaching and research-based activities. Many universities have been more aware with climate change issues and works together to minimize GHG emission and combat climate change impacts. There are many university networks on environment and sustainability that had been established at university, national, regional and global scale, such as Green Princeton, Sierranevada.com, Association for the Advancement of Sustainability in Higher Education (AASHE), Sustainable University Network (SUN) of Thailand, Campus Sustainability Network in Japan (CAS-Net Japan), International Sustainable Campus Network (ISCN), Sustainability Tracking, Assessment & Rating System (STARS), UI GreenMetric World University Rankings, etc. Through such specific network, universities from developed and developing countries can share their best practice on sustainability policy and the implementation.

Known as the best campus in Indonesia according to the QS World Universities Ranking 2017/2018 (1st ranked in Indonesia, 54th in Asia, 277th in the world) [2] and ranked >800 in the THEs ranking 2017/2018 [3], Universitas Indonesia (UI) kept its predicate being the nation's oldest, most prestigious, and greenest campus. The university has continuously promoted awareness and sustainability program internally to create a better and greener living. Geographically, UI campus is located in two areas: Salemba (Jakarta) and Depok (West Java).

* Corresponding author: nsuwartha@eng.ui.ac.id

https://doi.org/10.1051/e3sconf/20184802001

As the main campus, Depok campus covered approximately 320 ha, recognized around 70% of the area with astonishing greenery as a city forest and an ideal landscape for academic nuance of beautiful and tradition tranquil (Fig. 1). As an educational institution with a world-class university perspective, UI has a vision to create an environment-based 'green campus' to more than its 50,000 inhabitants.

**Fig. 1.** Landscape of Universitas Indonesia Depok campus as one of the greenest campus in the country

## 2 University policies on campus sustainability

UI has a strong commitment to achieve green and sustainable campus which can be seen from several Rector's Act issued on 2011, related to environmental and sustainability policy on green transportation, the use of bicycle and pedestrian, waste and toxic waste management, limitation on of toxic used for drinks and foods packaging, mitigation and adaptation of global climate change, reducing papers and plastic used, clean water conservation, and energy conservation program.

### 2.1. University landscape

In term of campus setting and infrastructure, UI maintains the present condition with continuous development and improvement. Based on the allocation of campus area plan, there are four land use type can be identified in UI Depok campus: physical building and green landscape buffer of 170 ha, water ecosystem of 30 ha, city forest area of 100 ha, and supporting facilities, infrastructure, including environment buffer area of 12 ha.

#### *2.1.1. Building and green landscape*

One of the UI commitments in creating green campus is represented by the new library building "Crystal of Knowledge" at Depok campus which was developed by adopting sustainable concept. Built on a 33,000-meter square area, the building is considered as the largest library in Southeast Asia. It is equipped with a cooling system water cooled chiller, solar panels, green-roof, and smoke-free area (Fig. 2). The library has the capacity to accommodate about 20,000 visitors per day and have collection of 1,500,000 books [4].

**Fig. 2**. The green-roof at new library building

Other building which adopted the green-building concept is the Faculty Club (FC) building. It was built in ±10 ha area as an integrated space for sports (national Olympic standard pool, tennis court, golf driving, gymnasium) meeting rooms, restaurant, etc. This building was equipped with thin film solar PV capacity of 25 kW peak, to supply university's electricity demand [5]. Since 2017, the integrated faculty club was cooperatively managed by UI and Relife Property Company (Felicity Festival).

As additional unique features to the existing green landscape and vegetation, there are about 9 big African Baobab trees aging over 200 years have been relocated and planted in UI Depok campus.

#### *2.1.2. Water ecosystem*

Apart from the beautiful and comfortable green landscape, there are series of lakes inside UI Depok campus aimed to function as the water storage and absorption/groundwater recharge area. Six lakes were built in cascade and called as KAMPUS (abbreviation of the first character of 6 species name of Indonesian trees), from upstream to downstream:

1) **K**enanga lake (28,000 $m^2$), located between Rectorate building and UI Mosque, built in 1992
2) **A**ghatis lake (20,000 $m^2$), located between Faculty of Mathematics and Natural Science and Jakarta State Polytechnic, built in 1995.
3) **M**ahoni lake (45,000 $m^2$), located aside the main street of South Boulevard (Eastern side of Faculty of Humanities and Faculty of Psychology, and western side of Faculty of Economics), built in 1996.
4) **P**uspa lake (20,000 $m^2$), located between Lake Ulin and Lake Mahoni, built in 1995.
5) **U**lin lake (72,000 $m^2$), located between Lake Puspa and Lake Salam, built in 1998.
6) **S**alam lake (42,000 $m^2$), located in lowest part series of Lake Ulin and Lake Puspa, built in 1998.

The lakes, as a natural water body are used as habitat for more than hundred species of fish and invertebrate (microorganism). It is also being used as an object for several researches on surface water quantity and quality, marine laboratory, as well as sports and sight-seeing spot.

#### *2.1.3. City forest*

Since the first stage of Depok campus development, UI inherited large area of tropical rainforest which is known as city forest. It has unique characteristics with three types of advanced ecosystem, they are: (1) Tree ecosystem from eastern part of Indonesia; (2) Tree ecosystem from western part of Indonesia; and (3) local original vegetation complex combined with Tectona Grandis/Teak Tree (so called Jati Mas in Indonesian) Forest [4].

Total area for reforestation is allocated approximately 100 ha. Beside reforestation, Sub-Directorate of Campus Environment Development (PLK) UI also promotes an enhancement planting program aimed to enriching trees variety. It is expected to reach the target of 50% of the 680 species of trees planted by 2017, since its first initiation of about 27,350 trees planted in 2013. Several events related to environmental sustainability regularly conducted, such as workshop on planting "Belimbing tree" (star-fruit tree) in vacant land inside Depok campus areas that organized by student association involving local community around UI, reforestation program by PLK in corporation with some CSR of private companies.

Since 2013 UI implemented a Green Campus program. The program aimed at creating campus as a centre of activity and empowerment of stakeholders and strategic partners in the conservation of the environment, prevent pollution and damage as well as the creation of clean and green campus. The program covers several activities supporting sustainable campus policy such as: water conservation and utilization; renewable energy and utilization; land utilization; and minimizing waste generation.

#### *2.1.4. Supporting facilities & infrastructures*

Several facilities and infrastructures that supporting the green-building and landscape inside the UI Depok campus are summarised as the following [4].

*Integrated Laboratory and Research Center* (ILRC) building was built with an aim to be the home for integrated laboratory and a center for research administration in UI. One of the examples is the development of the UI-Olympus Bioimaging Center, as the first bioimaging laboratory established in Indonesia. The ILRC building was designed in a half-circle shape, specifically aimed to receive the sun light equally distributed to each rooms of the building.

*The Makara & Art Center* building establishment is one of the UI efforts as an education institution to support the nurture of the soft skill competency of the students. The area of 11,600 $m^2$ building is located beside the UI Mosque and Kenanga lake. The building design concept is to create a magic box and it is planned to be a unique building that motivates and encourages the student's talent in arts world. This building is designed and equipped with the green chiller which is the 1st green chiller implements in Indonesia.

*Solar Air-conditioning System* are introducing to mitigate environmental pollution and reduce the greenhouse gas emissions of a building inside the university. A research collaboration in 2014 has resulted in a 239-kW solar air-conditioning system using a single-double effect combined absorption chiller. It was installed in a building at the Faculty of Engineering, Universitas Indonesia [6].

*Dormitory* is provided to first year students. The building covers an area of 25,534 $m^2$, surrounded by lakes and greenery, and accommodated 400 male students and 445 female students.

*Campus Bus* is available for free andi s dedicated to providing green public transportation which can reduce emission. In 2017, UI Depok campus maintain approximately 20 campus bus called "yellow buses" (*bis kuning*). They serve for students, staff, and public in regular routes inside the campus area from Monday to Friday, from 7:00 am to 9:00 pm and Saturday from 7:00 am to 2:00 pm.

*Campus Bicycle*, as an UI commitment and support in implementing "go-green" campaign, more than 300 bicycles were provided for free-of-charge in Depok campus. This facility is open for students and academic staff to travel from one faculty to another or buildings inside campus. The students need to show their UI student card to the rental officers who manage the bicycle shelter. The bicycle must be used on bicycle pathways and it must be returned to the bicycle shelters by 5:00 pm.

 https://doi.org/10.1051/e3sconf/20184802001

# 3 Towards green and sustainable campus

## 3.1. Managing of campus setting

Campus setting plays significant role for creating a more convenient and better academic atmosphere. On the other hand, it is undeniable that infrastructure also plays an important role in supporting various activities for any institution to function properly. Hence, better management on campus setting along with related infrastructure may significantly contribute to nurture the spirit of learning and intensifies connection of academic community to the surrounding society. The following section describes some main activities that has been carried out by Universitas Indonesia in order to create and provide open- and green-spaces, green public transport, and environmentally related programs

### *3.1.1. Maintenance of trees in the university's city forest*

UI City Forest is designated as a conservation forest area based on Jakarta Governor's Decree No. 869 in 2004. Besides the function as absorption area and research on biodiversity, the city forest also designated as a conservation area of germplasm and habitat of various wildlife. Thus, it is very much helps students who are studying tree identification. The UI Forest City UI stores approximately 186 species of trees that come from various regions in the archipelago. This makes 'UI City Forest' as the urban forest with highest variety of tree species compared to other urban forest in Jakarta. In addition, UI City Forest becomes a house for at least 30 species of birds.

However, due to the aging trees and the lightning, storm, and typhoon which are quite dominant during the rainy season, many fallen trees events were found around the campus of UI. Hence, maintenance activities and pruning of trees that prone to fall are conducted by Directorate of Facilities Maintenance and Management (DPPF) to minimize and eliminate the potential danger of fallen trees.

Identification of the potential hazard of fallen trees is carried out through inspection of the condition of existing timber trees in the main lane and surrounding building facilities. This activity results in 202 fallen-prone trees on the main lane were pruned and logged intensively.

In 2017, the maintenance activities of fallen-prone trees covers pruning (16%), logging (37%), and elimination (47%). Pruning is aimed at reducing the burden of trees and the risk of fracture of branches and twigs due to trees too dense. Logging is aimed to reduce the burden of trees and the risk of broken branches and fallen trees (broken at base). Elimination is aimed to eliminate the risk of fallen trees due to the condition of trees that severely suffered and damage due to pest diseases. Replanting and reforestation have been conducted in several locations to replace the fallen and eliminated trees.

### *3.1.2. Improvement of road and pedestrian highway*

Apart of the city forest maintenance, several improvements activities in setting and infrastructure were also been conducted (Fig. 3). The scope of this works include renovation work of road cleanliness, road markers, roads'median, pedestrian way, open and closed channels, bus stop, bicycle track, bicycle stop, parking lot, curb grass, park (all parks including those above pedestrian), pedestrian highway, drainage system (water channels), grass areas (parks), bicycle tracks, bus stops, bicycle shelters, ornamental plants throughout the campus area of UI Depok.

The idea of this improvements are to provide infrastructures that directly support green- and public-transportation which meet the standards quality. For example, the pedestrian highway were developed as much as possible to be convenience for all pedestrian including for person with disability.

(a) Renovation of the pedestrian highway

(b) Development of pedestrian road crossing pathways

(c) Rejuvenation of bicycle shelter and improvement of 120 bicycle track portal

(d) Development of additional bus stop

(e) Installment of freestanding signs at faculty's entrances

**Fig. 3**. Several improvements on public transport infrastructures

## 3.2. Developing green-building and infrastructure

In Depok campus, there are several buildings that dedicated as green-building model such as the Integrated Laboratory and Research Center (ILRC) building, new library building, Art and Culture Center, and the Integrated Faculty Club (IFC) building. As the UI commitment to save energy and support green environment, there is new regulation for any construction should adopt the green-building concept as much as possible.

### *3.2.1. Proposing LEED green building certification*

One more new green-building is being developed since 2017 known as the student activity center (Pusgiwa) as shown in (Fig. 4). The first Pusgiwa building was developed in 1989 served for only 15 students activities unit. The new Pusgiwa building within an area of 13,296 $m^2$ (GFA) is provided for 45 student activity units. The building was designated to be first building model acknowledged by Leadership in Energy and Environmental Design (LEED) green building certification.

**Fig. 4.** Development of the new Pusgiwa (students activities center) building

The new building will be used as a student spot center in addition to the existing spot center in UI new library building. The idea is to make this Pusgiwa as a place to gather, discuss, and engage students, both in the field of academic, sports, and arts. Hence, we want this new Pusgiwa area to be completely equipped in terms of facilities.

### *3.2.2. Promoting renewable energy development*

In line with the green-building concept, UI also promoting several renewable energy development. One of the massively developed is the installation solar PV at several buildings at faculties and university buildings. Those buildings are the MRPQ building Faculty of Engineering with the capacity of 2.5 kW peak and at the Integrated Faculty Club (IFC) with the of capacity 25 kW peak, etc.

In 2018, several solar PV will be developed at cluster of science of helath (RIK). The total production of solar PV installation in UI Depok campus by the end of 2018 is targeted to be approximately 140 kWP.

### *3.2.3. Monitoring energy consumption*

In 2017, the university developed a new website system aimed to monitor the energy consumption and sustainability. At this moment, the system can be accessed internally through the *sinergi1.ui.ac.id/ems_ui/*. (Fig. 5) shows the front page of the website.

By using this website, the UI management can monitor the energy consumption form each faculty in real-time (daily, hourly). For example, the red mark colour, shows the over-limit energy consumption, while the green mark colour indicates the normal energy consumption. It is expected with the new website system all information related to energy consumption and campus sustainability can be accessed by stakeholders in UI.

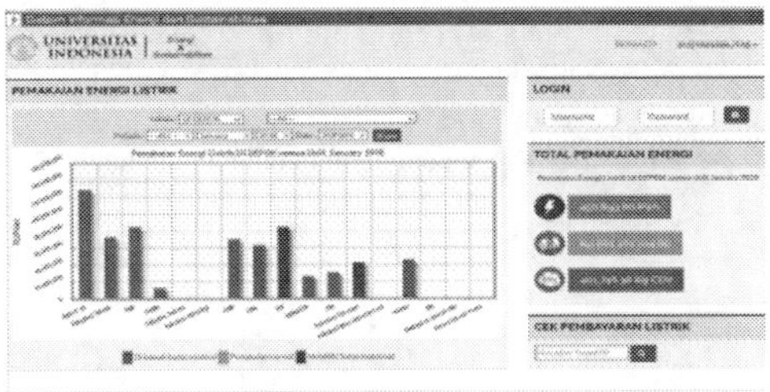

**Fig. 5**. Front page of the UI's Energy Information System.

In the sustainability program of energy, during 2017 several replacements of conventional lamps with LED type have been conducted. This will reduce the energy consumption to about 50% (the LED lamp has power consumption requirement of about 50% lower than conventional lamps with similar quality).

## 3.3. University achievement

Based on the UI GreenMetric 2017 ranking results, for setting and infrastructure category (Table 1), UI collects higher points from three main indicators; the first is the ratio of open space area towards total area; secondly, the ratio of open space area towards campus population; and the third is the university budget for sustainable effort.

**Table 1**. Universitas Indonesia' scores for setting and infrastructure based on UI GreenMetric 2017 result.

| Indicators | Score | Max. Score |
|---|---|---|
| SI1 | 295 | 300 |
| SI2 | 203 | 300 |
| SI3 | 122 | 200 |
| SI4 | 67 | 200 |
| SI5 | 70 | 300 |
| SI6 | 200 | 200 |
| Total score | 957 | 1,500 |

**SI1**: the ratio of open space area towards total area; **SI2**: the ratio of open space area towards campus population; **SI3**: area on campus covered in forest; **SI4**: area on campus covered in planted vegetation; **SI5**: area on campus for water absorbance; **SI6**: university budget for sustainable effort.

https://doi.org/10.1051/e3sconf/20184802001

With a total score of 957 for setting and infrastructure category (overall score for 6 categories of UI Greenmetric was 6,519), UI stays at rank #23 [7].

If we look more detail into this category, compared with the results in the last four years (Fig. 6), there are decreasing score for the area on campus covered in forest (SI3); the area on campus covered in planted vegetation (SI4); and the area on campus for water absorbance (SI5). These three decreasing scores is might be due to the effect additional of new participants of UI Greenmetric in 2017 (about 100 universities) which have better performance of setting and infrastructure in their respective campus. This is because the calculation algorithm for scoring of this category/indicators is by normalising the score based on the maximum score in regards to all participants [8]. In addtion, it obvious due to development of several buildings in UI Depok campus (university hospital, art center building, cluster for health and science building, integrated faculty club building, etc.) brought consequences to reduce the green open spaces and water absorbance area.

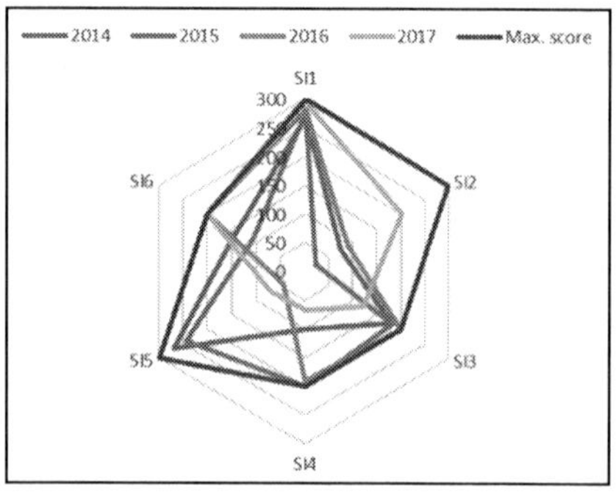


**Fig. 6.** The setting and infrastructure indicator's score of UI for the last 4 years.

However, for the other indicators (SI1 and SI6), UI has approximately reach the maximum scores. The setting and infrastructure category in UI GreenMetric contributes only 15% of overall score. Hence, the score for this category is not the merely significant factor in determining the final ranking result, because there are other categories such as energy and climate change (21%), waste management (18%), transportation (18%), education (18%), and water management (10%) that should be considered for improvement [9].

## 4 Concluding remarks

With regards to the Strategic Plan 2012-2017, Universitas Indonesia will continue to realizing the program and policy particularly in setting and infrastructure sector to reach the university vision and mission in creating sustainable campus. For sure, to realizing a green and sustainable campus with eco-friendly based education infrastructure (eco-science-park) required strong commitment and supports from all faculties members, staff and students. This commitment includes university budget for research and sustainability operation which are also a significant issue in the near future. It is time to put priority of university policy on sustainability of each development sectors which finally contribute to the global trends in making the SDGs university report.

**Acknowledgement.** We would like to express our thanks to Directorate of Facilities Maintenance and Management (DPPF) Universitas Indonesia and UI GreenMetric Rankings office for providing relevant data and information.

## References

1. The Climate Reality Project, 24 Hours of Reality: Resource Kit, Available online at https://www.climaterealityproject.org/content/24-hours-reality-resource-kit (1989)
2. QS World University Rankings 2016/2017, Available online at https://www.topuniversities.com/university-rankings/world-university-rankings/2016
3. The Times Higher Education World University Rankings 2015-2016, Available online at https://www.timeshighereducation.com/world-university-rankings/2016/world-ranking#!/page/0/length/25/country/16/sort_by/rank_label/sort_order/asc/cols/rank_only
4. Suwartha, N., Sari, R.F., Widanarko, B., Rarasati, A., Zakaria, A.A., *Towards Sustainable University through Setting and Infrastructure Development at the Universitas Indonesi, In*: Proceeding of the 2nd International Workshop on UI GreenMetric (IWGM 2016), Universitas Indonesia, Indonesia (1989)
5. Suwartha, N., Sari, R.F., Widanarko, B., Ilyas, T., *Cultivating Green energy at the Universitas Indonesia Towards Sustainable Campus, In*: Proceeding of the 3rd International Workshop on UI GreenMetric (IWGM 2017), Istanbul University, Istanbul, Turkey (2017)
6. Hajime Yabase, H., Saito, K., Lubis, A., Alhamid, I., Nasruddin, *Solar Air-Conditioning System at the University of Indonesia*, International Journal of Technology, **7(2)**, pp. 212-218 (2016)
7. UI GreenMetric, UI GreenMetric World University Rankings, Available online at http://greenmetric.ui.ac.id/overall-ranking-2017/ (2017)
8. Suwartha, N., Sari, R. F., *Evaluating UI GreenMetric as a tool to support green universities development: assessment of the year 2011 ranking*, Journal of Cleaner Production, **61**, pp. 46-53 (2013)
9. Lauder, A., Sari, R.F., Suwartha, N., Tjahjono, G., *Critical review of a global campus sustainability ranking: GreenMetric. Journal of Cleaner Production*, **108**, pp. 852–863A (2015)

# The challenges of adopting BIM for setting and infrastructure management of University of Minho

*Paulo J. S.* Cruz[1,*], and *Miguel* Azenha[2]

[1]Lab2PT, University of Minho, Pro-Rector, 4800-058 Guimarães, Portugal
[2]ISISE, University of Minho, Dep. of Civil Eng., School of Engineering, 4800-058 Guimarães, Portugal

**Abstract.** The University of Minho (UMinho) is aware of the relevant role of higher education institutions in the global challenge of sustainability. The integration of sustainability values into its strategy encompasses all the fields of its activity. Taking into account the growing importance of shared knowledge resources as a reliable basis for decisions during infrastructures life cycle, the UMinho is currently undergoing an important process of implementing a 'Building Information Modelling' framework (BIM). This paper briefly describes the implementation steps that are current towards such a goal, in particular in terms of facilities management aspects.

## 1 Introduction

UMinho is currently one of the most prestigious Portuguese higher education institutions and it is gradually becoming more prominent internationally. It was founded in 1973 and offers a wide range of courses corresponding to first, second and third cycles of studies leading to undergraduate, graduate or doctorate degrees. The UMinho has approximately 20.000 students, 1.300 professors and 900 non-teaching and non-research staff, which represents a population higher than 70% of the 308 Portuguese municipalities.

The sustainability policy adopted by UMinho is defined in the strategic plan of UMinho 2020. It contains the administration, management, operational management, educational activity, research activities, the dissemination of its performance and its impact which is based on a holistic vision, a fundamental element to ensure life quality levels in the campuses. UMinho's endeavours regarding Environmental, Social, Economic and Cultural dimensions are carried out, on a regular basis, according to the best international practices. This paper should also be, on the one hand, a tool for communicating strategy and the paths taken and an instrument for demonstrating and inducing new behaviours and, on the other hand, a way to promote new partnerships with other institutions that share the same view as UMinho's.

In 2017 the UI GreenMetric World University Ranking awarded the UMinho with the 1st position in Portugal and with the 48th world position among 619 universities from 76 countries. This recognizes UMinho's commitment in developing 'environmentally friendly' infrastructures and a relevant score in the six specific indicators that compose the ranking (setting and infrastructure, energy and climate change, waste, water, transportation and education).

In line with the sustainability goals stated above, this paper presents the strategic roadmap that the UMinho has established towards a more effective and sustainable process of infrastructure management, through the implementation of 'Building Information Modelling' techniques.

## 2 The UMinho infrastructures

Teaching and research activities take place in the three UMinho campuses: Gualtar Campus (Fig. 1), in the city of Braga, and the Azurém Campus (Fig. 2) and Couros Campus, both in the city of Guimarães.

The Schools of Sciences, Health Sciences, Law, Psychology and Nursing, as well as the Institutes of Human and Social Sciences, Education, Arts and Human Sciences, and part of the School of Engineering are established on the Gualtar campus, where there is also a set of services, such as a general library, several specialised libraries, a canteen, a university restaurant, seven cafeterias, a medical centre, a sports complex and two copy centres. It is the largest campus in terms of total implantation area (45 hectares), gross floor area (152.115 $m^2$) and number of buildings (21). It is also the campus where the largest number of students attend their classes or other academic activities (approximately 2/3 of the total number).

The Azurém Campus is located in Guimarães, next to the city's historical centre. The School of Engineering is located on this campus, where most of its study cycles are taught. The School of Architecture and some departments of the School of Sciences and of the Institute of Social Sciences are also on this campus. It also includes a set of services, such as a general library, several specialised libraries, a canteen, a grill, three

*Corresponding author: pcruz@reitoria.uminho.pt

cafeterias, a medical centre, a sports complex and a students' residence. It is the second largest campus in what concerns the total area of implantation (30 hectares), the gross floor area (89.231 m2) and the number of buildings (20).

**Fig. 1.** The Gualtar Campus

**Fig. 2.** The Azurém Campus

Couros is the southern area of the historic centre of Guimarães. In 1996 the Town Council of Guimarães initiated the rehabilitation of this industrial centre, and in 2001 the Multifunctional Complex of Couros was born. Within this context, and in order to strengthen UMinho's integration in the city, the Town Council of Guimarães and the University developed the Campurbis project (2006-2012). The intervention and rehabilitation of the former factories led to the creation of a new university space -the Couros campus- which, given its location on the city outskirts, emphasises the concepts of 'virtual campus' and 'university without walls', which are ingrained in UMinho's mission and strategy.

The infrastructure asset of the UMinho includes a significant number of buildings with quite diverse characteristics and ages and, consequently, significantly different maintenance needs.

## 3 Building information modelling: a brief context in EU, Portugal and UMinho

It is relevant to start contextualization with a definition of BIM, which briefly frames the topic for those not fully aware of its extent. According to the definition supported by UMinho [1], 'Building Information Modelling' is a methodology for information sharing and communication amongst all the stakeholders and during all the stages of the lifecycle of a given construction. It is normally supported by a digital model, accessible by software, which allows the virtual manipulation of the construction in concern. This digital model contains data about the geometrical features of the elements that compose the construction (e.g. beams, columns, electric plugs, installed equipment), as well as their properties and attributes, regardless of their nature, including physical properties, cost, construction timing, among others.

Even though BIM covers all stages of the lifecycle of construction, the scope of concern in this paper is the stage of operations and infrastructure management, normally termed as BIM-FM (Facilities Management) [2]. BIM-FM takes advantage of the extensive quantity and quality of information available in an as-built BIM model in order to assist operations and maintenance management of the building throughout its entire service life. The well-structured information available in the BIM Model can assist in several operational aspects simultaneously, such as: streamlined maintenance (e.g. through the integration of asset information and predictive/corrective work orders), improved space management, efficient use of energy (e.g. monitoring use of rooms/classrooms and even air conditioning or central heating), up-to-date available information for retrofits and renovations; enhanced lifecycle management (including improved prediction capacity of needs). Even though these aspects were already addressed by 'Computer Aided Facility Management' (CAFM) [3], their potential has been truly enlarged by BIM [4].

Nowadays, many world and chiefly EU countries have made BIM mandatory for their public procurement building projects, with private owners/investors following suit, the requirement for BIM usage and BIM experts is increasing dramatically. BIM progress can be felt strongest in northern and central EU countries which are rapidly developing their own BIM standards and specifications such as UK's BS and PAS documents while the European Committee for Standardization is promoting general CEN/TC 442 BIM standards. EU BIM awareness is further promoted by the EU BIM Task Group which is bringing together national efforts into a common and aligned European approach to develop a world-class digital construction sector.

Even though Portugal has not yet set out mandatory BIM use, there are many owners becoming aware of the inherent advantages and making BIM mandatory in their procurement processes. One of the main recent contributors for BIM adoption and spreading in Portugal is the National Technical Committee CT197 on BIM (www.ct197.pt, where UMinho has an active participation), which has successfully issued a BIM procurement guide (Employers Information Requirements) [5], and is currently finalizing the draft of a national BIM Execution Plan template.

It is also worth to mention the extensive experience of the Civil Engineering Department from UMinho in BIM, which was the first in Portugal to start teaching BIM as an independent Curricular Unit back in

2012/2013, and continues today [6], in several study cycles. The UMinho also leads the National Course of Building Information Modelling at the National Engineers Association since 2014, with 6 successful past editions and more than 250 professionals trained (www.cursobim.com). Furthermore, the UMinho has hosted the first National Conference of Building Information Modelling PTBIM in November 2016, and it now leads the scientific committee of the 2nd PTBIM to take place in May 2018.

## 4 A strategic roadmap for BIM-FM implementation at UMinho

The challenges and opportunities of bringing infrastructure management towards a fully digital BIM based approach are numerous. Firstly, it is essential to define information and details to be included in the models. After that, it is necessary to make the digital models of the buildings and infrastructures, potentially assisted by advanced techniques, such as laser scanning. This situation is being tackled with an ambitious full campus digitalization plan. The opportunities being brought by a database of materials used in the Campuses can also represent an important keystone at the level of the analysis of buildings from a lifecycle management point of view.

Therefore, bearing in mind the potential benefits and the hurdles to overcome in order to implement BIM in the operations of UMinho, a 36-month plan has been drafted, beginning during the first semester of 2018. The outline of the main stages of the process is given below.

### Stage 0 - Diagnosis and human resource assignment - M1-M3

- Work meetings with the current Infrastructure Management team of UMinho, as to establish an accurate description of current procedures, available information, further needs, human and material resources, etc.
- Definition of potential needs to hire a person for the implementation stage (e.g. with a grant), or alternatively basing efforts on the currently existing staff at UMinho (situation in which the exclusive dedication of one staff member with BIM competences is necessary).
- Clear definition of a 33-month implementation plan, according to the diagnosis of challenges and resources (human and financial). The plan to be drafted will correspond to an adapted version of the ideas forwarded in Stages 1-4 of the current planning (shown below).

### Stage 1 - Definition of modelling rules; pre-implementation of pilot model - M4-M6

- Careful study of the guidelines to be applied in terms of surveying and modelling, as well as modelling software and 'Common Data Environment' definition.
- Final definition of the information required in the models, as well as the protocols for their visualization (in the case of non-specialists) and updating. Attention will be also driven to 'maintenance orders' and management of 'predictive/corrective maintenance', as well as stock management for quick intervention (e.g. lamps, projectors, etc.).
- Creation of a small pilot demonstrating model to show to all potential users, with a workshop to be held at M6.

### Stage 2 - Modelling 3 buildings together with practical implementation of BIM-FM - M7-M12

- There will be selected three buildings to demonstrate the feasibility of the plans/discussions laid out in Stages 0-1.
- These models will begin to be used and tested since M8-M9, and thus difficulties and problems shall be identified. Consequently, improvements and adaptations will be made to FM procedures and the modelling rules themselves.
- Production of specific UMinho documentation for future reference:
  - Drafting of the strategic BIM Plan for UMinho, with larger extent that BIM-FM alone
  - Definition of relevant 'Employer Information Requirements' for procurement processes in new constructions to be built within UMinho (based on the documentation issued by CT197-BIM).
  - UMinho BIM Guidelines for surveying and modelling BIM Management protocols

### Stage 3 - Integral modelling of UMinho Campuses and extension of procedures - M13-M24

- Based on the rules and protocols defined in Stage 2, the modelling of all Buildings of Gualtar and Azurém Campuses will be promoted. Both internal and external resources will be deployed. The duration of this process will strongly depend on the human and financial resources that may be invested.
- Increase the possibilities of management and attaining added sustainability through pilot implementation of IoT sensor networks in key buildings (temperature, humidity, fire, presence) as to support the next steps of integrated digitization of the Campuses.
- Potentiate the possibilities of generalized use of the information contained in the BIM models through direct interfaces in webpages (e.g. management of room reservation, inventory inputs from users, etc.). This stage will request specific support of UMinho specialists in the fields of Computer Science and Information Systems.
- Internal training programmes to all relevant stakeholders within the UMinho community. Dissemination activities both within and outside UMinho.

**Stage 4 - Consolidation and integrating demonstrators - M25-M36**

- Stage of demonstration of the integrated concepts, with close intervention and monitoring by all the stakeholders in the process. Definition of adaptations and improvements, as well as their implementation.

## 5 Concluding remarks

Based on the consistent background and knowledge that the UMinho has in 'Building Information Modelling', a practical roadmap towards the full BIM digitalization of the infrastructure management process has been described it this paper. Indeed, the integration of databases that the digital BIM model provides is likely to bring significant advantages over traditional processes of infrastructure management (even those supported in digital databases that are difficult to integrate). In fact, the cross-linking of information will provide unique opportunities on both day-to-day operations (e.g. situation reporting), but also on strategic planning of predictive maintenance and other features of infrastructure management.

After an initial diagnosis stage, the process of BIM implementation formally involves four stages related to the definition of modelling rules/requisites (Stage 1), the preparation of three prototype buildings to test the concepts (Stage 2), the enlargement/generalization of the process towards the Campuses of the UMinho (Stage 3) and a final stage of consolidation and improvements (Stage 4). The entire process is expected to take 36 months, which is a duration considered feasible for the ambitious goals that have been set.

## References

1. Azenha, M., J.C. Lino, B. Caires, *BIM in Civil Engineering: design and construction*, Lecture notes (in Portuguese), Integrated Master in Civil Engineering, UMinho (2018)
2. Eastman, C., Teicholz, P., Sacks, R., Liston K., *BIM Handbook: a guide to Building Information Modelling for Owners, Managers, Designers, Engineers and Contractors*. John Wiley & Sons, Inc. Hoboken, New Jersey USA (2011)
3. Roper, K., Payant, R., *The Facility Management Handbook* ($4^{th}$ Edition), McGraw-Hill Education (2014)
4. Teicholz, P., *BIM for facility Managers*, Wiley (2013)
5. Costa, A.A., Matos, B., Drumond, D., Rodrigues, I., *Guia de Contratação BIM*, CT197 BIM (2017)
6. Azenha, M. J.C. Lino, Couto, J.P., *BIM Teaching Strategies in Civil Engineering at the UMinho, In:* the Proceedings of EDUBIM 2017, CESI Centre de Paris-Nanterre (2017)

# The University of São Paulo on the 2017's GreenMetric Ranking

*Patricia* Faga Iglecias Lemos*, *Fernanda* da Rocha Brando, *Paulo* Almeida, *Roberta* Consentino Kronka Mülfarth, *Tamara* Maria Gomes Aprilanti, *Luis* Otávio do Amaral Marques, *Nayara* Luciana Jorge and *Tadeu* Fabrício Malheiros

Superintendence of Environmental Management, University of São Paulo, SP, Brazil

**Abstract.** The University of São Paulo is a public University and the largest in Brazil. It is broadly organized into distinct campuses that are installed in different cities of the state of São Paulo, each one with its specific aspects. There are more than 95 thousand undergraduate and graduate regularly enrolled students, standing out as one of the best universities in Brazil and Latin America [2]. The performance of the University of São Paulo at the "UI GreenMetric World University Ranking" has improved considerably in 2017 (28th position) on the general ranking. The sustainability management process for USP is carried out considering the strengths each campus presents, but the USP governance process ensures standards and good practices are applied similarly to all areas under the university responsibility. As for example, the recent approval of the USP Environmental Policy, which process engaged the whole USP community, working together to understand the complex context of all campuses, what resulted at the end in very innovative arrangements for researching, teaching and practice for sustainability. Different of the past rankings, data provided for GreenMetric 2017 Ranking contemplated USP as a whole, because it is not possible to assess the university environmental sustainability performance in parts. This approach allowed showing the different enhanced efforts of the University as a whole. In addition, the USP's environmental policy-making process brought about more integration between the campuses and effective actions were put into practice, helping to consolidate USP as a sustainable university. The highlight in the score was "Setting and Infrastructure": there are 76.4km² of total area and only 2.6% from them are built-up. The University has been prioritizing the protection of green areas and biodiversity, in São Paulo city campus, totally surrounded by urban environment, but also in the other campuses in the countryside, some of them located in rural environment. This shows that USP's campuses are environmentally responsible, providing students and staff with enjoyable, healthy and in close touch with nature places. On the topic "Transportation", the University of São Paulo stands out also. Incentives to improve choice for public transportation and biking within the university community are among the constant concern of the University of São Paulo, which will be keeping up efforts for that.

## 1 Introduction

One of the main concerns of the generations living in the 21st century is called *sustainable development*. This term refers to the aim to meet the current society needs without damaging the wellbeing of future generations and ecosystems resilience. For that, it is essential that attitudes and policies in the social, economic and environmental spheres be implemented not only by the people in their daily lives, but also by institutions and companies.

In this context, it is important that each institution of higher education act in order to encourage their students, researches and employees to acquire and maintain sustainable behavior in their routines. An effective way to bring them up to this is by adopting exemplary and effective policies and practices that contribute to the environment and society.

One way to measure and provide transparency to the universities' efforts in this regard is through sustainability reports and rankings. This evaluation and classification, based on sustainable criteria, serves as an incentive for more specific analysis of the aspects in which it needs to get better as well as the maintenance of the current policies in their strengths aspects. In addition to that, the comparison between many universities around the world can result in successful actions being implemented in different countries and thus achieve greater range and efficiency in terms of sustainability.

The *UI GreenMetric Ranking* is an international ranking that annually shows the sustainability and environmental management efforts of higher education institutions around the world. In its eighth edition, 619 universities from 75 countries located on 5 continents were evaluated. The results can be seen by country, by region, by campus setting and overall.

In this context, this paper aims at presenting some of the good practices University of São Paulo has been put to work in the last years.

*Corresponding author: sga@usp.br

 https://doi.org/10.1051/e3sconf/20184802003

## 2 The sustainability in the context of the University of São Paulo

### 2.1 Description

The University of São Paulo is an international academic community that is broadly organized into 12 distinct campuses installed in 7 different cities of the state of São Paulo. Five of them operate in city of São Paulo and the other campuses operate in 6 other localities in Brazil, each one with its own specific aspects and colleges. The University has 2 more units on the coast (Santos and Ubatuba cities), where studies about marine biology and petroleum engineering are developed.

The total area of USP is about 76.40 km² and its built-up area is about 1.99 km². In 2017 there were almost 59 thousand undergraduate students enrolled in the 309 available courses; 37 thousand postgraduate students enrolled in the 265 available programs; 6 thousand faculty members and 16 thousand technical and administrative staff [1].

The University of São Paulo was responsible in 2016 for 13% of the total brazilian PhD students. Besides that, 77,813 Masters and 50,388 PhD have been awarded since the university was founded.

### 2.2 USP's performance in 2017

The University of São Paulo reached the 28th place overall, and the first among the Brazilian Universities in the 2017 GreenMetric Ranking. The highlight was in "Setting and Infrastructure": only 2.6% of its total area are built-up (1.99 km²). In addition, it also stood out in the themes "Energy and Climate Change", "Water" and "Transport", in which the University of São Paulo reached the six first places of the 64 participating universities from Latin America.

The combined efforts of the entire university community as well as the integrated approach of its several campuses were responsible for this good standing. Besides that, the current process of formulating USP's environmental policies brought more integration between campuses and effective actions were put into practice, helping to consolidate USP as a University that has Sustainability as its goal. In 2012, the Superintendence of Environmental Management - SGA was launched as a USP's agency responsible for the environmental and sustainability institutional relations of the University with federal, state and municipal public institutions [6, 10].

### 2.3 USP's environmental policy

In order to promote the integrated environmental management and aiming to guide and legitimize the environmental actions of the University of São Paulo, the University Environmental Policy Working Group was launched in 2014, pushing the process of the USP's Environmental Policy. Their main *modus operandi* included interdisciplinary approach, transparency, participation and respect for local specificities [7]. The USP's Environmental Policy was enacted in January 2018 and covers 11 themes such as: water and effluents, land use, sustainable buildings, energy, green areas and ecological reserves, mobility, greenhouse gases emission, fauna, administration, solid waste and environmental education [8]. The policy also provides a basis for the formulation of the Environmental Management Plan, and the Environmental Master Plans.

USP's Environmental Policy was elaborated in three main phases: the first, the thematic environmental policies were defined for all university campuses, each one with its different aspects. In this stage 150 volunteers (teachers, technicians and students) were involved, divided into several working groups. In the second (the current one), the environmental management plans are being drawn up with goals and indicators, with the engagement of 170 volunteers from the USP community. The third aims at the Environmental Master Plans to be developed for each of the campuses [9].

### 2.4 The USP recycling program

The USP Recycling Program (*USP Recicla*) is the oldest and main program, and targets Environmental Education and Waste Management. The program's principles are participation, belonging, empowerment, autonomy, environmental technologies and the 3Rs principle: reduction, reuse and recycling [5]. This program is responsible for the USP sustainability culture index improvement.

USP has a Disposal and Reuse Center for Electronic Waste (CEDIR), which redistributes part of the equipment that can still be used by assistance entities, through a loan. The remaining is disassembled, and its parts are separated and sent to recyclers [13]. There is also the Recicl@tesc Program, in São Carlos campus, launched in 2009, whose objective is the environmentally correct disposing of electronic waste [14].

### 2.5 USP infrastructure

Most of the campuses of the University of São Paulo are located in urban areas and the main campus (in São Paulo City) has the largest built-up area (almost 2 billions m²). The largest total areas belong to the Piracicaba Campus (49%), followed by the Pirassununga Campus (31%), which are incrusted in urban areas, but kept as rural campuses.

#### *2.5.1 Green areas*

The University of São Paulo established 2,312 ha of its total area as Ecological Reserves, aiming the conservation, restoration, research, teaching and extension. Most of these natural reserves are fragments of semideciduous forests and tropical savanna [12]. It is important to highlight the importance of the green areas in the USP campuses in the city of São Paulo, once today this city has a significant deficit of vegetation [11].

*Corresponding author: sga@usp.br

**Fig. 1.** City of São Paulo Source: Tadeu Malheiros

**Fig. 2.**. Green areas of USP - Campus Cidade Universitária - São Paulo.

### *2.5.2 Sustainable buildings and energy*

The University has invested in its lighting system, which now uses high-tech LED luminaire HP3, solar panels with excellent performance even in cloudy weather, with an autonomy from 1 day to 3 nights and the intelligent remote management system. The system of remote control of lighting consists of monitoring the points by means of communication antennas installed in each of them, which will transmit, through a georeferencing software, data regarding consumption and operation, that will allow the monitoring of the network at a distance, to control operation, to program drives and lighting levels by location or time.

**Fig. 3.** High-tech luminaire system

The system also provided a reduction of more than 30% in electricity consumption, even with almost double luminaires (from 3,200 to 6,113 units) [4], since USP's total energy consumption between October 2016 and September 2017 was 61,068 MWh, which represents a lower consumption compared to the previous months.

### *2.5.3 Mobility*

There are 50 available free shuttle buses operating at the University campuses and more than 5,000 bicycles transiting at USP daily. The campuses located at the city of São Paulo have easy access with bus and subway. The other campuses are mostly accessed by walking distances for the local students [3].

## 3 Concluding remarks

The management of a university with the size of the University of São Paulo is certainly a challenge, as it requires developing governance mechanisms that encompass teaching, research, and extension issues. The sustainable development movement of the second half of the last century brought new and urgent responsibilities for alignment with the principles of sustainability. USP was an active part of this movement, both inside and outside Brazil, and has progressively incorporated the culture of sustainability into its DNA. From speech to practice, today USP invests in a campus that dialogues with its community, with its open doors policy in the cities where it is installed is a key actor and contributes to opening opportunities to society. The environmental issue occupies a prominent place on its campuses, where biodiversity protection and significant amount of green areas contribute to local and global quality of life. Its environmental policy, built on a participatory basis, incorporates and translates its degree of maturity and commitment to the Sustainable Development Goals.

## References

1. USP - Universidade de São Paulo, Anuário Estatístico USP. São Paulo, Brazil (2016)
2. USP, Relatório de Gestão: Uma Universidade em Evolução | 2014-2017, São Paulo, Brazil (2017)
3. Superintendência de Gestão Ambiental, Iniciativas Ambientais na USP (2018) Disponível em: <http://www.sga.usp.br/acoes-da-sga/>. Acesso em 02/2018.
4. Alper Energia, USP. Disponível (2016) em: <http://www.alper.com.br/project/usp/>. Acesso em 02/2018.
5. USP, Portaria GR Nº 5438, de 22 de dezembro de 2011, São Paulo, Brazil (2011)
6. Superintendência de Gestão Ambiental, Áreas verdes e reservas ecológicas (2018) Disponível em: <http://www.sga.usp.br/grupos-de-trabalho-da-sga/politica-ambiental-na-universidade/>. Acesso em 02/2018.
7. Superintendência de Gestão Ambiental, Política Ambiental na Universidade (2018) Disponível em: <http://www.sga.usp.br/grupos-de-trabalho-da-sga/politica-ambiental-na-universidade/>. Acesso em 02/2018.
8. USP, 2018. Relatório de Gestão (2014-2017). São Paulo, Brazil.
9. USP, Environmental Policy University of São Paulo: Sustainability Masterplan for Policies, Plans, Goals and Actions. São Paulo, Brazil (2016)

10. USP, Resolução nº 6062, de 27 de fevereiro de 2012 (2012)
11. Prefeitura do Município de São Paulo, "Atlas Ambinetal" do Município de são Paulo: Fase I: Diagnóstico e Bases para a Definição de Políticas Públicas para as Áreas Verdes no Município de São Paulo. São Paulo, Brazil (2002)
12. Superintendência de Gestão Ambiental, Reservas Ecológicas da USP (2018) Disponível em: < http://www.sga.usp.br/acoes-da-sga/reservas-ecologicas-da-usp/>. Acesso em 02/2018.
13. USP, Projeto Recicl@tesc coleta lixo eletrônico (2012) Disponível em: <http://www.saocarlos.usp.br/index.php?option=com_content&task=view&id=9529&Itemid=171>. Acesso em 02/2018.
14. USP, Cedir da Esalq recebe descarte de informática da população (2013) Disponível em: < http://www5.usp.br/26894/cedir-da-esalq-recebe-descarte-de-informatica-da-populacao/>. Acesso em 02/2018.

# Green@Universiti Putra Malaysia: cultivating the green campus culture

*Ahmad Zaharin* Aris[1, *], *Zakiah* Ponrahono[1], *Mohd Yusoff* Ishak[1], *Nor Hazlina* Zamaruddin[2], *Nor Kamariah* Noordin[2], *Renuganth* Varatharajoo[3], and *Aini* Ideris[4]

[1]Faculty of Environmental Studies, Universiti Putra Malaysia, 43400 UPM Serdang, Selangor, Malaysia
[2]Corporate Strategy and Communications Office, Universiti Putra Malaysia, 43400 UPM Serdang, Selangor, Malaysia
[3]Office of the Deputy Vice Chancellor (Industry and Community Relations), Universiti Putra Malaysia, 43400 UPM Serdang, Selangor, Malaysia
[4]Office of the Vice Chancellor, Universiti Putra Malaysia, 43400 UPM Serdang, Selangor, Malaysia

**Abstract.** This article highlights the University's strategic plan and institutionalization commitment in cultivating the green campus culture encircling the Green@UPM initiative. It presents a range of insights, lesson learned and way forward of UPM green strategies in continuously nurtures the environmental-friendly campus atmosphere.

## 1 Introduction

A dynamic sustainability process in university require a holistic approach on university's teaching, research, operations and daily campus life. Campus community; individually and collectively roles in various environmental efforts are vital in engaging with the university's green systematic transformation fundamental. A comprehensive green campus mission aimed at achieving sustainability will indirectly create a new generation of socially and ecologically responsible citizens. In its role as a higher learning institution at the forefront of the sustainability movement, UPM is fully committed to alleviate and mitigate adverse environmental impacts. UPM is committed to provide services and to be developed based on the principles of sustainable development and responsible in raising the quality of life and ensure continuity of universal life (Fig. 1). In terms of campus environment, Sustainable Development Goals (SDGs) defines as fundamental that explains the importance of conserving ecological balance by reducing the depletion of natural resources. Comprehensively, SDGs adaptation into university management and operation, maneuver as the fundamental approach to addresses, involves and promotes, on the minimization of negative environmental, economic, societal, and health effects. It nourishes the transition of sustainable lifestyle of the campus society through the activities of teaching, research, outreach and partnership.

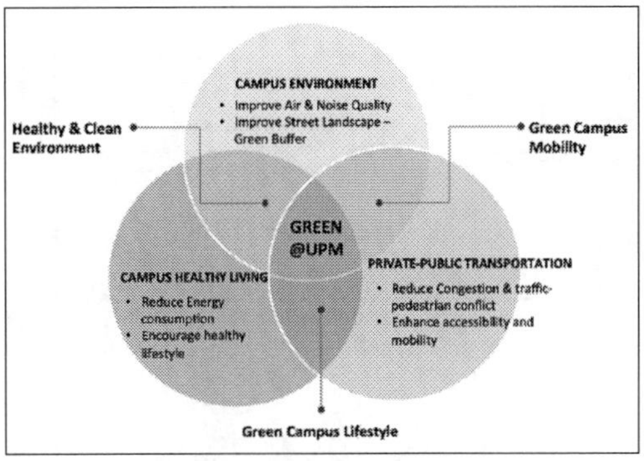


**Fig. 1.** Green@UPM fundamental goals.

The Green@UPM strategic plan to meeting SDGs is outlined to reflect on university contributions on achieving the sustainable campus environment. Innovative projects aligned with the fundamentals and guidelines of sustainability and in connection with the SDGs are comprehensively planned to build campus-wide green campus culture to addressing the issues arising from the imperative for university development, management and operation. Contribution of Green@UPM strategic plan to meeting SDGs is presented in Table 1 that evidently maximizes the survival and expansion of pioneering initiatives.

*Corresponding author: zaharin@upm.edu.my

**Table 1.** Contribution of Green@UPM strategic plan to meeting SDGs.

| Sustainable Development Goals (SDGs) | Contribution of Green@UPM Strategic Plan to Meeting SDGs |
|---|---|
| *Goal 1.* End poverty in all its forms everywhere | •Centre For Management of Waqf, Zakat and Endowment (WAZAN)<br>•MySadaqah™@WAZANdiHati<br>•Sichuan Rural Poor- Household Biogas Programme of Activities" (CDM PoA 2898, Gold Standard 1239)<br>•PadiU Putra Technology |
| *Goal 2.* End hunger, achieve food security and improved nutrition and promote sustainable agriculture | •e-Rakan Bakti Putra<br>•UPM Freemeals Program<br>•Sichuan Rural Poor- Household Biogas Programme of Activities" (CDM PoA 2898, Gold Standard 1239)<br>•Institute of Agricultural and Food Policy Studies<br>•Institute of Plantation Studies (IKP)<br>•Halal Products Research Institute<br>•Institute of Tropical Agriculture and Food Security (ITAFoS) |
| *Goal 3.* Ensure healthy lives and promote well-being for all at all ages | •Campus Car-Free-Zone Program<br>•Greener than Green program<br>•Cleaner than Clean program<br>•Carbon Footprint Reduction program<br>•Family, Adolescent and Child Research Centre Of Excellence (FACE)<br>•MALAYSIAN RESEARCH INSTITUTE ON AGEING (MyAgeing) |
| *Goal 4.* Ensure inclusive and equitable quality education and promote lifelong learning opportunities for all | •UPM Green Wave Roadshow<br>•Accepts students from IPTA who are interested to undergo industrial trainings, year-end projects (undergraduates) and research projects in Marine Science Centre<br>•Serdang Spirit Program<br>•Professors expert in environmental field<br>•Putra Science Park<br>•Faculty of Environmental Studies |
| *Goal 5.* Achieve gender equality and empower all women and girls | •Sichuan Rural Poor- Household Biogas Programme of Activities" (CDM PoA 2898, Gold Standard 1239)<br>•Women Bus Coach<br>•Putra Bakti Program<br>•Women Political Participation Research<br>•Cancer Education and Research Project (CaEd) |
| *Goal 6.* Ensure availability and sustainable management of water and sanitation for all | •Greener than Green program<br>•Cleaner than Clean program<br>•Carbon Footprint Reduction program<br>•Green landscape rainwater harvesting<br>•River of Life @ Serdang programme<br>•Environmental Forensics Research Center |
| *Goal 7.* Ensure access to affordable, reliable, sustainable and modern energy for all | •UPM Biofore Company<br>•Biocompost Pilot Plant<br>•Bioprocessing and Biomanufacturing Research Center<br>•Fermentation Technology and Bioreactor System |
| *Goal 8.* Promote sustained, inclusive and sustainable economic growth, full and productive employment and decent work for all | •Green Sustainability Steering Committee<br>•3R Urban Garden Concept<br>•PadiU Putra Technology |
| *Goal 9.* Build resilient infrastructure, promote inclusive and sustainable industrialization and foster innovation | •TN50 Dialogue Series: The Aspiration of Green Technology Prospects<br>•MyCOMS Electric Vehicle<br>•Institute of Bioscience |
| *Goal 10.* Reduce inequality within and among countries | •Rebrand all green activities at main campus and branch campus - including making existing program improvements<br>•International Green Research Collaboration |
| *Goal 11.* Make cities and human settlements inclusive, safe, resilient and sustainable | •Recycle@U<br>•Redcube Coca Cola Recycle Center |
| *Goal 12.* Ensure sustainable consumption and production patterns | •Putra Green Campus<br>•Green Car Parking<br>•MyCOMS Electric Vehicle<br>•PadiU Putra Technology |
| *Goal 13.* Take urgent action to combat climate change and its impacts | •UPM Waste Bank<br>•Greener than Green program<br>•Cleaner than Clean program |

 https://doi.org/10.1051/e3sconf/20184802004

| | |
|---|---|
| | •Carbon Footprint Reduction program |
| *Goal 14.* Conserve and sustainably use the oceans, seas and marine resources for sustainable development | •UPM Marine Science Center (COMAS)<br>•Produces authentic live feed stock (*Nannochloropsis oculata*, *Chaetoceros aureus*, Spirulina spp, *Chlorella vulgaris*) to be used in PSM and spawn of lobsters<br>•Held workshops regarding lobsters together with researchers from Universiti Malaya, Queensland University of Technology, Australia, Universiti Selangor, University of Fiji, Department of Fisheries and UPM |
| *Goal 15.* Protect, restore and promote sustainable use of terrestrial ecosystems, sustainably manage forests, combat desertification, and halt and reverse land degradation and halt biodiversity loss | •Maintain existing UPM Green Policy and make UI-GreenMetric policy and indicators mapping<br>•Coordinate green activities and agricultural visibility integratively strategically<br>•Sultan Idris Shah Forest Education Center (SISFEC)<br>•Institute of Tropical Forestry and Forest Products (INTROP) |
| *Goal 16.* Promote peaceful and inclusive societies for sustainable development, provide access to justice for all and build effective, accountable and inclusive institutions at all levels | •UPM Green Policy<br>•Green Sustainability Steering Committee<br>•3R Urban Garden Concept |
| *Goal 17.* Strengthen the means of implementation and revitalize the Global Partnership for Sustainable Development | •Green@UPM website<br>•Green Mandate<br>•Nature Education Camp & Tree Planting<br>•Sustainable Merit/Assessment of Excellence Evaluation of Administration |

# 2 Successful approaches of green initiatives, commitment and way forward

## 2.1 The green campus management and policy

Environmental management systems assist UPM in identifying and controlling the environmental impact of its activities and products, and in improving its environmental performance on an ongoing basis. Almost all of UPM's production sites, as well as wood-sourcing and forestry operations, have a verified environmental management system in place. UPM's aim is that all of its production units will have an environmental management system certification, including possible new units and acquisitions. The Green Mandate is an initiative of UPM to contribute towards environmental awareness in reducing carbon footprints among industries and create a more sustainable environment for our future. Echoing the spirit of 'Go Green, Reduce Carbon Footprints, Business Sustainability', UPM is promoting an environmental-friendly office atmosphere among industries in Malaysia. It is an effort to convey a strong message of awareness of the Green Nation to industries in Malaysia. In 2013, UPM has invited approximately 110 industries to join the The Green Mandate by registering their mandate [2]. It became the evidence of the industries' efforts in reducing the carbon footprint. Through participating the Green Mandate, the company will be classified as a Green initiative company. The Green Mandate 2013 shows that the industries in Malaysia are indeed concerned on the Green issues [2]. UPM has made significant improvements, particularly in the sectors of transport and infrastructure such as the size of available green space, in addition to the use of efficient energy-saving facilities and bicycles within the campus area.

## 2.2 The green governance and campus community

The university green governance is important to determine the direction of a green sustainability policy. UPM green governance is led by the vice chancellor of the university and supported by the green sustainability committees structured (Fig. 2).

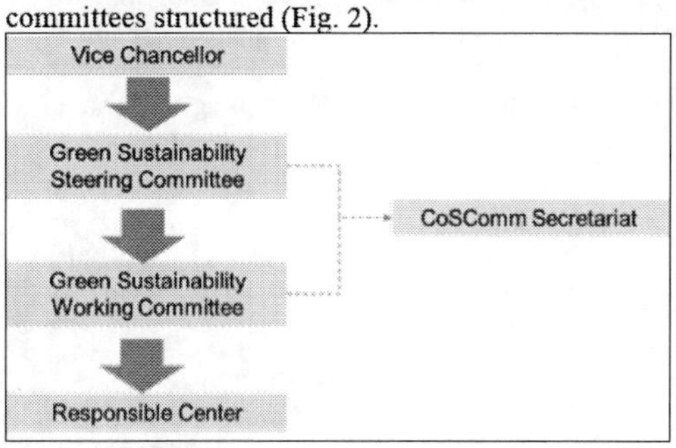


**Fig. 2.** UPM green governance structure.

The Green Sustainability Committee is responsible in formulating, planning and implementing all the activities/programs/projects according to UPM's Green Policy. Besides, the committee also reflects the commitment of UPM in forming network with various agencies to get funding or collaboration on green projects. It also ensures the continuous improvement and success of the green strategies implementation that requires a high competency in listening, communication, relationship building, vision development, responsiveness and continuous strategic adaptation from the university citizens. Right management framework will encourage more effective subcultures of decision-making style that promotes the top management-staff-student-community and industries partnership. The wide-scale participation effectively increases the impact in illustrating the university commitment to environmental

 https://doi.org/10.1051/e3sconf/20184802004

sustainable responsibility on regional and global level. This includes on direction to achieve the inclusiveness of SDGs main vision and aim towards shared large-scale participations on the program. The key characteristics of students, faculties' member and top management engagement that ensure the liveable and continuity of university are important.

### 2.3 The leading green campus initiatives

UPM has become a green sustainability leader among Malaysian institutions, as well as model that cultivate green activities in the teaching, learning, research and professional services ecosystem. The Green@UPM initiatives are our commitment to achieve environmental sustainability through strategic partnership with industries and the community as well as for higher campus quality of life. UPM Green Policy adopts five major pillars namely 1) to provide awareness and education, 2) to preserve biological diversity, 3) endeavor to reduce the release of greenhouse gases, 4) to reduce in way of waste management and motor vehicle dependency and 5) adopting the concept of sustainable development in the management and planning in the whole university's system (Fig. 3).

Additionally, this commitment has been translated into UPM's Green Sustainability Roadmap that is in line with Eleventh Malaysia Plan (RMKe-11) - "Pursuing green growth for sustainability and resilience" – 4 Focus Area (Fig. 4). The focus of Green@UPM direction is towards 1) strengthening the enabling environment for green growth, 2) strengthening resilience against climate change and natural disasters, 3) adopting the sustainable consumption and production concept; and 4) conserving natural resources for present and future generations.

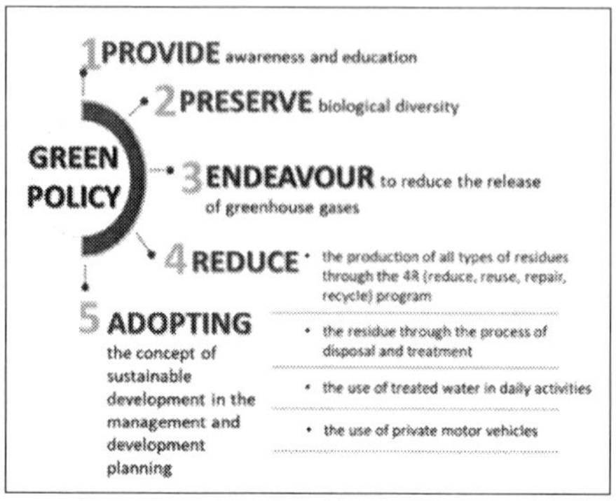


**Fig. 3.** UPM green policy [1].

Green@UPM way forward is focusing on five (5) strategic plans in engaging the campus community to involve with designated UPM green activities (Fig. 5). To heighten the UPM green profile, improvements to existing programs are made through action plan to rebrand all the green activities within the campus. Our strategic plan is also focus on maintaining the existing UPM green policy and mapping the policy with UI-GreenMetric indicators. The university also looks forward in disseminating the information of green activities through websites and other media in reaching the community. Moreover, the future undertaking of green commitment is strategically coordinated to enhance the green activities and agriculture visibility

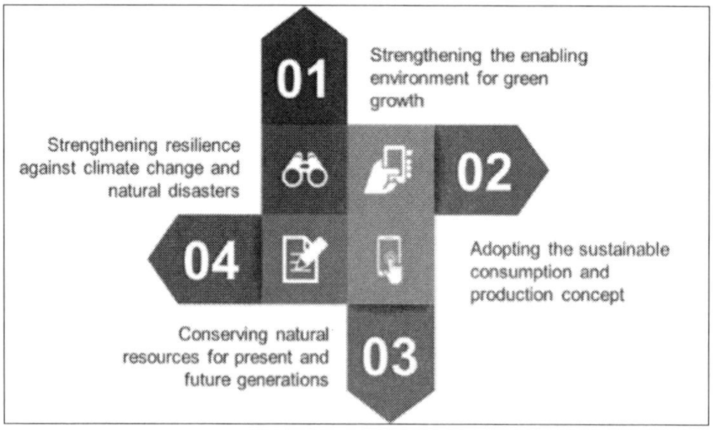


**Fig. 4.** UPM green sustainability roadmap.

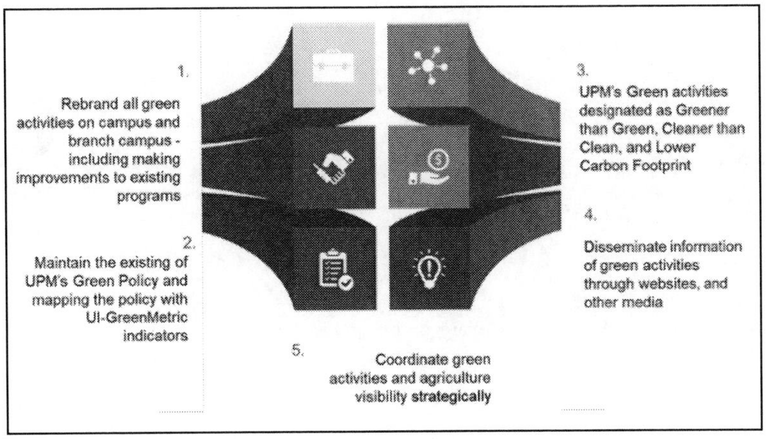


**Fig. 5.** UPM green strategy [1].

## 3 Conclusion

UPM is committed to conserve the environment through various eco-friendly and effective learning activities, research, co-curriculum and quality management systems based on sustainable and efficient environmental management. Starting with the inception of Green Policy in 2011, UPM has shown its green leadership capability which is reflected by the world recognition award, GreenMetric World University Ranking.

## References

1. Green @ Universiti Putra Malaysia, Available online at http://www.green.upm.edu.my/, accessed on 2nd of February 2018 (2016)
2. UPM Biofore Company, Available online at http://www.upm.com/, accessed on 2nd of February 2018 (2016)

https://doi.org/10.1051/e3sconf/20184802005

# Setting and infrastructure at North Carolina Agricultural and Technical State University

*Godfrey A.* Uzochukwu*

Interdisciplinary Waste Management Institute, Greensboro, North Carolina, USA

**Abstract.** North Carolina Agricultural and Technical State University (N.C. A&T) is an important intellectual and cultural center, and it contributes significantly to improving the quality of life globally. Campus setting and infrastructure provide basic information of the university's consideration towards green environment. Type of higher education institution, climate, campus site, campus setting, campus area, campus ground floor of buildings, campus buildings area, parking area, campus smart buildings, ground cover, drainage systems for reducing soil erosion and runoff, number of students including online students, number of academic faculty and staff and university budget for sustainability have made N.C. A&T attractive and the university of choice globally for 11,177 students in 2017. The University has one campus site and one university research and training site. Academic buildings occupy about 39% of main campus area. Buildings account for 40% of the energy and 72% of the electricity used in the United States. Buildings also account for 16% of United States water consumption, 40% of all material flows and produce up to 40% of waste in landfills depending on the region. N. C. A&T is committed to providing more spaces for greenery and safeguarding the environment, as well as developing of sustainable energy solutions.

## 1 Introduction

North Carolina Agricultural and Technical State University (N.C. A&T) is an important intellectual and cultural center, and it contributes significantly to improving the quality of life globally. North Carolina A&T is one of America's highly respected land-grant universities [5]. The University was founded on principles that embrace setting and infrastructure, student access and quality academic performance that integrates campus sustainability. N.C. A&T participated in the 2017 world ranking of university sustainability – UI GreenMetric and was delighted to showcase the university's programs and innovations in sustainability to the world. Developing a Green Campus relies heavily on setting and infrastructure. N.C. A&T has invested in cutting-edge sustainable activities and policies that aim to improve setting and infrastructure as well as protect the environment. Setting and infrastructure, a category of UI GreenMetric is a framework for constructing a green university [4]. Global circumstances demand new setting and infrastructure solutions for our science and technical fields, new enhancements in cultural competencies and social awareness, and an uncompromising campus sustainability expectation among students, faculty, staff and administrators. University setting and infrastructure helps to achieve uncompromising campus sustainability. N.C. A&T has developed important partnerships that impact the social and environmental challenges of regional, national and global communities. The University enhances intellectual environment through creative use of virtual and physical space. (Fig. 1) shows students enjoying the campus setting of N.C.A&T.

**Fig. 1.** N.C. A&T Campus

Campus setting and infrastructure provide basic information of the university's consideration towards

* Corresponding author: uzo@ncat.edu

https://doi.org/10.1051/e3sconf/20184802005

green environment. These indicators show whether or not NC A&T deserves to be called a Green Campus for environmental safeguard.

## 2 Methodology

### 2.1 Campus setting

Campus setting data (Table 1) was collected between August and October 2017 from Facilities Unit, Institutional Research Unit, Budget Office and Division of Research and Economic Development.

**Table 1.** Campus setting data

| **Type of higher education institution** | **Comprehensive** |
|---|---|
| Climate | Humid subtropical |
| Number of campus sites | 1 |
| Main campus setting | Urban |
| Total main campus area | 893,772 (meter square) |
| University research and training area | 2,428,114 (meter square) |
| Campus perimeter | 1,751 (meter square) |
| Total main campus ground floor of buildings | 97,647 (meter square) |
| Total main campus buildings area | 251,228 (meter square) |
| Total parking area | 74,332 (meter square) |
| Total main campus smart buildings | 10,347 (meter square) |
| Number of students | 11,177 |
| Number of online students | 564 |
| Number of academic faculty and staff | 1,863 |
| University budget for sustainability (%) | 15 |

The University has one campus site and one university research and training site. The total main campus area (893,772 meter square or 27% of total campus area) is devoted to educational activities (Fig. 2). The main campus ground floor of buildings (97,647 meter square or 11% of main campus area) and main campus building area (251,228 meter square or 28% of main campus area) are for instructional activities (Table 1). A large area (2,428,114 meter square or 73% of total campus area) is used for research and training). Academic buildings occupy about 39% of main campus area. The remaining area is used for parking, green space, sports stadium, etc. N.C A&T's setting helps to attract and train the next generation of state, national and global workforce.

### 2.2 Campus infrastructure

N.C. A&T Campus infrastructure highlights adjusting campus master plan to restructuring process of colleges; revamping and modernization of seven (7) existing buildings; creating green spaces and global village; constructing eight (8) new energy and water efficient buildings including Science, Technology, Engineering and Mathematics (STEM) facilities; reimaging sports stadium access; connecting campus to downtown greenway trails and creating a sustainable Information Technology (IT) backbone for N.C A&T. The University is proactive in the following areas: use of energy efficient appliances in buildings, smart building implementation (New Student Health Center), energy audits, energy monitoring, lighting upgrades, retro-commissioning process for buildings, repair of leaks to reduce metric tons of water renewable energy produced, electricity usage

### 2.3 Land use policy

The University uses land in the most appropriate manner to advance campus sustainability. The main campus area (893,772 meter square) is well managed and used to advance university's mission.

**Fig. 2**. N.C. A&T main campus perimeter and area

## 3 Summary/ concluding remarks

Type of higher education institution, climate, campus site, campus setting, campus area, campus ground floor of buildings, campus buildings area, parking area, campus smart buildings, number of students including online students, number of academic faculty and staff and university budget for sustainability have made N.C. A&T attractive and the university of choice globally for 11,177 students. N. C. A&T is committed to providing more spaces for greenery and safeguarding the environment, as well as developing of sustainable energy solutions. The University has invested in cutting-edge sustainable activities and policies that aim to improve setting and infrastructure. Buildings account for 40% of the energy and 72% of the electricity used in the United States. In addition, buildings account for 16% of United States water consumption, 40% of all material flows and produce up to 40% of waste in landfills depending on the region [1]. N.C. A&T campus is covered with planted grass and tree vegetation year-round for improving air quality, water quality, water absorbance. Drainage systems for reducing soil erosion and runoff are excellent. N.C. A&T's setting and infrastructure have a positive effect on student choices and institutional selection. Student enrollment increased from 10,852 in 2016 to 11,177 in 2017 (3% increase). Freshman enrollment is up by 11% for Fall semester 2018 [3]. The University was ranked $6^{th}$ in 2016 and $7^{th}$ in 2017 for setting and infrastructure. In addition, the University was ranked $14^{th}$ overall and $4^{th}$ in USA for campus sustainability in 2017 [6]. The ranking is an indication of N.C. A&T's commitment to campus sustainability. University rankings have been shown to affect student choices and institutional selection [2]. The sub-humid tropical climate (Table 1) and urban setting (Fig. 2) make N.C. A&T an institution of choice for many students globally. The university allocates about 15% of it's budget to sustainability including setting and infrastructure improvement. The main campus area (893,772 meter square) is well maintained throughout the year. The ground is covered with vegetation to prevent soil erosion and enhance drainage. The ratios of open space to total area and open space to campus population are important to the university for advancing sustainability. A good setting and infrastructure advances education and research. This paper supports the selection of setting and infrastructure as a sub-category for UI GreenMetric ranking because of the impact on energy and climate, waste management, water conservation and management, transportation and education.

N.C. A&T is striving to become a Green and Sustainable Campus for environmental safeguard.

UI GreenMetric is beneficial to N.C. A&T for the following reasons:

- Tool for self-assessment of campus sustainability.
- Strong commitment to sustainability issues.
- Help the university's efforts in internationalization and recognition by getting its sustainability efforts on the global map.
- Result in an increase of hits to the university website for campus sustainability
- Increasing awareness of sustainability issues.
- Membership in UI GreenMetric World University Rankings Network

**Acknowledgments.** The author would like to thank Chancellor Harold L. Martin for his support of campus sustainability, Provost Beryl McEwen and Vice Chancellor Robert Pompey for embracing sustainability. Thanks to N.C. A&T GreenMetric Team in Facilities unit, Institutional Research and Division of Research and Economic Development for contributing setting and infrastructure data. Special thanks to my colleagues who are advancing sustainability in their classes.

## References

1. Greening Sustainability, North Carolina Agricultural and Technical State University (2012)
2. E. Hazelkorn, *The Effects of Rankings on Student Choices and Institutional Selectio,* Dublin Institute of Technology (2012)
3. News and Record, Enrollment is up 11%, News and Record of Greensboro (2018)
4. Nyoman S. and R. F. Sari., *Evaluating UI GreenMetric as a tool to support green universities development: assessment of the year 2011 ranking* (2011)
5. Preeminence 2020, North Carolina Agricultural and Technical State University (2010)
6. UI GreenMetric, UI GreenMetric World University Rankings (2017)

# Greening campus experience: moving towards living laboratory action plan

*Wahid* Omar[1], *Azlan* Abdul Rahman[2*], *Mohd* Fadhil Md Din[3], *Shazwin* Mat Taib[3], *Santhana* krishnan[3], *Irina* Safitri Zen[4], *Norhisyam* Hanafi[3]

[1]Chancellor Office, Universiti Teknologi Malaysia, 81310 Johor Bahru, Johor, Malaysia
[2]Office of Deputy Vice Chancellor (Development), Universiti Teknologi Malaysia, 81310 Johor Bahru, Johor, Malaysia
[3]Campus Sustainability Office, Universiti Teknologi Malaysia, 81310 Johor Bahru, Johor, Malaysia
[4]Faculty of Built Environment, Universiti Teknologi Malaysia, 81310 Johor Bahru, Johor, Malaysia

**Abstract.** Sustainability is a growing priority for higher education institutions around the world. Universiti Teknologi Malaysia (UTM) through UTM Campus Sustainability Office ready to transform the campus as platform to inculcate the sustainability mind set of young people through education and empowering them to continue living the values in the future as stated in Transformasi Nasional 2050 (TN50) development plan. In 2010, among the first institution supporting UI GreenMetric auditing, UTM was positioned at rank 85, and then gradually improving the performance to position ourself to rank 66 in 2017. To boost this effort, sustainability elements have been implemented in Universiti Global Plan III (2018-2020) and UTM Master Plan 2017-2037. One of the main strength of UTM in UI GreenMetric evaluation components is Setting and Infrastructure that incorporates environmental and social considerations in addition to the traditional cost, time and quality. Improvement through transnational researches which is under Water and Energy Nexus projects such as biomass utilization and groundwater projects are escalating the efforts. For developing more robust approach, Sustainable Development Goals (SDG) has been incorporated in to six Living Labs (LL) initiatives. Positioning in global branding, integration across campus and regional level networking was aimed to leverage currents assets into further opportunities for research, teaching and institutional development.

## 1 Approach to sustainability

Universiti Teknologi Malaysia (UTM) as a leading innovation-driven entrepreneurial research university in engineering science and technology taking sustainability seriously by incorporate Sustainable Development Goals (SDGs) and Education for Sustainable Development (ESD) thorough a comprehensive framework of research integration, educational and operational subsidiary was constructed to balance of academic and non-academic implementation.

### 1.1 Introduction

The role of the university in advancing sustainability development has been widely recognized and the college campus is seen as offering an ideal setting for exploring and practicing sustainability. In universities, sustainability presents an opportunity to make education more problems based, more interdisciplinary and more applied [1]. Some value its broad based international political clout in bringing environmental issues back to the forefront. Others see in sustainability an opportunity to reflect on the role the university has to play in society and in the health and lifelong learning of employees and students. A sustainable university campus should be a healthy campus environment, with a prosperous economy through energy and resource conservation, waste reduction and an efficient environmental management, and promotes equity and social justice in its affairs and export these values at community, national and global levels [2].

### 1.2 Resolve for future sustainable campus

The university determination towards sustainability at campus reflected through detailed planning and UTM sustainable policies (UTM-P). Totally there are fifteen policies (UTM-P1 to UTM-P15) was introduced and being executed for sustainable practice inside campus [3]. Sustainability already incorporated into University Global Plan phase I and II from 2012-2017. From position 85 in 2010 UI Greenmetric ranking, UTM elevates by leaps and bound to positon 66 in 2017. Based on the excellent achievement, UTM has won two awards in 2017: Kelestarian Kampus Hijau Universiti Awam Malaysia (Malaysia Public University Green Sustainable Campus) from Ministry of Energy, Green Technology and Water (KeTTHA) and also Diamond Recognition Certificate LCCF Award Recognition Ceremony & Low Carbon Cities Framework (LCCF) Implementation Sharing Session [4].

To step up the game, UTM have executed the University Global Plan phase III (2018-2020) which consist of introduction of in-campus electric bus project, development of database of national and international sustainability awards and enhance "Waste-to-Wealth", "Recycle, Reuse, Reduce (3R)" programs. Universiti Global Plan III (2018-2020) is developed based on six main key focus areas namely: 1. Excellence in Learning and Teaching, and Transformative Campus Experience, 2. Research Excellence, Industry and Community Engagement, 3. Sustainable Campus, Infrastructure, Information and Communication (ICT) System, 4. Talent Transformation, Governance and High-Performance Delivery, 5. Advancement and Business Development for Financial Sustainability, and 6. Global Prominence

and Branding. The six main focus areas are translated into Strategic Objectives (SO) and Key Amal Indicators (KAI) of various initiatives and programs to be implemented. The success of PGU Phase III is dependent on four core values: Integrity, Synergy, Excellence and Sustainability towards universal well-being that strives for the good of the stakeholders, mainly the community and industry. The third phase also inculcates volunteerism among staff and students on its implementation. UTM Master Plan 2017-2037 for future administrator guideline [5, 6].

## 2 Setting and infrastructure of campus

A well-planned campus landscape supports the university's commitment to sustainability by enhancing ecosystem services such as clean water and air, regulating climate, soil building and erosion control, habitat enhancement, and improvement of human health and well-being. These benefits extend to the community surrounding the campus as well. UTM aims to develop an excellent infrastructure through the adoption of sustainable planning, design and construction principles. The basic setting and infrastructure information of the Universiti Teknologi Malaysia (UTM) towards sustainability is represented as Table 1. UTM Main campus is built in an 1145 hectare of land and total up to 612 buildings which consists of 14 faculties, 13 hostels, and 6 administrative buildings.

**Table. 1.** Setting and infrastructure at UTM.

| Criteria | Setting & Infrastructure |
|---|---|
| Total main campus area | 11,440,000 m$^2$ |
| Total main campus smart building area | 128,025 m$^2$ |
| Total area on campus covered in vegetation in the form of forest | 39% |
| Total area on campus covered in planted vegetation | 32% |
| Total area on campus for water absorption besides forest & planted vegetation | 29% |

UTM comprise a large-scale type of building complex development which making UTM responsible to a huge challenge in order to manage the whole resources and facilities with 33,000 populations of staffs and students. The total ground floor and building area comprises of 426,748.68 m$^2$ and 1,132,403.95 m$^2$, respectively. The campus maintains the 39% of green area in form of forest and 32% covered in planted vegetation. Total main campus smart building area consists of 128,025 m$^2$. UTM emphasises respect for natural resources, adapting to climatic conditions, and setting new standards for building performance through an environmentally-conscious setting.

### 2.1. UTM eco-home project

UTM Eco-Home is a project (Fig.1) that was initiated from UTM involvement in Solar Decathlon China 2013 (SD China 2013). The main purpose of an Eco-Home is to help disseminate, as well as become a showcase to all the technologies engaged for a green and sustainable home cum small office development. The development is to provide an exhibition space for UTM research products; to promote UTM Eco-Home to the public as an intelligent sustainable home; and to create medium for related researchers to conduct research works at this location. UTM Eco-Home designed and constructed to be a showcase of sustainable single storey three-bedroom bungalow located at UTM main campus in Skudai, Johor Bahru. The building is also an experimental and research venue for sustainable, green, low $CO_2$, low energy, smart building control, energy efficiency, building technologies and construction technique. One of the main objectives of the building is to demonstrate sustainable solutions in the design, construction and operation. That includes thermal comfort, low energy demand and low or zero carbon emissions building. Through this Eco-home project, the implemented UTM policies (UTM-P) include adopting green building and infrastructure design through a clear sustainable development framework to achieve cost effectiveness (UTM-P-1), optimising university assets and sustainable business opportunities to promote economic viability (UTM-P-2) and achieving efficiency in operational management of resources, facilities (UTM-P-3), and promoting low carbon practices among campus community (UTM-P10).

**Fig. 1.** UTM eco home.

## 3 SDGs initiatives reflection in sectoral action plan

Concept of Living Laboratories (LL) can be interpreted and used as a human-centric research and development approach. UTM LL approach not only focuses on involving students in development process, it also strives to facilitate the interaction among other relevant stakeholder, such as academia, researcher, and administrative personnel (Fig. 2). This interaction will generate a showcase of campus society as a translator of sustainable development and allows more creative and innovative conducive teaching and learning environment. A university would operate as a fully integrated community that models social and biological sustainability itself. In many cases, people think on education, research, operation as separate activities but they are not actually. Because students learn from everything around them, these activities form a complex of experience and learning. It is an interdisciplinary, trans-disciplinary and problem-based approach. The students need to overcome this larger, extensive form of learning, education experience must reflect on intimate connection among curriculum and research; understanding any negative ecological and social footprint of the institution and improve local and regional communities to healthier and environmental sustainable. Therefore, the concept of LL is good approach to enhance the sustainable development in UTM. The currently six LL in our campus are LL1 Sustainable Arcade, LL2 Green office, LL3 Sustainable Energy Management, LL4 Bio-Recycling Centre, LL5 Green School and LL6 Green Community. All of the initiatives direct and indirectly reflect few Sustainable Development Goals initiatives that have been ruled out by United Nations (UN).

### 3.1 LL1 sustainable arcade

It is one of strategic key initiatives of UTM to promote Green, Healthy and Safety Practice to the consumers and food operator in food arcade. It has main objective to reduce the waste generated in arcade (solid waste, food waste), practice recycling and resource recovery which reduce the carbon emission generated. The benefits not only contribute to the environment but also the campus society as a whole in combating zero hunger (SDG 2), good health and wellbeing (SDG 3) and responsible consumption and production (SDG 12).

### 3.2 LL2 green office

Conventional office provides lots opportunities for green practices. From minimize the use of the paper, adjusting the office lighting system to let the natural sunlight enter the room, recycle the used computer, ink toner and cartridge until the selection of electronic that save the energy. Affordable and clean energy (SDG 7), climate action (SDG 13) and responsible consumption and production (SDG 12) are applied SDGS on this effort.

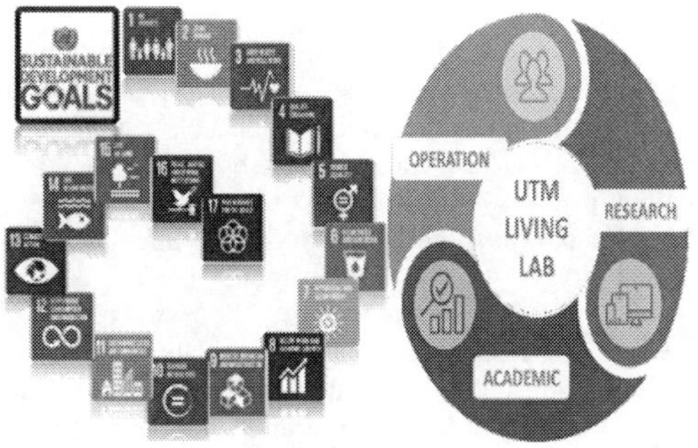


**Fig. 2.** UTM key living lab initiatives & SDGs.

### 3.3 LL3 sustainable energy management

Sustainable Energy Management (SEM) is a University-wide commitment program. The University will strive to efficiently manage the consumption of energy in a manner that is consistent with providing a conducive and sustainable campus environment for teaching, learning, research and intellectual development. Among SDGs served on this initiative are sustainable cities (SDG 11), climate action (SDG 13), responsible consumption and production (SDG 12) and communities and affordable and clean energy (SDG 7).

### 3.4 LL4 bio recycling centre

Food waste and kitchen waste from food arcades around the campus will be collected and mixed with shredded garden waste. Other conditioners and Effective Microbe (EM) will be employed to effect decomposition of the organic waste under controlled conditions. The composted products will be further treated and converted into soil enhancer, compost fertilizer or animal feeds. Chemical and physical analysis will be carried out to characterize the raw intermediates and final products of the recycling process (Fig. 3). These initiatives reflect good health and well-being (SDG 3) and responsible consumption and production (SDG 12).

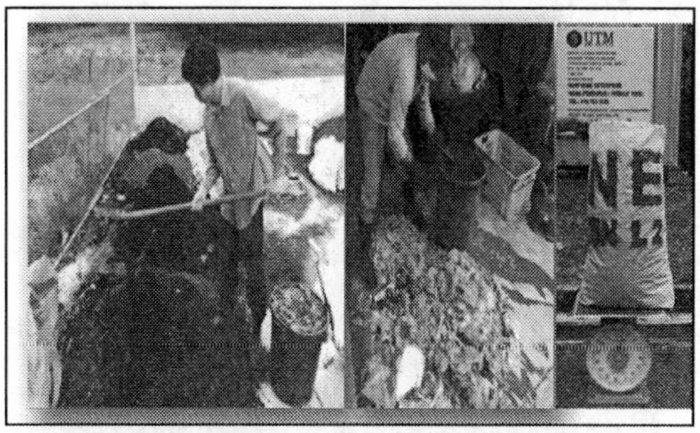

**Fig. 3.** Organic waste recycling facility in UTM.

### 3.5 LL5 green school

Green school is one of the sustainable campus initiatives which aim to educate high school and primary school students regarding sustainability concepts and issues. With various platform such as workshop, exhibition and sustainable activity, UTM experts reaching out to young generation building the capacity on the importance of sustainability. It's expected these pool of youth can act as ambassador communicating the concept and importance of sustainability at large and live a positive life appreciating the three sustainable element in the future, which is Human, Nature and Economy. Quality education (SDG 4), strong institution (SDG 16) and partnership for the gaols (SDG 17) are SDGs initiative can be fulfilled by this action. Encouraging activities that appreciate the existing natural environment (UTM-P14) is the element targeted by this initiative and promoting active and healthy lifestyle within the secure environment (UTM-P15).

### 3.6 LL6 green community

UTM contributing back to the community through LL6 Green Community which aim to help the society problems by providing a sustainable solution. Besides helping the community by using sustainable approach, green community initiative is also at the same time try to educate the society regarding the importance of sustainability environment and several issues related to it such as carbon and waste management. Various initiatives can cover various aspect of SDGs elements but mainly supporting by partnership for the gaols (SDG 17). The UTM-P components are maintaining healthy balance between developed and green areas to achieve campus eco system vitality (UTM-P8), augmenting the conservation of wetland features to support habitat and recreational water activities (UTM-P9).

## 4 Collaboration breakthrough showcases

### 4.1 Sustainable Energy Management Program (SEMP)

One of the successful achievements of UTM sustainability initiative is the Sustainable Energy Management Program (SEMP) which champion by the energy management group. The group is the unique combination of academic researchers which involves several faculties such as chemical engineering, electrical engineering, professionals and the operational staffs of Office and Asset Development (PHB). This program applies the holistic approach of effective and sustainable energy consumption and management which covers the inclusions of organizational structures and provides campus as a living lab. It is estimated that the renewable energy production inside campus is 81 KW as solar power. Concurrently, the ratio of renewable energy production towards total energy usage per year is 40%-60%. The programs apply the holistic approach of effective and sustainable energy UTM was officially awarded the Energy Management Gold Standard (EMGS) 3rd Star rating under the ASEAN Energy Management (AEMAS) in 2016. Based on the comparison of energy usage in 2009, which is the benchmark year used for the UTM Sustainable Energy Management Program (SEMP), there has been a total saving of 42.0 million kWh or RM21 million until April 2017. Some of the initiatives taken by the SEMP team to reduce Greenhouse gas (GHG) emission includes retrofitting the air-conditioning system by implementing variable refrigerant volume (VRV) in many faculties and buildings, lighting retrofits, which included de-lamping and installation of high efficiency T5 and super T8 lights, maximum demand management which involves shifting loads so as to reduce maximum demand on capitalizing off-peak tariff rate (OPTR) discount, utilizing its in-house Online Energy Billing and Management System (EBMS) to communicate and provide up-to-date information to all stakeholders on the campus. Other that, key elements in the formulation of policies and guidelines adopted at UTM to support energy-saving conservation such as ensuring the use of air-conditioner is at the setting rate that has been specified according to the needs of the room and weather. Air temperature setting recommendations are between 24 - 26°C; Air conditioner should be switched off for one hour beginning the lunch-hour between 13:00 and 14:00 daily; Indoor lighting or room may use natural lighting if the lighting rate exceeds 40%; Minimize the use of printed papers and ordinary meetings using a variety of printed materials; Using tumbler bottles or multi-purpose food containers in the office to reduce dependency on plastic bottles or plastic containers; All existing and constructed buildings at UTM should at least encourage and implement Smart Building and Green Building features such as devices and system of rainwater harvesting, energy and water conservation tools, green planting and etc. Encourage staff involvement in volunteerism planning related to energy conservation initiatives and GHG emission reduction commitments; Diversify healthy and vibrant lifestyle activities by practicing cycling, walking and strolling cultures instead of using vehicles and elevators. The implemented UTM-P segments are UTM-P10 and instilling integrity and ethical values through volunteerism and continuous commitment at all levels of the community (UTM-P13).

### 4.2 Food Waste Recycling Program (FWRP)

UTM leading by Office of Asset and Development (PHB) is promoting the Bio Recycling Centre to transform all organic waste into valuable products to promote the sustainable and environmental-friendly practices and various activities/programmers related on waste management towards Zero Waste Program. Most of the activities are encouraging behavior, motivation and awareness on reducing the wasted materials to be dumped at the landfills, which is currently the capacity of organic waste (mainly food wastes) occupied up to 55% of the total component in UTM. It is estimated from 140 arcades around campus could produce nearly two tons of food waste every day, which is the most difficult way to handle it by limited number of staff. Some of practices for the easiest way of separation at sources such as differentiate between organic and non-organic wastes in your plates/food container before throw in the bins; All contractor and catering unit inside UTM must allow the proper bins, in order to separate the food waste and non-food waste to ensure proper services by the Waste Unit, UTM; Ensure the bins at the building/section must equipped with the covered bins; Must alert the sticks/tissue and other related matters which always treated as non-organic features in our plates/food container; Never use plastic bag/straw in the same plates/food container to make sure we are not disposing it together; Always promote to other members on reducing wastes and separation at source campaign; and the concept of 3R which is only useful for mostly of the recycling part. The Implemented UTM-P elements are eliminating non-biodegradable food and beverage packaging (UTM-P11) and promoting community spirit and enhancing the quality of the life, responsive to local and global context in harmonious and conducive environment (UTM-P12).

### 4.3 Water Sustainability Program (WSP)

UTM has a clear vision and commitment towards the improvement of the policy and guidelines on sustainable water and water resources. Some of initiatives are; clean water sources such as Lata Jernih in Recreational Forests inside UTM campus are preserved and kept of pollution-free; All of the catchment area of the upstream and a number of fringe forests around UTM will be targeted to reduce the risk of contamination of suspended matters; All units in UTM are encouraged to use rainwater harvesting by installing wire shelf divider mosquito's breeds. Water usage is for the purpose of cleaning and urban farming spraying/fumigation available around the Office building; The use of tap water must be minimized with water saver tools installation, whether in the home or in the main piping building/laboratory unit; Stringent care and awareness about water resource rivers and lakes in the vicinity of UTM will be given to the added value, especially involving biodiversity, prevent water pollution and recreational/leisure; Exploration of groundwater are encouraged with additional information about the level of hygiene and safety to users; and use of non-sustainable detergents and soaps are not encouraged in the kitchen, and floor-type. The implemented UTM-Ps are enhancing sustainable consumption of available resources (Water and Energy) (UTM-P5), minimizing waste and pollution through effective waste management (UTM-P6), and introducing more local flora and fauna to protect and enrich biodiversity (UTM-P7).

## 5 Conclusions

Concept of sustainable campus is relatively new in Malaysia, and several public universities are still at the infant stage in adopting and implementing the sustainability in campus. The institutionalization of campus sustainability involves educational and awareness campaign, dialogue, series of meeting in identifying the target, establishes the system, mechanism and Key Performance Indicators (KPI's). The processes at the same time coupled with the Living Lab approach which allows the contribution three different groups; academic, students and professional and administrative groups of campus sustainability contributes to the campus sustainability campus contributes to the campus sustainability initiatives, efforts and projects. Apparently, the planning and implementation of these initiatives should be phased and prioritized accordingly so that they can affect optimally on sustainability in campus operations. There are various initiatives and sustainable practices that can be adopted and implemented by the top management to ensure that universities operate in a sustainable manner. However, more focus and attention should be given to the initiatives that can give positive impact on the environment, economic and social aspects of the campus community.

**Acknowledgements.** The authors would like to acknowledge the efforts and contribution of Pejabat Harta Bina, INSTEG group, Landscape Unit, Electrical Unit, Office of Vice Chancellor and Office of Deputy Vice Chancellor (Development).

## References

1 O. Saadatian, K. B. Sopian, E. Salleh, Sustainable Cities and Society, *Adaptation of sustainability community indicators for Malaysian campuses as small cities*, **6**, 40-50 (2013).

2 I. S. Zen, R. Ahamad, & W. Omar, Jurnal Teknologi. *The development and measurement of conducive campus environment for Universiti Teknologi Malaysia (UTM) of campus sustainability*, **68**(1) (2014).

3 http://www.utm.my/sustainable/our-policy/.

4 K. Tenaga, & T.H. Air, Low carbon cities framework & assessment system, ISBN-978-967-5893-06-3. (2011).

5 UTM Sustainability Report, *UTM press*, Towards Low carbon campus, ISBN-978-967-0194-43-1. (2013).

6 R. Zakaria, E. Aminudin, *UTM press*, Sustainable development campus, ISBN-978-983-52-1478-3 (2017).

# Making an urban university 'green': uniting administration and students towards synergy.

*Aleksandr* Fedorov, *Evgeny* Zakablukovskiy* and *Anna Galushkina*

Minin Nizhny Novgorod State Pedagogical University, 1 Ulyanov St., 603950, Russia

**Abstract.** The authors study a case of increasing the environmental sustainability of Minin University. Specific obstacles found on the path of the 'green' development of urban universities is analyzed. Objective reasons for difficulties include the geographical position of the university, absence of traditional campus system, buildings' age and condition, limited availability of financial resources and legal grounds for restructuring of buildings; centralized sewerage, heating and water supply as well as waste disposal; plus, the relatively low level of environmental culture of the population. Even when an urban university 'inherits' these issues from the city itself, the authors see a real opportunity to improve the environmental sustainability of the university by implementing a system that includes environmental projects by the university administration blended with student initiatives. The university administration prepares the application for the 'green' rating, administers a contest of student projects and the allocation of resources for those, as well as the renovation and restructuring of university premises. Students are responsible for planning relevant student associations' agenda, generating ideas, participating in project workgroups, planning and driving environmental events. The authors conclude that it is possible to obtain a synergetic effect of the upstream development of the environmental sustainability of the university.

## 1 Introduction

One of the main obstacles crossing the path of the "green" development of urban universities in Russia is the geographic location of a university and, therefore, mainly severe temperature conditions, peculiarities of precipitation as well as insufficient sunlight amount for most of the year. All above does not allow using many resource-saving technologies. For example, there are only 25 sunny days per year on average in Nizhny Novgorod, while precipitation falls 253 days per year: that limits the possibility of using the sunlight as an alternative energy source, and at the average annual wind speed of 2.4 mps it becomes economically costineffective to use industrial wind power plants [1]. In addition, Russian urban universities differ significantly in their format from traditional campuses located in Europe and in the US: there is no joint orderly infrastructure that would transform the university buildings into a single complex. Dormitories and educational buildings exist somehow 'separately' from each other, the utilities system forms part of the city network, and there is almost no green space. Thus, to a large extent, urban universities 'inherit' the problems of a settlement, and the situation is the same in many world countries [2]. The research performed under the auspices of the European Commission shows that in foreseeable future the population's desire to settle (and, accordingly, to receive higher education) within the urbanized territories will gradually increase: by 2020, up to 80% of the European population will live in the urban environment [3], so the urban universities, as a phenomenon, is something we have to consider further for many decades or even centuries.

Minin University (Nizhny Novgorod, Russian Federation) is an urban institution with rich traditions, founded in 1911 in the historical part of the city. Most buildings of the university are part of the country cultural heritage, they are strongly protected by authorities and are not subject to radical restructuring: in accordance with the federal law requirements, the university may not carry out any works that would change the appearance of the buildings. This, in turn, entails the inability to bring buildings into full compliance with environment-friendly standards [4]. In particular, it is not possible to perform completely 'green construction' works which would ensure that the natural environment pollution is significantly reduced thanks to the renewable energy, water and other resources. The university founder is the Russian Ministry of Education and Science whose subsidies are primarily designed to pay for staff salaries, utilities and maintenance. Those funds are not enough to cover the cost of introducing expensive 'green' technologies. Centralized sewerage, heat and water supply does not allow the University to make full use of technologies designed for cleaning and reducing the use of water. Similarly, the situation with waste removal is mediocre: separate collection of garbage is fully driven only in the cities of s.c. federal significance (Moscow and St. Petersburg). There is a tendency to introduce the program in other regions, but there are no large-scale positive results at the moment. The sorted garbage, as a rule, eventually turns out to be stored at the citywide landfill and not processed

*Corresponding author: ezakablu@mininuniver.ru

separately. For the same reason, the ecological culture of the population is at a relatively low level: one-off events raise the 'recognition rate' of some environmental issues to some extent, but do not represent a serious environmental 'vaccination' of the population. The situation could possibly change with the opening of a unique (for this country) waste sorting plant in Nizhny Novgorod in June 2018. According to the authorities, the enterprise will allow to reduce the amount of waste sent to the dump site by 30%. Initially, the sorting site will be working based on the Russian reality: individuals and legal entities lack systematic practice of separate garbage collection. In future, the system can be reconfigured, based on the changing market needs.

Despite the obstacles described, Minin University administration is making systematic attempts to improve the environmental sustainability of the school by building a system of interaction between the students and the management. In particular, we encourage implementing environmental projects suggested both by the university administration ('from above') as well as by the students ('from below'). It should be noted that the system works only when there are established internal communication channels and responsible supervision lines.

The university administration undertakes to develop and implement environmentally sound policies thought to minimize negative impact on the environment, including re-equipping premises, purchasing energy-saving equipment, offering environmental education to its employees, launching new environmental courses and specialties, starting competitions on social and environmental projects for students, allocating resources for implementing winner-projects as well as administering the application process for UI GreenMetric rating.

The University administration also oversees its own Department of Environmental Education and Rational Nature Management. The Department is one of the oldest in Russia. Research activities run within the priority areas of scientific research at the University [5] as well as the areas preferred by the Russian Foundation for Basic Research which issues relevant grants. The work results are published in monographs, textbooks, teaching aids and scientific periodicals. Over the past three years, the Department has published more than 200 scientific articles, including those in journals that are indexed by the Web of Science and/or Scopus. The Department supervises the education of bachelors and masters in the following areas: 'Environmental Management and Audit', 'Environmental Design and Expertise', 'Theory and Methods of Teaching and Education (Ecology)'. There are also additional short-term education programs such as 'Environmental Education for Sustainable Development', 'Ensuring Environmental Work Safety in the Field of Hazardous Waste Disposal'. The Department is responsible for carrying out specialized scientific events, e.g. an international scientific and practical conference 'Environmental Education for Sustainable Development: Theory and Pedagogical Reality', an interregional conference for undergraduate and graduate students 'Environmental problems and how to solve them: natural and socio-cultural aspects', regional student Ecology Olympics, city contests for school students on ecology. There is an association of educators teaching ecology in the Nizhny Novgorod region, an environmental analytical laboratory for monitoring and environmental protection and the 'Ecodesign of the urbanized megalopolis environment' research laboratory functioning on the basis of the Department. The laboratory is a technical complex consisting mainly of advanced physical and chemical devices used in modern environmental and analytical practice. The laboratory has implemented all stages of environmental objects analysis, from sampling and sample preparation to direct environmental analysis and data desk processing. The main environmental objects explored during the educational process and scientific work, are soils, plant biomass, water, food raw materials, waste and air masses. 'Ecodesign of the urbanized environment of the megalopolis' laboratory develops a strategy for designing and mastering the environment of a big city, providing a synthesis of cultural, ecological, humanitarian, artistic, scientific and technical directions of culture, in order to assist in harmonizing relations within the s.c. 'human-nature' system.

One of the significant achievements of the Department has been the active involvement of students who study ecology. As a result, two environmental student associations were created.

However, the key performance indicators of these student associations have been quite modest due to students having no or little experience in project activities. It was also difficult to obtain large volumes of input data for research from the economic services of the University, despite the existing initial experience of network cooperation and the usage of advanced ICT formats [6]. The University administration has decided to consolidate its efforts and the work of students, in particular for the implementation of the 'Positioning of Minin University in the UI GreenMetric Global Green Rating'. A steering committee consisting of the employees and the students of the University was created. The initial stage involved an in-house kick-off project session, where the project team distributed roles and responsibilities, drafted the project documents and discussed the global rating methodology. The R&R distribution was organized as follows: students are responsible for the planned work of relevant student associations, for generating 'green' ideas, participating in working groups for the project implementation, organizing and driving student environmental activities, while the University administration takes care of collecting data and filling up the application for UI GreenMetric rating. At the second stage, an offline creative session was held to generate solutions: as a result, an initial pool of proposals was developed to improve the position of Minin University in UI GreenMetric rating. The third stage included conducting a full-time creativity and sorting session to discuss and sort out ideas. The events were driven in accordance with the s.c. design thinking methodology, adapted to the project situation by the students themselves [7] on the basis of the PMBOK. The solutions were discussed, clustered and carefully selected on the basis of the open in-class voting as a result of persistent three-hour long work. The fourth stage involves submitting projects to the intra-university student initiatives competition and receiving funds within the internal university grant system. The fifth stage consisted in organizing the interaction of the project team and relevant university administrative staff with the aim of collecting design data and developing preliminary feasibility study documents in order to implement selected solutions. At the end stage of the project, priority solutions will be presented to the senior management for approval. The presentation will include a feasibility study as well as calculation results for potential number of rating points, provided that the proposals are fully implemented. In addition, based on the interim report by the project managers, a number of decisions should be made regarding the appropriate adjustments needed for the University ecostrategy as a whole. Proceeding from the above written material, it is possible to draw a well-grounded conclusion about the possibility of obtaining a synergy effect of bottom-up development in the direction of the environmental sustainability of an urban university by combining efforts and goodwill of all parties even in difficult objective conditions.

## References

1. The climate in Nizhny Novgorod. Weather and climate monitoring in Russia, CIS and the world. Available online at: www.pogodaiklimat.ru/climate/27459.htm (accessed on 19.02.2018).
2. D. J. Edelman, M. Schuster, J. Said, 2017. Urban Environmental Management in Latin America, 1970-2017. Current Urban Studies. **5**, pp. 305-331.
3. Making our cities attractive and sustainable. How the EU contributes to improving the urban environment. Luxembourg: Publications Office of the European Union, 2010 — 36 pp. doi: 10.2779/42720
4. Federal Law No. 73-FL of June 25, 2002 'On the Objects of Cultural Heritage…'. Available online at: www.consultant.ru/document/cons_doc_LAW_37318/ (accessed on 19.02.2018).
5. The main directions of scientific activity at Minin University. Available online at: mininuniver.ru/scientific/scientific-activities/schools (accessed on 19.02.2018).
6. E. Patarakin, S. Shustov, *Digital Ecology: Social Networks and Informational Ecosystems*. Vestnik Mininskogo universiteta. **3**, pp. 13. (2013)
7. S. Goldman, Z. Kabayadondo, A. Royalty, M.P. Carroll, B. Roth. *Student Teams in Search of Design Thinking*. pp. 11-34 // Design Thinking Research. Building Innovation Eco-Systems. Heidelberg: Springer, (2014).

E3S Web of Conferences **48**, 03001 (2018) https://doi.org/10.1051/e3sconf/20184803001
*IWGM 2018*

# Creating a low-carbon campus in Chaoyang University of Technology (CYUT)

*Tao-Ming* Cheng[1], *Hwa-Nan* Chang[2], Clive Yen[3,*], *Ming-Hsiung* Chen[4]

[1]President, Chaoyang University of Technology, 168, Jifeng E. Rd., Wufeng District, Taichung, 41349 Taiwan, R.O.C.
[2]Assistant professor, Department of Golden-Ager Industry Management, 168, Jifeng E. Rd., Wufeng District, Taichung, 41349 Taiwan, R.O.C.
[3]Postdoctoral Researcher, Department of Environmental Engineering and Management Innovative Center of Sludge Recovery, 168, Jifeng E. Rd., Wufeng District, Taichung, 41349 Taiwan, R.O.C.
[4]Director of Officer for Environmental Safety and Health, 168, Jifeng E. Rd., Wufeng District, Taichung, 41349 Taiwan, R.O.C.

**Abstract.** Founded in 1994, the campus of Chaoyang University of Technology (CYUT) is located in the suburban area of Taichung, Taiwan. In 2009, the president of CYUT signed the Talloires Declaration to show his commitment for promoting environmental sustainability on campus. In 2012, CYUT and many other universities in Taiwan cofounded the Green University Union of Taiwan (GUUT) to collaboratively promote the concept of environmental sustainability. Following the announcement of the Sustainable Development Goals (SDGs) set by the United Nations in 2015, CYUT has been putting a lot of effort into converting itself into a green university. Four different fields of management, which are energy and resources management, accident prevention and rescue management, occupational safety and health management, and environmental management, have been promoted in the campus of CYUT for this cause. Furthermore, four management practices including implementing management systems, organizing green courses, promoting green activities and creating green environment, have been applied to improve the effectiveness of the campus management. In the case of energy and resources management in particular, not only ISO 50001 and 14001 management systems were implemented but also an intelligent energy network (iEN) was established for maintaining effective usage of the campus energy. For years of striving in creating a green university, CYUT had several remarkable accomplishments. The green open space ratio in campus is 94.43%. Water-saving equipment was installed in the whole campus and reclaimed water is collected for urban reuse. Garbage reduction and classification have been enforced to make ease for later treatments. For issues related to energy and climate change, strategies, such as renewal of high energy consuming facilities, installation of green roofs, utilization of recycle energy and education of staffs were enacted and the consumed energy in campus gradually decreased in recent years. For example, the electricity and oil (including gasoline and diesel) consumption in campus of 2017 were 7.59% and 14% lower than those of 2016, respectively. In order to overcome the challenges from climate change, CYUT will continue on its process of creating a low-carbon campus focusing on energy sustainability.

## 1 Introduction

Since the Industrial Revolution, fossil fuel has been greatly used by humans which resulted in huge amount of greenhouse gas emissions. From 1880, the global average surface temperature has increased by about 0.85 ℃ [1]. To date, the Intergovernmental Panel on Climate Change (IPCC) from the United Nations (UN) has published 5 assessment reports. In their 5th assessment report published in 2013, the term used to indicate the likelihood of an outcome between human activities and global warming was "extremely likely". It was upgraded from "very likely" which was stated in the 4th assessment report published in 2007. In other words, 95% of global warming is related to the outcomes of human activities. According to the IPCC report, the global average surface temperature could increase 0.3～4.8℃ from the years of 2081 to 2100. During that time, the crisis of global ecosystem collapse could unfold and human survival would probably be severely challenged. As for Taiwan, due to environmental pollution and highly urbanization, the phenomenon of urban heat island was formed. Therefore, during 1911 to 2009, the mean annual temperature of Taiwan has increased 1.4℃, that is 0.14℃ for the rate of temperature increase of every 10 years [2]. Also, the rate of the increasing amount of greenhouse gas emissions in Taiwan is nearly the highest of the whole world. Furthermore, since Taiwan is located on the western Pacific typhoon basin, the Climate Risk Index ranked Taiwan as one of the ten countries who suffers most from extreme weather events.

According to the Paris Agreement, which was negotiated in the 2015 United Nations Climate Change Conference (COP 21) held in Paris, the global temperature rise for this century is aimed to be kept under 2℃ and even further to 1.5℃ by pursuing with great efforts. But from the current circumstances, it

* Corresponding author: pres@cyut.edu.tw

would be an uphill battle to achieve the goal. From 2006, the annual Global Risks Report published by the World Economic Forum enlisted climate change as one of the top 10 global risks. Therefore, the extreme weather events that were accompanied by climate change, such as Hurricanes Katrina and Sandy in US and Typhoon Morakot in Taiwan, have created a great survival crisis for human beings. In order to face the challenge of climate change, universities should take more social responsibility.

## 2 About CYUT

Chaoyang University of Technology (CYUT) was newly founded in 1994. Currently, CYUT comprises 5 colleges including management, science and engineering, design, humanities and social sciences, and informatics. Within the colleges, there are 22 departments (e.g., finance and business administration) that offer 22 master's programs (e.g., environmental engineering and management) and 5 doctoral programs (e.g., architecture and urban design). The currently enrolled students are about 16,000, and the faculty and staff members are over 600. CYUT is situated on the hills in the southeastern part of Taichung, Taiwan (Figure 1). The total campus area is 66.4 hectares and is divided into two campus locations according to the development time (Figure 2). Campus I is the main campus for student activities and the carbon footprint and greenhouse gas emissions in this area could be lowered. Campus II is located to the southeast of Campus I with a distance of 3 km. It has an area of 39 hectares which is currently covered with trees and its carbon sequestration is about 291 metric tonnes $CO_2e$ per year. At present, the green open space ratio in campus is 94.43% [3].

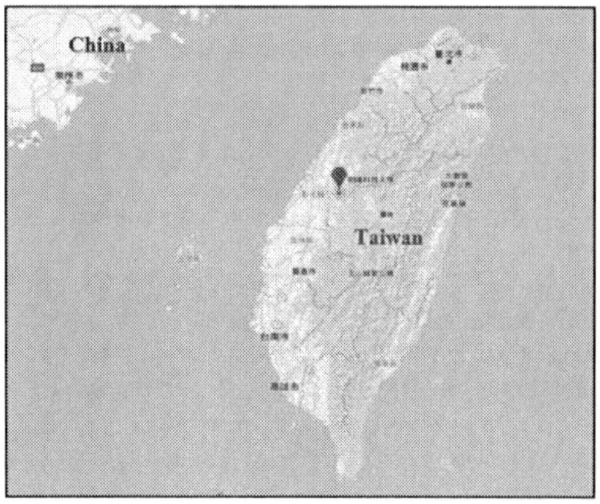


**Fig. 1**. Location of CYUT.

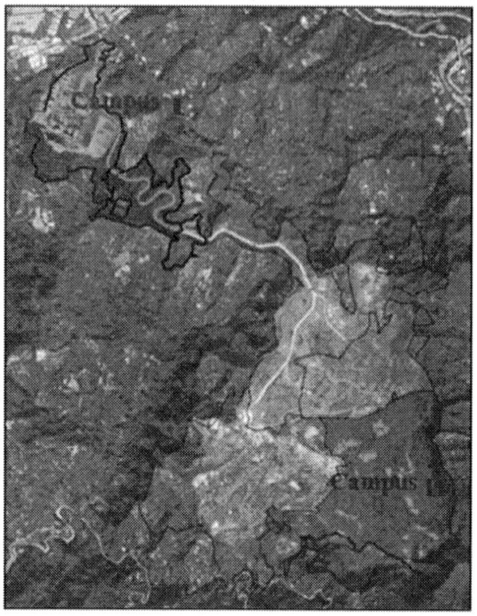

**Fig. 2.** Campus locations of CYUT.

## 3 Factors for green university promotion

### 3.1 Setting up a dedicated organization

The Office of Environmental Safety and Health of CYUT is the dedicated organization for the promotion of green university. Together with the Office of Academic Affairs, Office of Student Affairs, Office of General Affairs, General Education Center, and other units or departments, the green university promotion committee was formed to act as the principal consultant. Referring to the Sustainable Development Goals set by the United Nations [4] and the environmental characteristics of Taiwan, four different fields of management, including energy and resources, accident prevention and rescue, occupational safety and health, and environmental, have been promoted in the campus of CYUT for this cause. Furthermore, four management practices including implementing management systems, organizing green courses, promoting green activities and creating green environment, have been applied to improve the effectiveness of the campus management (Figure 3). By learning, living and experiencing, the students and staff members can adapt the concept of sustainability in their daily lives and hope to achieve the goal of creating a sustainable and low-carbon campus [5].

https://doi.org/10.1051/e3sconf/20184803001

**Fig. 3.** CYUT's tactics for green university promotion.

### 3.2 Participating in green university organizations

Besides the operation of the in-campus organization, CYUT is aggressively linking other universities and organizations in Taiwan for promoting environmental sustainability. Currently, CYUT is participating in the Green University Union of Taiwan (GUUT) and has partnership with the Central Environmental Education Regional Center and the Environmental Education Facility and Field by the Joint Regional Planning Panel of Chung-Chang-Tou-Miao (Table 1). The main reason for CYUT to join GUUT is for having better connections with other universities in order to promote the works of green university in Taiwan and interact with international green universities academically and practically. On the other hand, having partnership with the Central Environmental Education Regional Center and the Environmental Education Facility and Field by the Joint Regional Planning Panel of Chung-Chang-Tou-Miao could assist the local government to promote environmental education and raise the public environmental awareness for developing a sustainable society.

**Table 1.** Environmental sustainability organizations with CYUT participation

| Name of organization | Time of participation | Remarks |
|---|---|---|
| Green University Union of Taiwan | 2012.01 | 1. Cofounded by CYUT.<br>2. The president of CYUT is one of the executive directors. |
| Central Environmental Education Regional Center (Partnership) | 2014.12 | |
| Environmental Education Facility and Field by the Joint Regional Planning Panel of Chung-Chang-Tou-Miao (Partnership) | 2016.04 | 1. Cofounded by CYUT. |

### 3.3 Implementing environmental and energy management systems:

Since CYUT implemented ISO 14001 Environmental management in 2007, more management systems, such as ISO 14064-1 Greenhouse gases and ISO 50001 Energy management, were implemented in the following years due to the improvement in effectiveness of campus management. By following the PDCA in each system, the energy consumption, environmental risks, and greenhouse gas emissions all could meet the management goals.

**Table 2.** Environmental and energy management systems implemented in CYUT

| Management systems | Time of implementation | Time of certification | Certification organization |
|---|---|---|---|
| ISO 14001 | 2007.07 | 2008.01.14 | Lloyd's Inspection Limited Taiwan |
| ISO 14064-1 | 2008.06 | 2009.12.30 | SGS Taiwan Ltd |
| ISO 50001 | 2010.07 | 2012.03.22 | SGS Taiwan Ltd |

### 3.4 Establishing iEN management and monitoring system

In order to construct smart buildings, CYUT has been establishing management and monitoring systems for water and electricity consumption since 2006. Due to the good results, other management systems including curfew, surveillance, fire, and indoor air quality were added year by year into the system and the combination is called iEN (Intelligent Energy Network) management and monitoring system. The current iEN can function in systems including electricity, water supply and sewerage, air conditioning, lighting, and environment monitoring. The framework and control interface of the iEN system are shown in Figure 4 and Figure 5, respectively.

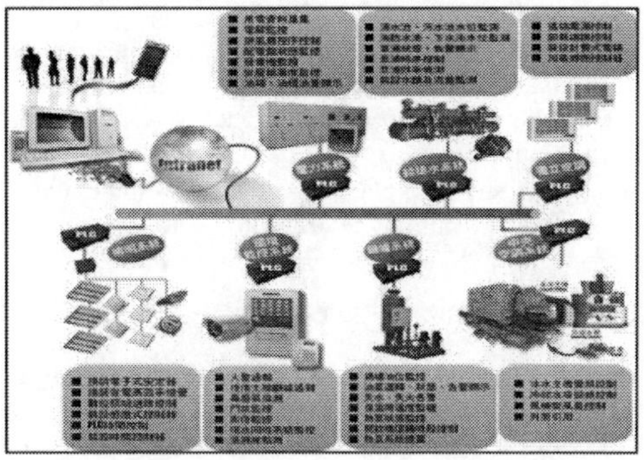

**Fig. 4.** The framework of the iEN management and monitoring system in CYUT.

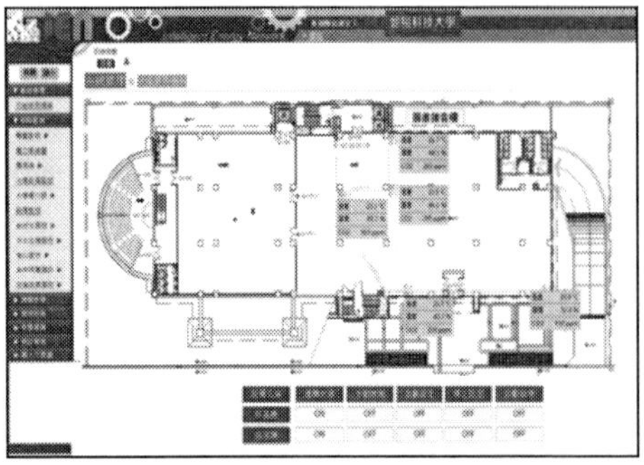

**Fig. 5.** The control interface of the iEN management and monitoring system in CYUT.

The functions of the iEN management and monitoring system in CYUT are described as follows:

1. Electricity system: including managing the campus demand, collecting and analyzing data from electricity consumption and monitoring relays. Over 95% of electricity usage in CYUT is controlled by this system.
2. Water supply and sewerage system: including maintenance of water reservoirs in campus (tap water, ground water, waste water and firewater), timing of pumping, recording and monitoring of water usage. The iEN system can control the water pump to function during off-peak periods and it can reduce the electricity consumption during peak hours. Also, by monitoring the water usage, leakage of water pipes can be detected.
3. Air conditioning system: functions for central air conditioning system include water cooled chiller and cooling tower with variable frequency drives, and infiltration control; functions for individual air conditioner include remote power control, billing meter and unloading valve circuit operation. Customized air conditioners are purchased with a temperature controlling chip (set at 26℃) built-in. Each classroom follows its own schedule for air conditioning. In other words, only if there is class in the classroom, the air conditioner may function, and vice versa.
4. Lighting system: using time as a factor to switch on/off the outdoor lightings.
5. Environment monitoring system: including curfew, surveillance, fire alarm and detector central control. Also, following the Indoor Air Quality Act of Taiwan, environment monitors, such as carbon dioxide sensors and hygrometer, are placed in the library and in the hall of the Design Building in order to assure air circulation.

### 3.5 Building green data centers

With proper planning in campus spacing, the computer facilities were designed as green data centers. In other words, cold and hot aisles were separated in the facilities. After renovation, the power consumption decreased from 34kw to 25kw, that is a 26.5% saving ratio and the power usage effectiveness (PUE) is in the gold level.

### 3.6 Replacing old chillers with active magnetic bearing chillers

Old and less effective water-cooled chillers (720RT) in the library were replaced with new active magnetic bearing chillers (400RT) and the power saving ratio was as high as 51.16%.

### 3.7 Replacing power draining lightings

In order to improve lighting efficiency and reduce power consumption, indoor lightings were all replaced with T5 fluorescent and LED lightings and the power saving ratio was at least 30%. The lightings in the Athletes Building were replaced with electrodeless lamps and the power saving ratio was more than 50%. The lightings in the athletic fields and parking lots were replaced with ceramic discharge metal-halide lamps and the power saving ratio was also more than 50%.

### 3.8 Installing green roofs

According to the data from Taiwan Power Company, the roof of a building is one of the main indoor heat sources since it has the most solar irradiation. Therefore, in order to reduce the heat load and energy consumption of a building, lowering the temperature of the roof is an effective way. Installing green roofs in CYUT not only effectively lowered the indoor temperature but also increased the green open space ratio in campus. Currently, green roofs have been installed on the Design Building, Teaching Building, and Dormitory (Figure 5). From real measurements by the members of the architecture department of CYUT, the indoor temperature could be lowered by more than 2.5℃ which could save the electricity consumption of air conditioners by more than 15%. Also, CYUT promoted this result to the middle and elementary schools, which are counseling partners of Green University, and helped them install sustainable insulation roofs in order to create low-carbon schools hand in hand.

**Fig. 6.** Green roofs in CYUT

Besides roof installation, vine planting is also aggressively conducted in CYUT since vine plants can reduce the heat radiation absorbed by building walls. Currently, vines are planted on the walls of several buildings including Science and Engineering Building, Humanities and Technology Building, Administration Building, Information Building (feature wall), and the retaining walls on the athletic fields (Figure 6).

**Fig. 7.** Green walls in CYUT

## 4 Results

Due to global warming and the passing of the Sustainable Energy Policy Framework by the Taiwan government, CYUT has been promoting the reduction of greenhouse gas emissions since 2008. After years of striving, the results are fairly good. Using 2008 as the base-year, the electricity consumption decreased from 13,484,000 kwh in 2008 to 10,769,600 kwh in 2017. The energy use intensity (EUI) also decreased from 85.66 kwh/m$^2$ to 68.41 kwh/m$^2$ with a reduction ratio of 20.1% (Table 3). For vehicle and hot water boiler, the oil consumption including gasoline and diesel decreased from 51,093 Liter in 2008 to 46,758 Liter in 2017, and the reduction ratio is 8.5% (Table 4). For greenhouse gas emissions, comparing to the emissions in the base-year of 2008 which is 9,104.82 metric tonnes CO2e, the emissions in 2017 was lowered to 6,268.41 metric tonnes CO2e which is a decrement of 2,836.41 metric tonnes CO2e and the reduction ratio is 31.2% (Table 5).

**Table 3.** Electricity consumption in CYUT.

| Year | Electricity Used(kwh) | EUI(kwh/m$^2$) |
|---|---|---|
| 2008 | 13,484,000 | 85.66 |
| 2009 | 13,782,800 | 87.55 |
| 2010 | 13,485,200 | 85.66 |
| 2011 | 13,229,600 | 84.04 |
| 2012 | 12,518,600 | 79.52 |
| 2013 | 12,106,200 | 76.90 |
| 2014 | 11,860,400 | 75.34 |
| 2015 | 12,107,200 | 76.91 |
| 2016 | 11,653,600 | 74.03 |
| 2017 | 10,769,600 | 68.41 |

**Table 4**. Oil consumption in CYUT

| Year | Vehicle | | Hot water boiler |
|---|---|---|---|
| | Gasoline (Liter) | Diesel(Liter) | Diesel (Liter) |
| 2008 | 25,951 | 1,294 | 23,848 |
| 2009 | 22,615 | 4,968 | 21,463 |
| 2010 | 23,646 | 4,317 | 21,781 |
| 2011 | 21,543 | 9,263 | 29,600 |
| 2012 | 12,738 | 11,359 | 29,800 |
| 2013 | 17,998 | 11,193 | 26,000 |
| 2014 | 18,652 | 11,176 | 29,200 |
| 2015 | 17,539 | 10,529 | 32,600 |
| 2016 | 17,210 | 9,562 | 27,800 |
| 2017 | 16,034 | 10,524 | 20,200 |

**Table 5.** Greenhouse gas emissions in CYUT

| Year | Emissions (metric tonnes $CO_2$e) | Reduction from base-year | Remarks |
|---|---|---|---|
| 2008 | 9,104.82 | -- | base-year |
| 2009 | 9,173.48 | -68.66 | |
| 2010 | 8,855.43 | 249.39 | |
| 2011 | 7,707.68 | 1,397.14 | |
| 2012 | 7,449.75 | 1,655.07 | |
| 2013 | 6,991.79 | 2,113.03 | |
| 2014 | 6,897.23 | 2,207.59 | |
| 2015 | 7,018.10 | 2,086.72 | |
| 2016 | 6,775.35 | 2,329.47 | |
| 2017 | 6,268.41 | 2,836.41 | |
| Total reduction (metric tonnes $CO_2$e) | | 14,806.16 | |

## 5 Conclusion

According to the assessment reports by IPCC, it is getting more obvious that global warming and climate change are caused by massive usage of fossil fuel by humans and the consequences are huge loss of lives and properties. It is inevitable to reduce the use of fossil fuel and improve the efficiency of energy utilization. Since the operational logic behind campus life is a kind of pseudo-socialization process [6], universities should also take responsibility for leading people to face the challenge from climate change.

For creating a green campus in CYUT, the management goals for energy consumption and greenhouse gas emissions have been achieved in recent years by implementing energy management systems, replacing high energy consuming facilities, installing green roofs, and changing the behaviors of students and staffs. For improving in the future, ISO 14001 Environmental management and ISO 50001 Energy management will still be implemented to reduce the energy consumption and increase the efficiency of energy utilization. Last but not least, environmental education will be emphasized to raise the environmental awareness of the students and staffs for creating a low-carbon and sustainable campus in CYUT.

https://doi.org/10.1051/e3sconf/20184803001

## References

1. Intergovermental Panel on Climate Change (IPCC), Available online at http://www.ipcc.ch/
2. Huang-Hsiung Hsu, Chia Chou, Yi-Chao Wu, Mong-Ming Lu, Cheng-Ta Chen, Yung-Ming Chen, *Climate Change in Taiwan: Scientific Report 2011(Summary),* National Science Council, Taipei, Taiwan, ROC (2011)
3. Executive Yuan Agriculture Bureau Forestry, Available online at http://lifetree.forest.gov.tw/cp05.asp
4. United Nations, Available online at http://www.un.org/sustainabledevelopment/sustainable-development-goals/
5. Chaoyang University of Technology Officer for Environmental Safety and Health, "2017 Campus Environment Report" (unpublished), Chaoyang University of Technology, (2018)
6. Cortese, A, *The Critical Role of Higher Education in Creating a Sustainable Future, Planning for Higher Education.* March-May 2003: pp.15-22, (2003)

E3S Web of Conferences **48**, 03002 (2018) https://doi.org/10.1051/e3sconf/20184803002
*IWGM 2018*

# The Unibo energy saving and climate change approach

*Mirko Degli* Esposti[1*], *Olivia* Bernardi[2], and *Luciana* Sacchetti[3]

[1]Alma Mater Studiorum – Università di Bologna, Deputy Rector, Professor of Mathematical Physics, Italy
[2]Alma Mater Studiorum – Università di Bologna, Sustainability Office, Italy
[3] Alma Mater Studiorum – Università di Bologna, Research and Third Mission Division, Italy

**Abstract.** For the university of Bologna, the core of sustainability combines both environmental and social dimensions: enhancement of the territory, ensuring its protection and the renewal of its natural resources and the ability of the parties involved, encouraged by the concerted efforts of the various institutions, to work together effectively. The commitment of the university is to reduce the impact on the environment deriving from its activities and to encourage the community to assume responsible and environmentally respectful behaviours. Since 2016, the university integrates its Strategic Plan with a re-reading of the 17 UN SDGs and reports the direct and indirect impact produced in the dimensions of teaching, research and third mission.

## 1 Introduction

The Alma Mater Studiorum - Università di Bologna founded in 1088 and recognized as the oldest university in the Western world, today is one of the most important institutions of higher education throughout Europe with more than 84,000 registered students, 2850 academics, 3100 administrative employees and over 3700 young scientists.

Managing more than 1,000,000 sq. m. of modern and historical buildings of the university citadels located throughout the region and beyond, the University of Bologna plays a decisive role within the urban, regional and national dynamics.

For the Alma Mater it is natural to be a promoter of the principles of sustainability such as the enhancement and protection of the territory, the improvement of community welfare, the economic development, the social equity and the capacity of the subjects involved to work together effectively for the common good.

To achieve objectives in terms of sustainability the university of Bologna takes a project-based approach towards pragmatic results. That involves adoption of innovative methodologies for the measurement of results and control of the various operational processes, defined in the 2016-2021 Sustainable Multicampus Project [1].

## 2 The university strategic plan 2016-2018 and the UN SDG's

Since 2016 the university's governance foresaw to develop the university Strategic Plan [2] following the 2030 Agenda and the sustainable development goals of the United Nations (SDG's) for tackling the challenges posed by contemporary reality and in particular for a sustainable development.

This innovative approach has led to re-read the university's activities to allow the insertion in everyday life of the values and the communicative and inspiring power of the SDGs, with a real impact on the university's community and its stakeholders.

Unibo has integrated its Strategic Plan with a full re-reading of the 17 basic sustainable development goals. Every basic target in the Strategic Plan is associated with one or more of the sustainable development goals (Fig.1).

**Fig. 1.** the strategic Plan's basic targets and the related UN sustainable goals.

To the area of the Third Mission is associated the strategic goal "to promote cultural development plus economic and social innovation". In this framework, the basic target C.1.3 recites as "to promote processes that enhance the environmental and social sustainability of the university's buildings, facilities and community, while also revitalizing the area in synergy with all local

* Corresponding author: mirko.degliesposti@unibo.it

institutions". The actions planned to reach that target are aligned with three SDGs: 5 Gender's equality, 10 reduce inequality, 11 sustainable cities and communities.

# 3 The report on U.N. sustainable development goals 2016

Consistent with the perspectives adopted in the Strategic Plan 2016-2018, the University has an innovative way to report on the contributions generated by its institutional activities, aimed at the achievement of the 17 SDGs. The report [3] presents 17 sections, each one dedicated to a UN goal, reporting the impact generated by the university of Bologna through its diverse activities. Each goal is reported in relation to the four academic dimensions of training, research, third mission and the institution.

The report and its metrics were formulated by the Scientific and Technical Committee for Social Reporting at the University of Bologna. Data refers to 2016 as well as the Academic Year 2016/17 for training data. The original metrics match and integrate other official document of the Alma Mater, a methodological note on page 62-65, gives a full description of each metric cited in table 1.

**Table 1.** The reporting metrics

| Publications in Scopus | "Cited by" in Scopus | International benchmarking | National benchmarking |
|---|---|---|---|
| Course Unit | Students | Collaborations, teaching mobility | FP7, H2020 research projects |
| Cooperation & social engagement research projects | Patents (including plant varieties) | | |

# 4 Take urgent action to combat climate change and its impact – SDG 13

The impact of the university on the sustainable development goal 13 (Fig. 2) is described in the report from three different perspectives: teaching, research and from the viewpoint of the "institution". Evidences for the teaching engagement on climate change are the 54 course units and the 4,794 students who could choose the course unit.

From the research perspective, 6 publications during the period 2007-2016 could be fully identified in Scopus as related to climate change and its impact, generating 164 citations. The bibliometric outputs are responsible for ranking Bologna below the international benchmark (100) for publication per capita (Unibo index number 30), but above the national benchmark which consider the universities of Padua, Rome La Sapienza, Milano Statale (Unibo index number 130). More relevant is the impact of the transportation actions taken by the University as an institution, such as the use of public transport on favourable terms, an electric fleet management and the promotion of bicycle. As certified by GreenMetric 2016 on 2015 data, the Alma Mater CO2 emission over the last 12 months is about 512 Tons.

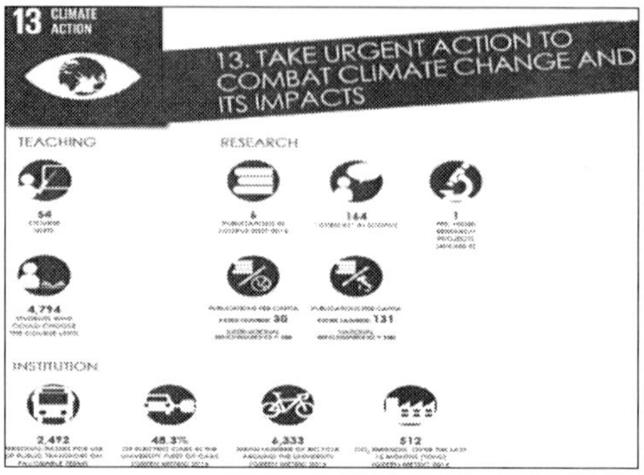


**Fig. 2.** the climate action at the University of Bologna.

# 5 Actions for sustainability

## 5.1 Energy

In order to reduce energy consumption, the University of Bologna strives to make technological and architectonic enhancements and encourages the University population to use energy more rationally. The measures identified seek to respond to the many requirements stemming from the constant improvement of buildings and to facilitate the rationalization of their differing energy needs.

The commitment of the University in the area of energy saving is designed to:

- o Improve awareness, as well as the management and control of building energy consumption, by adopting measures to obtain and manage energy data in real time, and to control heating and air-conditioning installations on a remotely.
- o Improve architectonic efficiency, by adopting measures that increase the energy efficiency of buildings and their heat insulation.
- o Improve the efficiency of systems by implementing technological upgrades and optimizing the configuration of the heating and cooling systems installed.
- o Encourage and spread the rational use of energy, as a key element in the reduction of waste.

The principal actions undertaken are:

*Renewable energy* – The university harnesses the renewable energy sourced from photovoltaic installations: Unibo has more than 8,500 sq. m. of photovoltaic surface in its buildings.

*Remote monitoring and energy management* – The University of Bologna monitors the consumption of utilities on a remote basis. The GECO project (Energy Management and Remote Control) seeks the maximum energy saving of buildings, by controlling energy flows and environmental parameters.

*Smart building implementation* – The university has applied a program of smart building implementation, providing several energy efficient appliances and tools. In almost all buildings is installed a network technology, in order to control facility functions, such as thermal power plant management system, internal volumetric sensor installed for physical security, remote control of an external lights system, automatic light and Fire Alarm sensor system, dashboard of the integrated energy management system.

*Replacement of lighting* – The University of Bologna guarantees the constant upgrade of internal and external lighting. Moreover, we are experimenting innovative remote management and remote-control systems for lighting bodies and progressively subsitute obsolete lighting fixtures with high-energy consumption with innovative systems.

### 5.2 Climate change

As a contribution to the protection of the environment, the university promotes the enhanced use of green areas, a more responsible use of water and waste, to help improve the quality of life in University areas. Via the rational use of natural resources in its buildings, the university pursues the objective of harmonizing the essential trade-off between built-up areas and the surrounding environment.

In addition to encouraging students and staff to dedicate greater attention to environmental matters, the approach adopted seeks to highlight the value of our territory and landscapes. Moreover, in order to reduce atmospheric emissions, the University of Bologna promotes a low impact mobility both when at work and when travelling between home and work/place of study.

Our commitment to mobility is designed to:

- Encourage public transport by adopting measures to reduce the number of private vehicles used.
- Lower atmospheric emissions by promoting the use of cycle paths and other forms of low-impact transportation.
- Offer innovative and sustainable mobility services aimed at the shared use of vehicles.

The actions taken to implement our commitment are the following:

*Reduce and recycling waste* – we manage the various types of waste produced by its structures with a view to proper separation, safety and recycling. Unibo has established formal policy to reduce the use of paper and plastic and separate waste for recycling. The university has provided suitable containers (about 10.000) for paper, plastic, glass and aluminium, batteries and toner cartridge, located in all the buildings. Moreover, the toxic waste are completely contained, inventoried and handled.

*Green roof* – The university installed approximately 1,000 sq. m. of green roofs and hanging gardens on the horizontal roof plans of the university buildings.

*Green areas* – The university increases the amount of green by transforming the use of existing areas, renovating green ones abandoned and developing the use of greenery in urban settings.

*Water* – The university strives to make best use of the available water resources, in order to contain the amount of water wasted not only with water efficient appliances but also with automatic irrigation systems.

*Public transport* – The university encourages public transport enabling students and personnel to purchase passes for the use of public transport on favourable terms agreed with the transport companies and municipal administrations.

*Electric cars and car sharing* – The University of Bologna has rationalized and renewed its fleet of cars, which are all-electric and used in car sharing mode.

*Bicycle* – we encourage the use of bicycles by providing each building with bicycle parking and bicycle paths. Moreover, in order to disseminate a culture of sustainable mobility the university has signed an agreement with the Bologna bike station that enables the university community to use its various services on advantageous terms: bike rental, secure parking and repair/self-repair facilities.

## 6 Concluding remarks

The Alma Mater Studiorum is committed to spreading a real culture of sustainability both as National coordinator of the UI GreenMetric Network and in collaboration with the RUS – University Network for Sustainable Development [4], which principal goal is to increase the environmental, ethical, social and economic impact of the actions implemented by the Italian universities.

## References

1. 2016-2021 Sustainable Multicampus Project available online at http://www.sostenibilita.unibo.it/unibo-sostenibile/progetto-multicampus-sostenibile
2. University Strategic Plan 2016-2018 available online at http://www.unibo.it/en/university/ who-we-are/strategic-plan
3. Report on U.N. Sustainable Development Goals 2016, page 67 available online at http://www.unibo.it/it/ateneo/chi-siamo/reporting-on-united-nations-sustainable-development-goals
4. RUS – University Network for Sustainable Development available online at https://sites.google.com/unive.it/rus/

# University of Turin performance in UI GreenMetric Energy and Climate Change

*Marcello* Baricco[1,*], *Andrea* Tartaglino[2], *Paolo* Gambino[3], *Egidio* Dansero[4], *Dario* Cottafava[5] and *Gabriela* Cavaglià[6]

[1] Vice-Rector, Department of Chemistry and NIS, University of Turin, Italy
[2] Energy Manager, University of Turin, Italy
[3] Department of Physics, University of Turin, Italy
[4] Sustainability Delegate, Department of Culture, Politics and Society, University of Turin, Italy
[5] Department of Physics, University of Turin, Italy
[6] UNESCO Chair in Sustainable Development and Territory Management, University of Turin, Italy

**Abstract.** At the University of Turin (UniTo) the first attempt to reduce the ecological footprint dates back to 2006. In 2013 UniTo participated, for the first time, to the GreenMetric World University Rankings and it launched its most ambitious sustainable initiatives by publishing the first annual University Sustainability Report. Since 2013, several efforts have been made to reduce carbon emissions, to improve the energy efficiency of buildings, as well as for reducing water consumption, improving waste management, promoting sustainable mobility, and increasing ecological purchases. The most recent achievement was in 2016, when the University Green Office (UniToGO) was established. Concerning the energy management, UniTo may be considered as a "city within a city": it counts about 70,000 students and 4,000 academic and administrative-technical staff studying and working in about 120 buildings, with an annual primary energy cost of over 10 M€. Thanks to UniToGO, UniTo adopted an Energy Plan with the aim to reduce primary energy consumption, to improve buildings energy efficiency and to increase the energy production from renewable energy. As a result, several actions relevant for GreenMetric were performed: the renovation of old chiller and substitution of new energy efficient LED, the implementation of Smart Building Systems (BEMS) for HVAC plants and the adoption of an OpenData policy for energy consumption, the increase of renewable energy production, mainly due to three cogeneration plants, and the adoption of a university policy in order to buy only renewable energy from the current Distribution System Operator. Moreover, UniTo took several efforts to improve or to design elements of green buildings in a partecipatory way. Finally, during 2017, the Environmental Sustainability Action Plan was set-up to plan future actions related to five sustainability fields: Energy, Food, Green Public Procurement, Mobility and Waste.

## 1 Introduction

Since the Rio de Janeiro Earth Summit in 1992, universities – considered as place of training of future generation - are called to assume their social responsibility and the important role of "agents of change" through targeted choices in their three spheres of intervention:

1. education, to foster the development of a culture of sustainability and the widening of the educational offer of green curricula and courses [1];
2. research, to stimulate the development of inter and multidisciplinary research projects with a holistic and innovative approach to deal with climate change and, more in general, with challenges facing 21st century society;
3. "third mission", to develop targeted networking actions with local authorities, stakeholder engagement and active citizenship involvement [2].

The goal of reducing the university ecological footprint needs the involvement and the commitment of the entire community, from the governing bodies to students, from academia to technical- administrative staff.

Over the years, numerous universities all over the world have started a reflection on the role that the universities can play in the challenge of environmental sustainability. Many campuses have undertaken a process of improvement of their environmental performance by planning policies, strategies, projects and precise interventions. Moreover, several university networks have been developed to share experiences, ideas, solutions and best practices [3-5].

The first attempt made by the University of Turin (UniTo) to reduce its ecological footprint dates back to 2006 with the project *Towards an Agenda 21 of the University*, a students' initiative supported by some professors. It led to the creation of the University Environmental Sustainability Commission with the aim of proposing structural interventions on campus

* Corresponding author: marcello.baricco@unito.it

https://doi.org/10.1051/e3sconf/20184803003

sustainability [6]. Some experimental pilot projects, aimed to improve mobility and canteen management, were carried out with a participatory approach promoting the creation of Local Action Groups, open to the participation of all members of the university community. Although the initiatives were carefully monitored, the lack of dedicated human and economic resources, together with the lack of strong political will on the part of the university's top management prevented the progression of the projects, which however represented the first systematic attempt to design a university environmental policy.

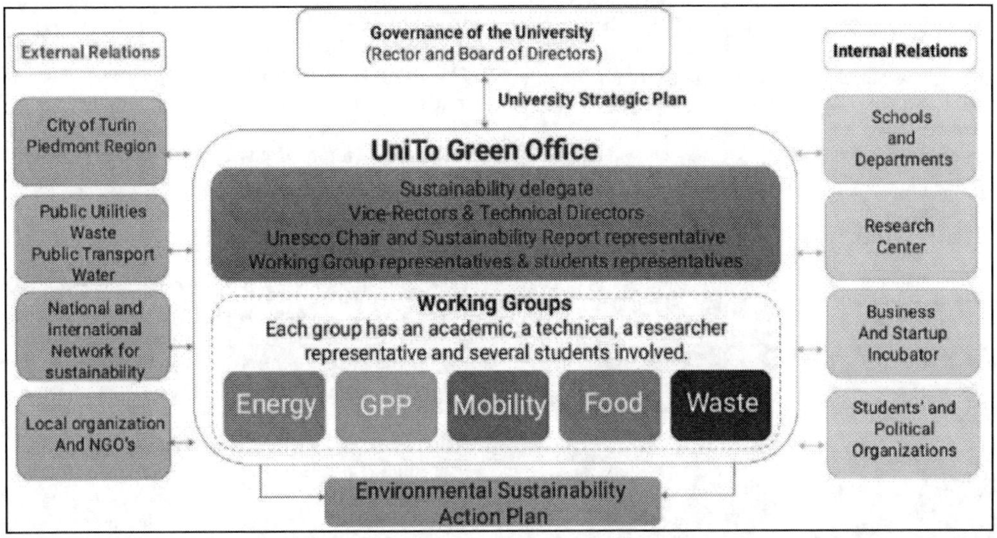


**Fig. 1.** Structure of the UniTo Green Office (UniToGO). The core of UniToGO is based on five working groups and a horizontal board composed by students, researchers, academic, technical and administrative staffs.

The following year, the *UniTo Programming Plan for the period 2007-2012* [7] explicitly mentioned, for the very the first time, sustainable development as a reference for the UniTo's activities.

In 2009 UniTo joined the Green Public Procurement Project (APE) promoted by Turin local government [8], signing a memorandum of understanding aimed at promoting purchases with reduced environmental impact and, in 2014, the commitment to respect some minimum environmental criteria in the procedures for the purchase of goods and services. Since then, UniTo settled an internal APE working group with the aim of monitoring every year UniTo's purchases and stimulates concrete effects in terms of purchasing choices, starting from the energy field.

Since 2014 a particular effort was made for drawing up and implementing the first *Energy Saving Plan* [9] aimed at improving energy efficiency, reduce waste, increase the share of energy production from renewable sources and improving building efficiency and user comfort. A multidisciplinary working group, *RE@UniTo*, was specifically settled [10]. The activity of this group and the results achieved are described thoroughly in the second part of this paper.

The focus on environmental sustainability strengthens in the *Strategic Lines of the University 2013 - 2015* by recognizing UniTo social responsibility and, even further, in the *2016 - 2020 University Strategic Plan* by introducing a specific reference to the environmental dimension – "Increasing the social, economic and environmental responsibility of the University of Turin" - , expressed in terms of objectives and indicators.

A fundamental step is registered with the decision to draw up and publish annually - starting from 2013/14 academic year - a *Sustainability Report* [11], with the purpose of analyzing and communicating publicly UniTo's performance in terms of economic, social and environmental sustainability, following the international methodology of the *Global Reporting Initiative* (GRI) [12]. The editing of this report is not just an assessment of UniTo's environmental impact, but also the opportunity to analyse policies and interventions carried out, to present the progress reached and to identify the steps still to be taken. The report constitutes a fundamental element in the formulation of a university environmental policy and it launched the creation of the UniTo Green Office (UniToGO).

UniToGO was created in the spring of 2016, gathering thrusts from the top (Rector and General Manager) and from the bottom (students, academia andadministrative staff), around the experience of the RE@UniTo and the APE working groups, in order to systematize UniTo knowledge and experience to promote initiatives on environmental sustainability, supported by the university community. UniToGO represents now the link between the planning addressed by UniTo governing bodies and five thematic working groups appointed to concretely develop research and actions in the various areas of intervention identified (above energy and green public procurement also, food, waste and mobility). It is a mixed and interdisciplinary entity made up by academic and administrative staff, together with students coming from many different disciplines within UniTo, as shown in Fig. 1. UniToGO has three main goals and lines of intervention:

- "Knowing" by building a shared knowledge base both on the initiatives launched within the UniTo and on good practices, with particular attention to the experiences of other universities in Italy and abroad;
- "Communication and Engagement" by intensifying networking actions within the university community and externally between UniTo and the relevant actors on a local, national and international scale;
- "Change" by designing and implementing interventions of various nature aimed at effectively reducing the environmental impact of the university in various areas.

**Fig. 2.** Renovation of the heating plants

**Fig. 3.** Trigeneration power plant at Campus Luigi Einaudi.

### 1.1 Local and international networks

2013 was not only the year in which UniTo published its first Sustainability Report but also the first time it took part to the *UI GreenMetric World University Rankings* with the idea to start a comparison with other universities committed on reducing their ecological footprint. Every year, UniTo succeeded in improving its score: from a total score of 4.372 in 2013 it reached 6.064 in 2017 moving its position in the world ranking from 211 to 55 (second on the Italian rank). GreenMetric represents for UniTo not just a reliable tool for national and international comparison but also a guide to internally identify sphere of intervention.

Up to now, in addition to the participation to GreenMetric, UniTo extended the opportunity for comparison with other sustainable campuses network both at national and international level. UniTo joined the *International Sustainable Campus Network* (ISCN) [3] in 2015; it brings together over than 80 colleges and universities in more than 30 countries in a forum aimed at the exchange of information, ideas and best practices to improve environmental performance of the universities and to integrate themes related to sustainability in research and higher education.

At European level, UniTo joined the *Green Office Movement* [5], promoted by the *Rootability* association [13], which counts 17 participating universities located in 7 European countries (and 25 in the process of joining). The Green Office Movement aims to facilitate the creation of formalized working groups, made up by students supported by technical and administrative staff, to guide and coordinate the realization of ecologically sustainable and economically feasible projects. UniTo was the first Italian campus to take part in the movement, and UniToGO received the formal membership precisely for the attention dedicated to the students' involvement.

At a national level, the *Network of Universities for Sustainability* (RUS) was established in 2015 within the *Conference of Italian University Rectors* (CRUI) [14]. It counts 50 Italian universities that agreed to commit themselves to the issues of sustainability and social responsibility.

## 2 Energy and climate change

### 2.1 Renovations and interventions

The real estate assets of UniTo include around 120 locations dating back to different periods (from 18th to 21st centuries), for a total area of about 800,000 square meters. Since 2000 a series of restorations were carried out in the university buildings, both to renovate the facilities and to improve energy efficiency. In 2007, these efforts resulted in an important project whose total cost exceeds € 12 millions. The project led to the following renovation works:

- installation of three trigenerators, with a power of 1,154 kWhe which enable the self-production of electricity and supply heating and cooling (Fig. 2);
- transformation of all the heating plants from diesel oil to methane, with the installation of condensation boilers (Fig. 3);
- creation of a remote management system for air conditioning systems, to allow complete management of the heating and conditioning plants (Fig. 4).

https://doi.org/10.1051/e3sconf/20184803003

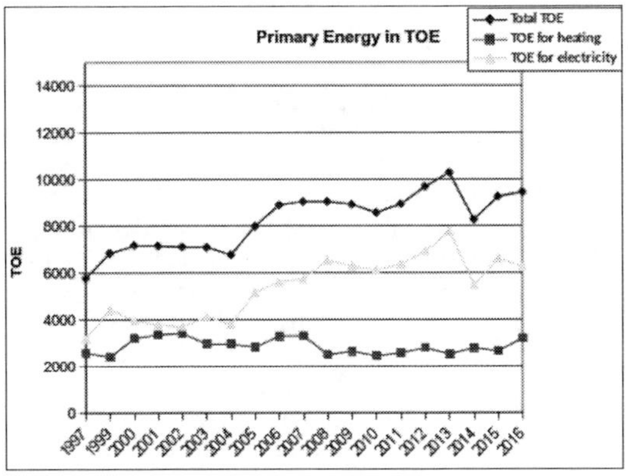


**Fig. 4.** Annual energy consumption trends from 1997 - 2016 of the University of Turin

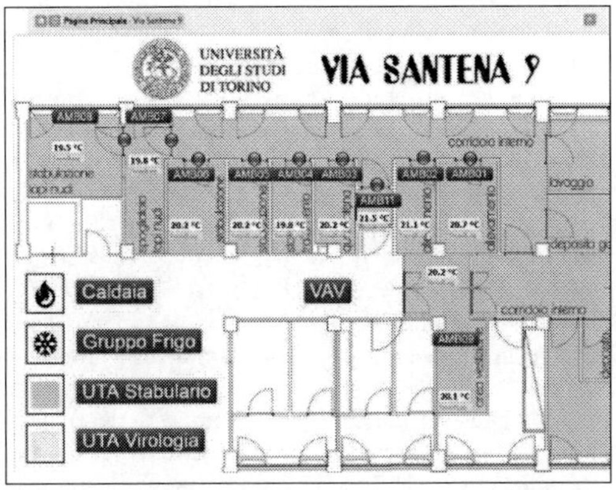


**Fig. 5.** A screenshot from the building energy management system

All these actions led to an annual saving of around 1.8 M€ and 1,600 TOE, bringing the total annual cost of primary energy for UniTo to around 10.5 M€. Currently, UniTo has an annual consumption of about 10,000 TOE and the consumption trend is summarized in Fig. 5.

Thanks to the activities of RE@UniTo first and UniToGO later, in 2016 UniTo adopted an *Energy Plan*. This plan includes, in particular, *long-term structural renovations* characterized by high cost from which high long-term savings are expected, *soft renovations*, with limited cost and fast return, *awareness-raising activities*, to raise users' awareness over a conscious and rational use of energy and *administrative interventions*, to optimize primary energy bills. As a result, several actions relevant for GreenMetric were performed:

1. numerous renovations and energy efficiency improvements have been made with LED technology and the use of energy efficient appliances and lighting fixtures, as well as the standardized the design of new spaces (EC 2.1);
2. all the old refrigeration systems have been censured to give life to a plan to replace the older chillers with new ones with energy class at least A++ (EC 2.1);
3. the implementation of Smart Building Systems (BEMS) for HVAC plants and the adoption of an OpenData policy for energy consumption (EC 2.2);
4. the increase of renewable energy production, mainly due to three cogeneration plants, and to the adoption of a University policy in order to buy only renewable energy from the current Distribution System Operator (EC 2.3, 2.5).

Moreover, UniTo took several efforts to improve or to design elements of green buildings in a partecipatory way (EC 2.6), and there is an attempt to create sustainable design standards for university building designers.

## 2.2 Community engagement

Several initiatives have been organized during the past years in order to engage and involve both the academic community and external stakeholders in the rational use of energy.

Firstly, several round tables have been organized during the past two years in order to engage all type of stakeholders, from students to professors, from local non-governmental organizations (NGOs) to energy providers. A round table, titled *UniToGO Community Engagement* is organized each year for each working group – i.e. food, energy, waste, GPP and mobility – in order to collect direct feedbacks from UniTo's stakeholders on particular topics [15].

Secondly, seminars were organized on different arguments, such as smart metering systems, behavioural change, the Sustainable Development Goals, green roofs and crowdmapping, in order to explore and test new approaches, to understand psychological implications, opportunities, barriers and threats of new technologies and participatory processes. For instance, during a two days-event, titled *"United for Sustainable UniTo"*, a workshop was run about *"Internet of Thing for Energy Monitoring and Energy Efficiency"* where students and participants could learn how to monitor temperature, humidity, energy consumption and other environmental measurements in real time - thanks to two open source projects: Scatol8 [16], based on Arduino, and OpenEnergyMonitor [17], based on Raspberry Pi - and how to build a basic circuits based on Arduino board [18].

In order to explore behavioural change, a workshop, titled *"Co-Design of the local Sustainability Ecosystem"* [19], was organized by engaging more than 40 local stakeholders from NGOs, public administrations, private enterprises, as well as citizens and students. The goal of the 4 hours workshop was to identify and represent the local ecosystem (i.e. 205 nodes and more than 300 links for the City of Turin), starting from four topics – energy, food, waste and mobility – highlighting the interrelations among them. In particular, related to energy topic, more than 60 relevant actors have been identified. Results from the workshop are publicly available [20] and are syncronized in real time with google drive spreadsheets [21]. As a final example, the *"HOME"* seminars, organized within the framework of a research project, *"Human Observations Meta Environment"* [22], based on a smart metering system for heating, ventilation and air conditioning (HVAC) systems, where buildings users may interact with buildings through natural interactions (gesture, voice and movements) or through social networks (Facebook, Instagram and Twitter) by writing feedbacks with a precise hashtag.

Above workshops and seminars, other type of actions and initiatives have been promoted in the past two years. For instance, *"On Hot and Cold"*, an art installation, during the Contemporary Art Week in Turin, and *"Hunt for kWh"*, a flashmob to promote a rational use of energy where participants, split in teams, walk around the university buildings to switch off light, heating, desktop computers, printers and other electrical loads. "Hunt for kWh" was an initiative organized by the RUS network within the framework of *"M'Illumino di Meno"* (I enlighten me less) an annual Italian campaign [23] about energy saving. It is noteworthy also to mention the students' thesis on detailed energy audit which have helped the UniTo energy management in identifying many energy wastes [24-26].

### 2.3 Research Projects

Various innovative projets related to smart buildings and smart metering systems, as well as OpenData, have been lead within UniTo. A strategic research project, *"ComfortSense"* [27], started in 2014 and aiming to decouple human comfort from energy consumptions. ComfortSense is based on a smart metering system, which allows to measure *objective variables* (temperature, humidity, $CO_2$ concentration and light intensity), and on a mobile app, which allows to measure *subjective variables*, i.e. direct feedbacks from buildings users. The project has been applied on three main buildings, engaging more than one thousand students and pointing out how it is possible to reduce energy consumption and, meanwhile, increasing human comfort.

Finally, an online dashboard for energy management has been developed [28], based on Open Source javascript library as D3.js [29] in order to help the analysis of energy consumption of a large buildings stock – as previously mentioned UniTo counts around 120 buildings – and to improve the efficiency of large energy audit. Further researches allow to develop and test multidimensional analysis tools in order to quickly recognize buildings depending on their main function and to identify and highlight outliers [30]. These analysis tools have been developed following the ASHRAE recommendations for a high quality and appropriate analysis method – i.e. accuracy, sensitivity, speed, reproducibility, ease of use, level of detail, availability of required data and quality of the output [31].

## 3 Conclusions & future planning

### 3.1 Education for sustainable development programmes

The United Nations constituted 2005–2014 as the decade for educational sustainable development [32]. Since then universities start to debate about how they can form students capable of taking on the 21st century challenges of global warming and climate change, social inequities, unsustainable lifestyles and the need to switch to renewable energy. UniTo started this process of greening curricula in 2015 with a brief research project on Green Jobs. Among others, one of the main project goal was to identify the green course already offered by the university, i.e. courses strictly related to environmental sustainability.

UniTo is one of the most ancient and prestigious Italian universities that virtually covers every field of knowledge (all disciplines, except for Engineering and Architecture) and offers a huge quantity of courses (around 8,000 different courses) every year. However, from the courses database it was not possible to simply identify the green ones. The project on green jobs run a qualitative analysis by interviewing several professors from different disciplines coming out with a result of 359 courses. Since then many green curricula have been developed in UniTo (Ph.D. Program in Innovation for the Circular Economy, Master's Degree Program in Economics of the Environment, Culture and Territory, Master's Degree Program in Socio Environmental Sustainability of Agrifood Networks).

Moreover, to answer the GreenMetric Education 6.2 indicator, UniTo developed a semiautomatic procedure to identify green courses by checking the occurrence of special keyword in the course and curricula descriptions. The testing of this procedure is still running comparing different sets of keywords.

### 3.2 Environmental Sustainability Action Plan

The last fundamental step towards the sustainability of the University of Turin has been achieved last October 10th, 2017, when the Academic Senate and the Board of Directors approved the *"Environmental Sustainability Action Plan" (ESAP)* [33], written in a collaborative and participatory way by UniToGO. The Action Plan was developed engaging dozens of UniTo stakeholders during the *"UniToGO Community Engagement"*. During these round tables, internal and external stakeholders proposed possible actions and initiatives to improve the environmental sustainability of the University based on the three guidelines of UniToGO – knowing, engaging, changing – and the five topics – Food, Energy, Waste, Mobility and GPP. Within the ESAP of the University of Turin, dozens of actions, divided in structural, soft and communicative interventions, have been planned for the period 2017 - 2020.

## References


1. L., W. Petronella, *Green curriculum: sustainable learning in higher education,* The International Review of Research in Open and distributed learning **14**, 1 (2013).
2. R., Pinheiro, P., V., Langa, and A., Pausits, *One and two equals three? The third mission of higher education institutions*, European Journal of Higher Education **5**, 3 (2015).
3. International Sustainable Campus Network (ISCN): Available online at https://www.international-sustainable-campus-network.org/
4. Unica-Green, Available online at http://www.unica-network.eu/page/unica-green
5. rootAbility gUG and the UNESCO Chair Higher Education for Sustainable Development at Leuphana University Lüneburg, Available online at http://www.greenofficemovement.org/
6. Commissione Sostenibilità Ambientale Del Senato Degli Studenti Dell'università Degli Studi Di Torino, Anno Academico 2007/2008, Available online at https://goo.gl/hXgEFP
7. Divisione Organizzazione e Finanza Staff Sistema di Programmazione e Controllo, Università degli studi di Torino Available online at https://www.unito.it/sites/default/files/fatti_e_numeri_del_piano_di_programmazione_aa2009.pdf
8. The Green Public Procurement (GPP) Working Group, Available online at http://www.green.unito.it/en/Green_Public_Procurement
9. University of Turin, Available online at http://www.unito.it/ateneo/strutture-e-sedi/piano-energetico-di-ateneo
10. The Energy Working Group, Available online at http://www.green.unito.it/en/Energy
11. University of Turin, Rapporto di sostenibilità 2015/2017, Available online at https://goo.gl/eDbPVV
12. Global Reporting Initiative (GRI), Available online at https://www.globalreporting.org
13. rootAbility, Available online at http://rootability.com/
14. RUS - Rete delle Università per lo Sviluppo sostenibile, Available online at https://sites.google.com/unive.it/rus/
15. UniToGO Community Engagement, Available online at http://www.green.unito.it/it/node/273
16. Scatol8, Available online at http://scatol8.net/
17. OpenEnergyMonitor, Available online at https://openenergymonitor.org/
18. GreenTo, Available online at http://www.unitixunitosostenibile.it/workshop/1/index_en.html
19. Workshop di Co-Design dell'ecosistema della sostenibilità, Available online at http://www.green.unito.it/it/Workshop_CoDesign_Sostenibilit%C3%A0
20. GreenTo, Available online at https://goo.gl/V83JDR
21. Stakeholder Sostenibilità Torino, Available online at https://goo.gl/QE6bVV
22. Human Observation Meta Environment, Available online at http://www.green.unito.it/it/home
23. Caterpillar, Available online at http://caterpillar.blog.rai.it/milluminodimeno
24. UniToGO, Sul Caldo e Sul Freddo, Available online at https://goo.gl/e1Es13
25. GreenTo, Innovation in the food sector From urban food production to the management of the university Available online at https://goo.gl/ns39QV -
26. UniToGO, L'Efficienza Energetica al Campus Luigi Einaudi: studio degli sprechi e idee per il risparmio, Available online at https://goo.gl/VPdHUU
27. ComfortSense, Available online at http://www.green.unito.it/en/ComfortSense_Project
28. D., Cottafava, P., Gambino, M., Baricco, and A., Tartaglino, Energy efficiency in a large university: the Unito experience. *Sustainable Built Environment. Towards Post Carbon Cities*, 92-101 (2016)
29. M., Bostock, V., Ogievetsky, and J., Heer, D3: Data-Driven Documents. *IEEE Transactions on visualization and computer graphics*, 92-101 (2011).
30. D., Cottafava, P., Gambino, M., Baricco and A., Tartaglino, Multidimensional analysis tools for energy efficiency in large building stocks. *12th SDEWES conference* (2017).
31. ASHRAE, *Fundamentals American Society of Heating, Ventilating and Air- Conditioning Engineers* (1997).
32. Shaping the Future, We Want - *UN Decade of Education for Sustainable Development*, UNESCO (2014) - Available online at https://goo.gl/q9PMmH
33. UniToGO, Approvato il Piano d'Azione per la Sostenibilità Ambientale, Available online at http://www.green.unito.it/it/ESAP_Unito

# Design and operation experience of zero-carbon campus

*Pan* Xu *, *Zebang* Jin, *Yuzhu* Zhao, *Xiaobo* Wang and *Haiquan* Sun

Shandong Normal University - Lishan College, Qingzhou, Shandong, China

**Abstract.** Shandong Normal University - Lishan College is a zero-carbon campus, using the technical scheme "multiple sources of energy complementing each other, 100% utilization of renewable energy, combining concentrated demonstration with popular extension". It has built 5 distributed energy stations, including roof PV power station, solar heating water system, biomass vacuum hot water unit, natural gas (straw pyrolysis gas, biomethane) CCHP demonstration project etc. Renewable energy can provide power, heating, air conditioning and hot water for 10,000 teachers and students. The zero-carbon campus save 3,650 tce/a, and the $CO_2$ emission reduction is 9,490 t/a. It offers great experience and is a model for the regional clean and low carbon energy usage and the realization of sustainable development.

## 1 Introduction

Energy shortage and environmental pollution are the two major problems facing the world today. The consumption of fossil fuel causes not only the emission of greenhouse gases such as $CO_2$ but also pollutants such as dust, $SO_2$ and $NO_x$. Renewable energy is an important means for the development of zero carbon [1,2,3]. Large-scale application of renewable energy can reduce carbon emissions, solve air pollution and achieve sustainable development. University should be the leader of social development and construction of zero-carbon campus plays an important leading and demonstrating role [4,5]. Zero-carbon campus is a test platform for technological innovation. It is a training base for students with sustainable consciousness and a model for leading society to achieve green, low carbon and sustainable development.

Shandong Normal University - Lishan College is an undergraduate university approved by the Ministry of Education of China. Its goal is to build a famous applied university. The school-running model is integration of production, education and research, integration of school and enterprise. The school has 400,000m$^2$ buildings with good insulation, floor radiation heating, and PV panels or solar collectors on roofs. The school has made breakthroughs in the field of renewable energy technology, energy saving and environmental protection, and it has established close cooperation with dozens of universities both at home and abroad. Lucy New Energy Technology Co., Ltd set up by Lishan College is a high-tech enterprise focusing on new energy technology. Its products were sold well in more than 40 countries or regions. Lishan College and Lucy New Energy Company are pioneers of R & D and utilization of new energy technologies in China and have built the zero-carbon campus of large-scale application of renewable energy. There are thousands of visitors from all over the world every year.

## 2 Technical schemes of zero-carbon campus

### 2.1 Design Concept

Shandong Normal University - Lishan College located in Yunmen mountain scenic spot, undertaking the responsibility of protecting environment. As for school construction, we strictly followed green development concept and abided by the law of developing circular economy, rejecting highly polluting fossil fuels such as coal, and applying renewable energy to build a zero-carbon campus.

The design concepts of the zero-carbon campus include:

(1) The application of renewable energy on large-scale to achieve zero carbon emissions.
(2) According to local conditions, using distributed energy technologies to achieve energy self-sufficiency.
(3) Energy cascade utilization and efficient utilization of low grade heat energy.
(4) Using the whole life cycle to analyze and design.
(5) Not to store energy as far as possible and seamless docking with the grid.
(6) To realize the closed loop chain of ecological cycle.

### 2.2 New energy technology for campus

The campus is densely populated, and the energy consumption is large [1]. Owing to stable and concentrated demand of electricity, heat, air conditioning and hot water, it is quite suitable for developing distributed energy system. Campus buildings are basically multi-storey, suitable for the application of solar energy. Generally speaking, the campus is rich in biomass resources, including agricultural and forestry waste, kitchen waste and so on.

* Corresponding author: xpan51@126.com

The rational use of biomass energy can not only change waste into treasure, but also protect the environment. According to the geographical location, energy demand and energy price of Lishan College, the renewable energy and new energy for application are analyzed as follows.

1) Solar energy, including solar absorption air conditioning technology, solar PV air conditioning technology, solar hot water, solar street lamps, and so on.
2) Biomass energy, including direct combustion of straw pellet fuel for heating, biomass pyrolysis gas CCHP, biomethane CCHP and so on.
3) The cascade utilization of energy. Lishan College has built natural gas (straw pyrolysis gas and biomethane) CCHP demonstration project.
4) Geothermal energy. In order to solve the problem of imbalance between heat and heat load of ground source heat pump, geothermal energy plus solar thermal storage technology can be adopted, which can not only realize geothermal energy recycling, but also store solar energy.
5) Low grade heat energy utilization. With continuous deepening of energy saving work, effective utilization of low grade heat energy resources has become the key and difficult point of energy saving work. The low-grade heat utilization technology suitable for the school includes the sewage source heat pump, the air source heat pump and the high temperature flue gas depth recovery technology.

### 2.3 Energy solutions

Lishan College adopts the technical scheme of "multiple sources of energy complementing each other, 100% utilization of renewable energy, combining concentrated demonstration with popular extension" to build a zero-carbon campus. Electricity, heat, air conditioning, and hot water are all from clean renewable energy

**Table 1**. The utilization of various new energy

| | Energy Utilization Equipment | Specifications | Energy Use Form | | | |
|---|---|---|---|---|---|---|
| | | | Electricity | Heating | Cooling | Hot water |
| Solar | Solar absorption air conditioning system | 5180[m²] | × | √ | √ | √ |
| | Solar PV Panel | 2[MW] | √ | √ | √ | √ |
| | Solar bath hot water system | 9600[m²] | × | × | × | √ |
| | Solar boiled water system | 220[m²] | × | × | × | √ |
| Biomass | Pellet vacuum boiler | 2×7[MW] | √ | × | × | × |
| | Pyrolysis gas CCHP | 1[MW] | √ | √ | √ | √ |
| | Biomethane CCHP | 35[kW] | √ | √ | √ | √ |
| Wind | Wind power generator | 20[kW] | √ | × | × | × |
| Low grade heat | Air source heat pump | 22240[kW] | × | √ | √ | × |
| | Wastewater source heat pump | 350[kW] | × | √ | × | × |
| | Ground source heat pump | 700[kW] | × | √ | √ | × |
| Natural gas | CCHP | 200[kW] | √ | √ | √ | √ |

## 3 Construction process of zero-carbon campus

### 3.1 Application of solar energy air-conditioning technology

Solar air conditioning's seasonal matching is good. The hotter the temperature, the greater demand for air conditioning, and the better condition of the solar radiation and the greater capacity of the solar refrigeration. Solar air conditioning includes solar thermal absorption air conditioning system and solar PV air conditioning system. These two forms of solar air conditioning are used in the Lishan College. Lishan college has installed 5800m$^2$ solar collector system to drive LiBr absorption chiller to achieve three kinds of energy supply, such as cooling in summer, heating in winter, hot water in four seasons. The system saves more than 60% energy consumption. It has installed 2MW PV power station. The air source heat pump driven by PV power is used to heat the students' apartment in winter and to provide cool air in summer. During theholiday period, solar PV power is sold to grid. In addition to the solar air conditioning system, it has other forms of solar applications. 9600m$^2$ solar hot water system provides bathing hot water for the students. 2000 metal heat pipe solar vacuum tubes provide 5 t/d boiled water, and the biogas as an auxiliary energy. The school has also installed more than 200 solar street lamps and so on.

### 3.2 Demonstration project of crop straw heating

2×7MW biomass vacuum hot water units heating 250,000m$^2$ building. The heating equipment is the patent product of Lishan College. Crop straw is used as fuel for the biomass vacuum hot water unit, and the thermal efficiency is up to 95%, and the emissions conform to China's ultra-low emission standards. Biochar slow-release fertilizer produced with straw ash to promote crop growth. It has realized the comprehensive utilization and ecological cycle chain of "straw - fuel - heating - fertilizer - crop (straw)".

### 3.3 Demonstration project of biogas CCHP

Lishan Colleg has undertaken and completed the national high-tech R&D program of China (863 program) "the demonstration project of the biomethane CCHP". There are three tasks in this project: (1) The organic wastes are efficiently converted into biogas. (2) Research and development of high efficiency biogas generator. (3) Optimize the control system of CCHP. This project realizes the organic waste harmless treatment, biogas energy cascade utilization, ecological utilization of fermentation residue. In addition to the CCHP with biogas as fuel, the school also has 1MW biomass pyrolysis gas CCHP, 200kW natural gas CCHP and 50kW diesel CCHP system. The energy utilization rate of CCHP is above 85%.

### 3.4 Air source heat pump

7,000kW air source heat pumps have been installed in 100,000m$^2$ of buildings for cooling, heating and hot water. The heat pump combined with the small temperature difference heat transfer end can further improve the COP and improve the comfort of the room.

### 3.5 Distributed energy stations

At present, Lishan College has built 5 major distributed energy stations, including 2MW roof PV power station , 1,5000 m$^2$ of solar hot water system, 2 sets of 7MW biomass vacuum hot water unit, 200kW natural gas CCHP demonstration project, 1MW straw pyrolysis gas CCHP demonstration project, 35kW biomethane CCHP demonstration project, and 200 solar street lamps, dozens of air source heat pumps, solar absorption air conditioning system, combined compression and absorption type chiller, ground source heat pump + solar seasonal storage system etc. The school's electricity, heat, air conditioning and hot water are all from renewable energy and new energy.

## 4 Operational analysis

### 4.1 Energy saving and environmental benefit analysis

The advantages of renewable energy are low carbon and clean. Lishan College attaches great importance to energy-saving work, taking technological, management, and other measures to achieve energy-saving, so the overall energy consumption is small. In 2016, the overall number of people in school is 9,640, the power consumption was 2,151,355kWh. The school's PV power plant generates more than 2,000,000kWh electricity per year. Supply heating for 400,000 m$^2$ buildings depends on crop straw, generator waste heat and air source heat pump. The bath hot water is all provided by the solar heat collector system and the air source heat pump. According to the calculation, the renewable energy and energy saving equipment of the school save 3,650 tce/a, and the $CO_2$ emission reduction is 9,490 t/a.

### 4.2 Economic analysis

Economic benefits are the biggest restrictive factors in the application of new energy and renewable energy. The initial investment in large-scale application of renewable energy is large. A total of $14.4 million is invested in renewable energy and energy saving equipment in Lishan College. Among them, the investment of 15,000m$^2$ solar collector and 700kW absorption chiller is $4,500,000. The 2MW PV power station and 200 solar street lamps were invested $2,900,000. The total investment was $2,810,000 of biomass energy utilization equipment. Air source heat pump investment of $650,000, sewage source heat pump investment of $260,000, ground source heat pump investment of $580000, low grade heat utilization of a total investment of $1,490,000. A total of $1,800,000 is invested in the CCHP. Wind power, EV and LED are invested in $900,000. In recent years, the Chinese government has made great efforts to support the new energy industry. Generally speaking, the investment recovery period of renewable energy project is within 8 years.

 https://doi.org/10.1051/e3sconf/20184803004

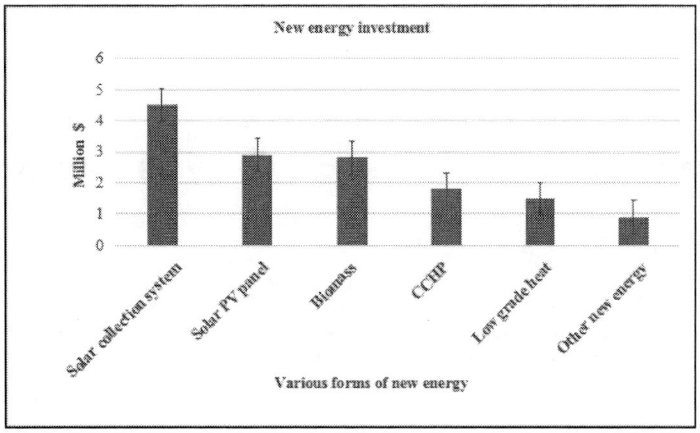


**Fig. 1.** New energy investment of Lishan College

### 4.3 Problems in operation

Renewable energy provides comfortable life for 10,000 teachers and students. But there are still some problems in the actual operation of each energy station, and it needs to be improved in the later period. The solar energy collector system needs to further reduce the energy consumption and optimize the operation of the circulating pump. In the spring and autumn seasons, part of the heat energy of the solar absorption air conditioning system is still not used. The price of natural gas is high, the electricity price of the school is low, and the application of the gas CCHP on campus is not good. At present, each energy station in the school is relatively independent, and the complex multiple energy flow network needs to be further optimized to improve the economic efficiency. We intend to establish an Integrated Energy Management System (IEMS) to achieve "multiple energy complementary, source, network, and load can be coordinated", and in the end it can realize maximized benefits.

## 5 Conclusion

The zero-carbon campus built by renewable energy, energy saving and environmental protection technology in Shandong Normal University - Lishan College is running well, providing clean and low-carbon energy for all teachers and students, basically realizing energy self-sufficiency, and continuously improving the distributed energy system in actual operation. The zero-carbon campus offers great experience and is a model for the regional clean and low carbon energy use and the realization of sustainable development.

## References

1. O. Opel, N. Strodel, *Climate-neutral and sustainable campus Leuphana University of Lueneburg. Energy.* **141**, pp. 2628-2639 (2017)
2. Hongbo Liu, Xinghua Wang, *The ecological footprint evaluation of low carbon campuses based on life cycle assessment: A case study of Tianjin*, China. Journal of Cleaner Production. **144**, pp. 266-278 (2017)
3. Dimitrios Hasapis, Nikolaos Savvakis, *Design of large scale prosuming in Universities: The solar energy vision of the TUC campus,* Energy and Buildings. **141**, pp. 39-55
4. Hongwei Tan, Shuqin chen, *Development of green campus in China*, Journal of Cleaner Production, **64**, pp. 646-653
5. Magnus Bonde, Jill Ramirez, *A post-occupancy evaluation of a green rated and conventional on-campus residence hall.* International Journal of sustainable Built Environment. **4,** pp.400-408 (2015)

# Industrial revolution 4.0: Universiti Malaysia Sabah perspective

*D Kamarudin* D Mudin[1], How *Siew Eng*[2,] *Md Mizanur* Rahman[3], *Pungut* Ibrahim[4], *Marcus* Jopony[5*]

[1]Vice-Chancellor, Universiti Malaysia Sabah, Jalan UMS, 88400 Kota Kinabalu, Sabah, Malaysia
[2]EcoCampus Management Centre, Universiti Malaysia Sabah, Jalan UMS, 88400 Kota Kinabalu, Sabah, Malaysia
[3,4]Faculty of Engineering, Universiti Malaysia Sabah, Jalan UMS, 88400 Kota Kinabalu, Sabah, Malaysia
[5*]Assistant Vice-Chancellor, Universiti Malaysia Sabah, Jalan UMS, 88400 Kota Kinabalu, Sabah, Malaysia.

**Abstract.** Industrial Revolution 4.0 or IR 4.0 is getting the attention of Higher Learning Institutions throughout the world. In the case of Universiti Malaysia Sabah (UMS), "transformation towards University/Industry 4.0" has been identified as one of the nine key result areas (KRAs) in the Strategic Plan 2018-2020. The transformation framework focusses on three areas namely Teaching and Learning 4.0, Smart Eco-UMS 4.0 and Research 4.0. Various initiatives have been planned for each area, some of which are currently being implemented. Previously, under the Strategic Plan 2013-2017, UMS gave great attention on developing and promoting UMS as an eco-campus in line with UMS aspiration to be the reference of eco-campus or green campus in the region. This will be further enhanced under Smart Eco-UMS 4.0 in line with the United Nation sustainable development goals. One of the related initiatives is Smart Energy. Through the initiatives under the UMS Ecocampus Plan 2013-2017, a total reduction of 44.50 % in energy consumption was recorded between 2014-2017 compared to 2013 and with 29 % reduction in carbon footprint from 2014-2017. The commitment towards an ecocampus has significant impacts on lowering energy usage and reducing carbon footprint. We believe that the smart energy initiative will further contribute to another level of energy saving and carbon footprint reduction towards reducing the impacts of climate change.

## 1 Introduction

Universiti Malaysia Sabah (UMS), established in 1994, currently has a total enrolment of about 18,500 students involving 10 faculties in three campuses. Some 12.3 % of students are postgraduates while 4.2 % are international students from 28 countries. The number of fulltime academic staff is 1078, of which 10.8 % are international. The total number of UMS alumni, so far, is about 66,500.

During the past five years, the progress of UMS was guided by the UMS Strategic Plan 2013-2017 [1]. One of the focus areas (key result areas) is Committing Towards Environmental Sustainability and one of the key initiatives that have been successfully implemented was the establishment of Ecocampus Management Centre under the EcoCampus Blueprint 2013-2017 [2]. Consequently, UMS has been aggressively promoting the university as an EcoCampus and has achieved significant inroads both nationally and internationally. For example, UMS is ranked at no.37 in the world according to UI GreenMetric World University Ranking 2017. Comparatively, UMS was ranked at no. 224 in 2014.

## 2 UMS and industry 4.0

The term "Industry 4.0" or the Fourth Industrial Revolution started to become a global buzzword only recently despite the concept of Industry 4.0 was initially proposed in 2011 [3]. Many definitions, ideas and scenarios on Industry 4.0 have been presented, some of which are difficult for the layman to understand. Nevertheless, due to continuous publicity given on Industry 4.0 lately, industries, government agencies and the general public are progressively becoming more aware of it.

In general, everyone and every industry will be impacted by Industry 4.0 but the good news is that all can benefit from it. Application or incorporation of digital technologies have wide ranging benefits such as increased productivity, increased competitiveness, increased flexibility, increased quality, increased innovation and lower staff cost. Increase in automation, however, will give rise to disappearance of some jobs while some tasks would change significantly [3].

### 2.1 Transformation towards university/ industry 4.0

Universiti Malaysia Sabah (UMS), likes other universities, needs to embrace Industry 4.0 to maintain relevance to prospective students and other stakeholders. UMS need to deliver future generations with the right set of skills and knowledge (future proof graduates) while at the same time needs to change and adapt for the digital transformation. For this, UMS has formulated a transformation framework known as Transformation Towards University/Industry 4.0. This covers several

* Corresponding author: marcj@ums.edu.my

https://doi.org/10.1051/e3sconf/20184803005

phases, the first phase being for the period 2017-2020. This transformation has been included as one of the nine focus areas of the UMS Strategic Plan 2018-2020 [4].

This Transformation Towards University/Industry 4.0 covers three principal areas, namely Learning and Teaching 4.0, Smart Eco-UMS 4.0 and Research 4.0. Learning and Teaching 4.0 focusses on the undergraduate curriculum, learning and teaching process and lifelong learning. To enable systematic and effective implementation of Learning & Teaching 4.0, two new centres have been created, namely Centre for Teaching Excellence and Academic Quality and Centre for E-Learning. To further enhance the teaching skills of the academic staff, a specific training programme on teaching excellence have been formulated and implemented. Additionally, an existing centre which is the coordinator of lifelong learning programmes, have been rebranded as Centre for Flexible Education. Among others, this centre offers Industry 4.0 relevant courses or training to UMS staff as well as to the external community (industries, government agencies and general public).

### 2.2 Smart eco-UMS 4.0

Smart Eco-UMS 4.0 focusses on digitalization of UMS. This agenda is very much in line with the expectation of the Ministry of Higher Education of Malaysia that public universities in Malaysia need to transform into smart campuses. In fact, the ministry targets all public universities to start implementing Smart Campus by June, 2018 [5]. The UMS Strategic Plan 2018-2020 retains "Committing Towards Environmental Sustainability" as one of the nine focus areas. Therefore, UMS will continue this EcoCampus agenda while at the same time pursue the digitalization agenda. Hence, UMS has branded this digital transformation as Smart Eco-UMS 4.0.

Smart Campus has a wide range of definitions. In general, a smart campus harnesses digital technology and tech-enabled solutions to enhance the performance (productivity, efficiency, etc.) of the university as well as for better learning and teaching, as well as research. Smart Eco-UMS 4.0 adheres closely to this definition whereby the emphasis is the digitalization of UMS. Smart Eco-UMS 4.0 has a number of initiatives, including digital ID, digital payment, integrated information system, smart administration, smart classroom, smart e-Learning, smart energy, smart security, smart facility and asset management, and smart hospital. The main enablers of all these initiatives are connected campus, financial resources, and governance, leadership and management.

The implementation plan for the respective initiatives has been formulated. Nevertheless, implementation will be in stages and according to priority. Some of the initiatives, namely University Integrated Information System, digital ID and digital payment have produced quick wins. In the case of smart energy, the university expects this initiative to significantly improve energy management including reducing electricity bills, in addition to the commitment in aligning itself to the national and international policies in sustainable development especially the Sustainable Development Goals 7 (ensure access to affordable, reliable, sustainable and modern energy for all) and 13 (take urgent action to combat climate change and its impacts). Meanwhile, Research 4.0 focusses on developing new research dimensions as well as intensifying university-industry collaborations for research and innovation.

### 2.3 Smart energy: sustainable energy management for sustainable development

UMS is committed to achieving a world-class reputation for sustainable energy management through high efficiency and cost-effective practices with the highest regard for the environment. The University recognises the importance of effective energy management and shall endeavour to continuously provide a conducive environment through the implementation of sustainable learning, teaching, research and human capital development.

By establishing a sustainable energy management system, there is an opportunity for an effective, continuous and extensive implementation of energy conservation, efficient practices and renewable energy throughout the University. To implement the concept of Industry 4.0 as well as for sustainable development, Universiti Malaysia Sabah has identified several energy issues in our buildings and open spaces in which our smart energy strategy will be implemented based on automated metering system and building management system and supported by Building Information Modelling (BIM) to monitor the energy consumption of buildings as well as reduce energy usage without undermining the comfort and safety of the campus community.

The significant challenge in this initiative is getting funding to support the campus wide system as the main campus area is 4.04-million-meter square with 47 buildings. UMS is currently in the process of engaging with a registered energy service company (ESCO) under Malaysia Energy Commission to establish energy performance contracting (EPC in order to support the initiative. Meanwhile, UMS is in the preparation stage of establishing a 20 MW solar project as a source of renewable energy.

## 3 Our achievement

UMS has identified electricity usage as one of the key areas which have a significant effect on the operation cost. Through preliminary energy audits in 2014 and 2017, it was identified that most of the electricity usage came from the operation of air-conditioning chillers. As an initial initiative of reducing energy usage, UMS has prepared a schedule for the operation and maintenance of the chillers to reduce the daily energy peak demand. Many other initiatives such as awareness program, quarterly utility usage report during the UMS development meeting chaired by the top management with the attendance of all the heads of department, stringent control of outside office hours usage, usage of energy efficient appliances, usage of solar energy best office practices based on the "Guidelines for Energy Saving Methods in Government Offices and Premises" [6] and the "Guidelines on No-Cost and Low-Cost Measures for Efficient Use of Electricity in Buildings [7] have been implemented.

By implementing these strategies, a significant reduction of electricity usages between the years 2014 to 2017 has been observed (Fig. 1) with a tremendous 35.32 % reduction reported in 2014 (compared to 2013) and further 11.51% reduction in 2017 (compared to 2016). A total reduction of 44.50 % was recorded between 2014-2017, the achievement has been in line with aims of the establishment of UMS Ecocampus Management Centre in 2013 and the EcoCampus Blueprint 2013-2017 [2]. To understand the impact of energy usage reduction, the total carbon emissions are calculated based on the electricity and transportation usage in the campus and presented (Fig. 2). Since the electricity is the major contributor to $CO_2$ emission in the campus, therefore the reduction of electricity usage has significant effects on $CO_2$ emission (estimated a total 29% dropped from 2014 to 2017).

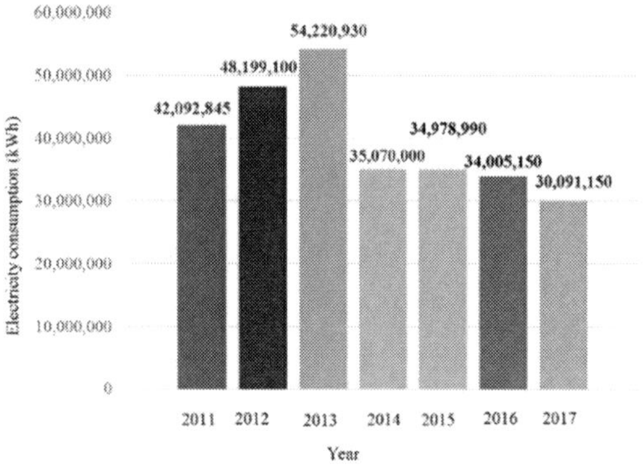


**Fig. 1.** Electricity consumption (kWh) of UMS main campus in 2011-2017

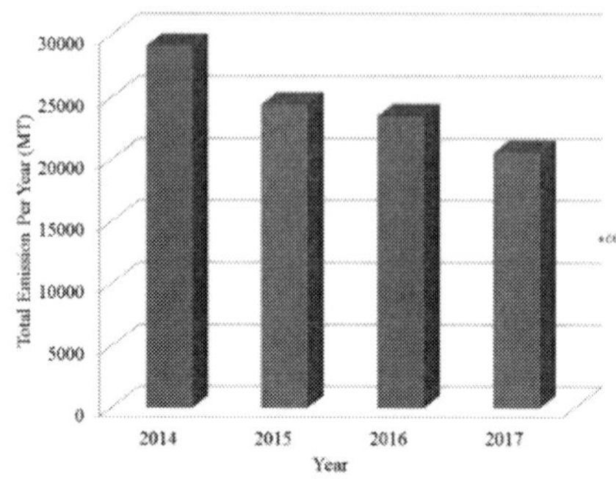


**Fig. 2.** Total estimated $CO_2$ from electricity usage and transportation

## 4 Summary

UMS has been aggressively promoting the university as an EcoCampus and has achieved significant inroads both nationally and internationally. The focus of the industries has changed due to the impact of the fourth industrial revolution that creates significant pressure on teaching and learning in the universities. The university needs to embrace Industry 4.0 to maintain relevance to prospective students and other stakeholders. By establishing a smart energy initiative supporting sustainable energy management system; there is an opportunity for an effective, continuous and extensive implementation of energy conservation, efficient practices and renewable energy throughout the university. UMS will continue to cultivate environmental citizenship among the students and staff through university activities, practices and operation towards the achievement of the sustainability goals as well as the Smart Eco-UMS 4.0.

## References

1. Universiti Malaysia Sabah, *Universiti Malaysia Sabah Strategic Plan 2013-2017: Sustaining Excellence Towards International Prominance.* Universiti Malaysia Sabah, Kota Kinabalu (2013)
2. Universiti Malaysia Sabah, *UMS EcoCampus Blueprint: Transforming Ideas into Reality.*Universiti Malaysia Sabah, Kota Kinabalu (2013)
3. H. Kagermann, Wahlster, W., Helbig, J., *Recommendations for implementing the strategic initiative Industrie 4.0.* Frankfurt/Main: acatech (2013)
4. Universiti Malaysia Sabah. *Universiti Malaysia Sabah Strategic Plan 2018-2020.*Universiti Malaysia Sabah, Kota Kinabalu (2018)
5. Embracing Industry 4.0. Available online at https://www.nst.com.my/education/2018/01/325914 /embracing-industry-40. Accessed on 8 March 2018. (2018)

6. Prime Minister Office, *Guidelines for Energy Saving Methods in Government Offices and Premises*.Prime Minister Office, Kuala Lumpur. (2014)

7. Energy commission, *Guidelines on No-Cost and Low-Cost Measures for Efficient Use of Electricity in Buildings*.Energy commission, Kuala Lumpur (2016)

# Expansion of renewable energy resources and energy-conscious behaviour at the University of Szeged

*László* Gyarmati*

Management, University of Szeged, Study and Information Centre, Szeged 6722, Hungary

**Abstract.** At the University of Szeged, as the greenest University of Hungary, the sustainability project is built on two pillars. One of them is based on events and communication campaigns held regularly for the University citizens to prompt environmental-conscious behaviour, whereas the other is built on technological developments and on the extensive use of renewable energy resources. Thus the development of built environment and social responsibility both support the adequacy to sustainability requirements. The spreading of the effective solutions to making more and more buildings of the University energy efficient, numerous investments using renewable energy are also responsible for the decrease of the natural energy use of the institution contrary to the fact that the number of the buildings of the University of Szeged is continually increasing. It can be stated that the University of Szeged is committed to using renewable energy which is taken into consideration of each investment planning. The following examples confirm it: using geothermal cascade system for heating and cooling of five university bulidings, solar panels on 24 builidings and a unique technology of using the heat of wastewater to cool and heat one of the main bulidings of the university, namely the Study and Information Centre.

## 1 Introduction

The University of Szeged is one of the biggest universities in Hungary as it welcomes more than 6000 freshmen every September. The new students become members of an institution that regards the establishment of sense of vocation, the practical way of teaching and shaping of the students' approach as priority. The University is proud to have dozens of titles, some of which are in connection with conservation and sustainability.

Meeting the aims of the Third Mission of the Development Strategy of the University of Szeged, the Institution makes every effort each year to be more and more environmental conscious and 'green'[1].

Selective waste collection, use of solar energy, grey water and the heat of the waste are only some of those solutions owing to which our University was among the top 20 universities in the UI GreenMetric World University Ranking at 2014 [2] [Table 1.].

**Table 1.** Achievements of the University of Szeged in GreenMetric

| Year | Place of the University of Szeged/number of Universities taking part in the ranking | Proportion of European Universities in the ranking behind the University of Szeged | Difference in points between the University in the first place and the University of Szeged |
|---|---|---|---|
| 2010 | 54./95 | 44% | 2972 |
| 2011 | 67./178 | 70% | 2232 |
| 2012 | 52./215 | 82% | 1738 |
| 2013 | 35./301 | 88% | 1192 |
| 2014 | 19./360 | 92% | 885 |
| 2015 | 29./407 | 90% | 1038 |
| 2016 | 107./516 | 90% | 2751 |
| 2017 | 88./619 | 79% | 1179 |

## 2 On the way to sustainability

At the beginnings, the green efforts of the University of Szeged got realized by individual initiatives and motivations. Seeing the success of these efforts, more and more university members and university units have joined these initiatives. It has all been started by the Management of the University of Szeged József Attila Study and Information Centre (henceforth SZTE TIK) (Fig. 1). Being the so called 'Heart of the University',

* Corresponding author: gyarmati.laszlo@tik.u-szeged.hu

the Centre has been a combination of a Studying Area, a Teaching Area, a Service Area, a Conference Site and a Meeting Place for its daily 3000-3500 visitors for more than 13 years thus being the ideal place for showing examples of social responsibility to be followed. As a first step, the Management of SZTE TIK organized the "Earth Day" with tradition forming intention in 2008. Then in 2009, they won "The office taking the most steps towards being green" Award by following the guidelines of KÖVET Association for Sustainable Management. Seeing the success of this initiative, three other university units have joined the invitation the following year, The Rector's Office, The Student Service Office and the Dean's Office of the Faculty of Economics and Business Administration all of which were awarded with Creative Special Award.

Considering the visitors' needs, such developments had been carried out in the Centre, owing to which it was awarded the "Bicycle Friendly Workplace" title in 2010 and in 2011. Also in 2011, the Management of SZTE TIK launched the so called 'Green Commando' initiative which aimed at helping other university units become environment-friendly according to the guidelines of the 'green office' and by taking environment-friendly viewpoints into account. 2011 was again the year when the Green Commando initiative won the Public Award at the CSR Market event organized by KÖVET Association for Sustainable Management, which meant that a voluntary initiative of a university got national acknowledgement. Besides expanding the number of programmes which aimed at shaping the visitor's approach (e.g.: Earth Day, Earth Hour, Green Film Club, World Water Day, Fighting against Global Warming series of events, Furnace Tour, Charity events), technical developments and optimization processes to decrease the environmental impact of the Centre have also started. Furthermore, the regulation system of the first green Congress Centre of Hungary has also been established [3].

**Fig. 1.** Study and Information Centre – University of Szeged (SZTE TIK)

## 2.1 Technical solutions

Besides shaping the university members' and citizens' approach, the second pillar of the sustainability efforts of the University of Szeged is the continuous infrastructural development. The spreading of the effective solutions to making more and more university buildings energy efficient, (applying motion sensor switches and LED bulbs, optimizations of building management systems, shaping the environment-conscious behaviour of the employees, etc.), numerous investments using renewable energy are also responsible for the decrease of the natural energy use of the institution contrary to the fact that the number of the buildings of the University of Szeged is continually increasing (at present there are about 300 buildings of the University of Szeged scattered around in Szeged) [4]. It can be stated that the University of Szeged is committed to using renewable energy and it is taken into consideration at the beginning of each investment planning. The following examples confirm it: By using geothermal cascade system the heating and cooling of five university buildings are guaranteed [5]. Grey water system is applied at the central building of the University of Szeged owing to which rainwater has its secondary function. In the summer of 2015, the project of establishing the unique technology of using the heat of wastewater to cool and to heat the SZTE TIK was carried out (Fig. 3). With the help of this system, the heat energy gained from the wastewater having permanent temperature is used for heating and cooling of the building and it will be also possible to reduce the $CO^2$ emission more than 300 tonnes [6].

**Fig. 3.** Heat pump for usage of wastewater heat

## 2.2 Widespread use of the solar panels

In 2012, a new infrastructure development was launched owing to which solar panels were placed on top of some of the university buildings to produce energy with the help of solar energy. At that time 2 754 pieces of solar panels were placed on top of 24 buildings of the University (Fig. 4). Besides the energy production of the solar panels, the emission of carbon dioxide also decreases as well as the proportion of sunlit surface, which results in energy consumption savings in the process of cooling, too. This way approximately 85 700 Euros are saved each year.

**Fig. 4.** Solar panels at the roof of SZTE TIK

The preparation of this project took long years as the first measures, calculations and site visits were carried out already in 2008. The reason for this was that the buildings of the University of Szeged can be found at every part of the city and because they were built in different times and styles so each of them had unique characteristics as for energetic aspect. The experts of the University examined more than 50 buildings from which the 24 most suitable for infrastructure development were chosen. At the process of choosing, one of the criteria was the appropriate building siting while the other was the 35-degree angular offset of the roof. In addition, the bearing capability of the roof had to be taken into consideration because of the weight of the solar panels and the extreme weather conditions.

Three years after this project was finished, the University successfully competed for a new project of establishing photovoltaic system. In this case again, The University of Szeged competed for an Environment and Energy Operative Programme to decrease its carbon-dioxide emission, and its energy costs as well as to harmonize the relation between people and environment.

With the help of this project 2 040 pieces of solar panels have been placed on top of 18 buildings of the University that again results in saving 69 841 Euros each year [7].

### 2.3 Our companions, the university citizens.

The popularity of the SZTE TIK ensures perfect opportunities for us to make the visitors get to know the green efforts of the University as well as it gives us the opportunity to get their feedback on it. Decreasing the environmental impact of the University depends just as much on the environment-conscious behaviour of the university students, citizens and event participants as on the commitment of the university employees.

Therefore, we use several communication channels to call the attention of our visitors to the environmental aspects in every spring and autumn semester. We plan a series of 'Get green during the registration with SZTE TIK' events for students registering for the following semester in order to draw their attention to the importance of environment-conscious behaviour (Fig. 5). We can reach more than 10.000 followers through the official Facebook site of the Centre. Furthermore, there are information panels placed all around the Centre to call the attention of its visitors to the conscious use of resources.

Besides raising the awareness of the visitors, it is also essential to inform the thousands of university professors, researchers and employees working in other fields, therefore a web-based internal social forum (so called coospace scene) was established in August 2014: Via coospace about 5.000 university citizens can get direct access to the information on the sustainable efforts of the university.

The Management of SZTE TIK has always considered it important to get direct feedback from the visitors on their expectations. According to the results of the visitors' satisfaction survey in 2012, the university citizens consider it extremely important that a state owned institution belonging to the Higher Education such as the SZTE TIK should be involved in social responsibility and should popularize the environment-conscious behaviour (78% of the more than 1.000 respondents regarded this topic essential) [8].

**Fig. 5.** Earth Hour event at the SZTE TIK (2015.)

## 3 Summary

The University of Szeged is committed to find and successfully compete for further projects which are to fund the use of renewable energy sources so that it can operate in more and more environment conscious and sustainable way. The University regards it important to find solutions to spreading the use of renewable energy in its buildings as well as to optimize its present technologies and improve their efficiency.

To keep the already achieved results, the University of Szeged has to take further steps in the following areas:

- defining the new short-term, mid-term and long-term strategies
- decreasing the greenhouse emission of the university within a set period of time

- establishing a continuous feedback channel on the ecological footprint of each Faculty of the University of Szeged together with suggestions for further improvement
- implementing new investments by which further old institutions of the University of Szeged can get greener
- ensure that the goals set in the strategies are reached by continually monitoring and evaluating the figures and the results
- raising students' awareness of the importance of sustainability in workshops and community events
- establishing and monitoring an online forum for the teachers and students working and studying in the field of sustainability thus helping their co-operation
- co-operation with the representatives of civil services and local suppliers

## References

1. University of Szeged Institution development plan, Available online at http://www.u-szeged.hu/egyetemrol 141002/minosegfejlesztesi/szte-ift-2016-2020-170124, accessed on 27th, February 2018 (2016)
2. UI GreenMetric, 2010-2017. UI GreenMetric world university rankings. Available online at http://greenmetric.ui.ac.id/, accessed 27th, February 2018. (2010)
3. Green institution introduction, Available online at http://www.u-szeged.hu/tik/zoldintezmeny, accessed on 27th, February 2018, (2018)
4. Press conference about GreenMetric results of University of Szeged 2013, Available online at http://www.u-szeged.hu/sztehirek/hirarchivum/2013-januar/magyarorszagon-szte, accessed on 27th, February 2018 (2013)
5. Press release about geothermal cascade system 2014, Available online at http://www.u-szeged.hu/sztehirek/ sajtoarchivum-2014-141105-8/sajtokozlemeny-ujszegedi, accessed on 27th, February 2018 (2014)
6. Project introduction homepage, Available online at http://www.u-szeged.hu/ fejlesztesiprojektek/projekt-bemutatasa/keop-2012-4-10-0-12-2013, accessed on 27th, February 2018 (2013)
7. Press conference about solar panel installations at the University of Szeged 2015, Available online at http://www.u-szeged.hu/sztehirek/2015-augusztus/2040-napelem-18-egyetemi, accessed on 27th, February 2018 (2015)
8. SZTE TIK visitors' satisfaction survey – research report 2012. Available online at https://www.u-szeged.hu/download.php? docID=20692, accessed on 27th, February 2012 (2012)

# Evaluation of electricity consumption and carbon footprint of UI GreenMetric participating universities using regression analysis

*Alfan* Presekal[1,*], *Herdis* Herdiansyah[2], *Ruki* Harwahyu[1], *Nyoman* Suwartha[3], and *Riri* Fitri Sari[1]

[1]Department of Electrical Engineering, Faculty of Engineering, Universitas Indonesia, Kampus UI Depok, Depok 16424, Indonesia
[2]School of Environmental Science, Universitas Indonesia, Kampus UI Salemba, Jakarta 10430, Indonesia
[3]Department of Civil Engineering, Faculty of Engineering, Universitas Indonesia, Kampus UI Depok, Depok 16424, Indonesia

**Abstract.** UI GreenMetric as sustainability-based university rankings has received a worldwide acceptance since its initiation in 2010. One of the criteria for this ranking is the annual electricity consumption of participating Universities. There are some challenges in evaluating the overall data, i.e. some electricity consumption information is missing or may not accurately represent the real condition. There is various information that can be used to calculate the university rank associated with electricity consumption. On the other hand, some external data sources from World Bank on the annual electricity consumption per capita for every country is highly correlated with the electricity consumption in every University. This paper aims to show our evaluation and prediction of the annual electricity consumption from participating university using regression analysis based on the available data of UI GreenMetric and relevant external information. This is conducted using regression analysis on the data submitted in 2017 and the predicted KWH based on the number of full-time student and staff in the university. The result shows that some universities are consuming more electricity than the average KWH used per-capita in their country. The result also shows that the prediction cannot be used accurately, especially for the carbon footprint. This evaluation may help universities to improve their policy in reducing the electricity consumption and the greenhouse gas emission reduction policy, and mainly helps UI GreenMetric to speed up the verification process when necessary

## 1 Background

UI GreenMetric has already been running for almost 10 years. The number of participating Universities already reach more than 600 in 2017. There are various parameters used by UI GreenMetric for scoring. One of the parameters is university's annual electricity usage. In this ranking, this parameter belongs to the Energy and Climate Change category. Basically, university with more efficient electricity consumption will get better score which at the end determines the rank. Unfortunately, data about electricity consumption sometimes are incomplete or inaccurate. In UI GreeMetric's system, there are two causes that made the data incomplete of inaccurate. The first cause is that the university might have left the field blank or filled it with zero valye. The second reason is that the data can be illogically large or small, i.e. outliers. With ever-increasing UI GreenMetric's member and received data, this incomplete and inaccurate data may decrease the accuracy and validity of the generated ranks.

Related with electricity, UI Greenmetrics also perform analysis into carbon footprint. Participating Universities required to calculate their own carbon footprint. Electricity is one key aspect to determine quantity of carbon footprint. Data the carbon footprint also has same problem with electricity. This work aims to find relation between electricity consumption and carbon footprint output.

This paper consists of 5 parts. The first part is background of this work which related with UI Greenmetric University Ranking. Second part of this paper explain about ecological footprint perspective for electricity consumption and carbon footprint. Third part elaborate methodology of collecting data and data analysis from this work. This is followed by fourth part which contains result and discussion, and the last part consist of concluding remarks from this works.

## 2 Ecological footprints

There are various measurements instrument and the concept of 'green' which in uses. Ecological Footprint (EF) was introduced in 1990 by Mathis Wackernagel and William Rees in the book "Our Ecological Footprint" [7]. Ecological Footprint was to convey the method for calculating EF and Ecological Footprint Analysis (EFA). In this book, based on the former activity by humans on earth, namely measuring trace of human level of interaction with nature. Wackernagel and Yount explained that the analysis of the EF is the assessment

* Corresponding author: apresekal@gmail.com

tool to calculate the level of human interaction and disposal [3]. Data collection and calculated of greenhouse gases (GHG) emitted from the atmosphere during a certain time period (the US Environmental Protection Agency, 2009), studied by Mieko A. Ozeki to serve as the basis of mitigation and adaptation to climate [6]. The difference between GHG inventory EF and EF is that it does not fully take into account climate change due to greenhouse gases. Climate mitigation and adaptation planning are also studied by Jeongho Kyoung-Sik Woo and Choi [8], they introduced concept called energy saving action program. The study determines the standard model in the university laboratory for implementation energy saving action program to implement the study guide. The result of research shows that energy saving action program to saving of power consumption significantly.

Carbon Footprint to illustrate the total carbon emissions from all everyday activities. An example is the use of fossil fuels [4]. There are two types of carbon footprint that includes traces of primary and secondary trail. The main carbon footprint represents the amount of CO2 emissions that occur directly from use of fossil fuels, while the secondary carbon footprint accumulated indirect carbon source and as an example is electricity. Another definition of carbon footprint is estimated individual in contributing to global warming which is calculated based on the unit that is equivalent to the CO2 [5].

In this context, electricity consumption and carbon footprint are closely related as electricity is considered to be the source of secondary carbon footprint. UI GreenMetric collected both the electricity and carbon footprint data. In this work, we evaluate the electricity consumption and carbon footprint data in UI GreenMetric database. A regression analysis is proposed such that we can guess the electricity usage of certain universities which do not fill the data or fill the data with zero. By this means, the accuracy of the produced rank can be retained up to some degree.

Performing calculation of carbon footprint can increase the awareness of how much carbon emission one has produced during certain periods, which may lead to carbon emission reduction. According to a study by Lambrechts dan Liedekerke (2014) 0f Leuven University College (KHLeuven), mobility from transportation contributes to 3.362 ton (44.22%), nearly half from total carbon footprint. UK Higher Education Funding Council for England (HEFCE) encourage higher education institution to reduce carbon emission (HEFCE, 2010a). After conducting the measurement and calculation of its greenhouse gas emissions, which of course include $CO_2$, Yale University pledged to reduce its greenhouse gas emissions 43% below 2005's levels by 2020. Currently, Yale University has achieved a 12% reduction, despite a 14% increase in campus size [6]. The ecological footprint from carbon footprint is very important, UI GreenMetric already use question to evaluate this. This paper will present method and analytical aspect of data related with ecological footprint.

## 3 Methodology

In this work we evaluate the university's annual electricity consumption and carbon footprint. For the electricity consumption evaluation, mainly we utilize the information of annual electricity consumption per-capita from the World Bank which has been specified for every country. Subsequently we match and compare it with the data from UI GreenMetric, i.e. the campus population, and available annual electricity consumption. Using university's annual electricity consumption (AEC) data from UI GreenMetric, we performed some regression analysis. Regression analysis is often used to predict future electricity consumption [1] [2]. In this work we show AEC along with the carbon footprint from universities to see its relation. We valuated some specific targeted countries as the sample to show how the prediction can be done for the blank data.

While the AEC value is taken from UI GreenMetric database, the predicted AEC is calculated as:

$$AEC\ est. = \ num.of\ population \times AEC\ per\ kapita \quad (1)$$

where the number of population is taken from UI GreenMetric database. Subsequently, regression is performed to find the trendline of the AEC vs predicted AEC. To calculate prediction of AEC from universities we are using formula 1.

## 4 Result and discussion

In the following, the result of AEC vs Predicted AEC is plotted. Each dot represents one university. The trendline is drawn based on linear regression against the scattered dots. Notice that in this 2D plots, the preferred situation is when all dots concentrate in one area or close to the trend-line. When it happens, it means the data is good as predicted, and also confirm the accuracy of the prediction.

### 4.1 Electricity consumption

Based on regression analysis we can create function to predict annual electricity consumption from certain country. Figures 1,2,3,4,5, and 6 show the annual electricity consumption from universities in Indonesia, Taiwan, Colombia, Italy, Malaysia and Russia.

 https://doi.org/10.1051/e3sconf/20184803007

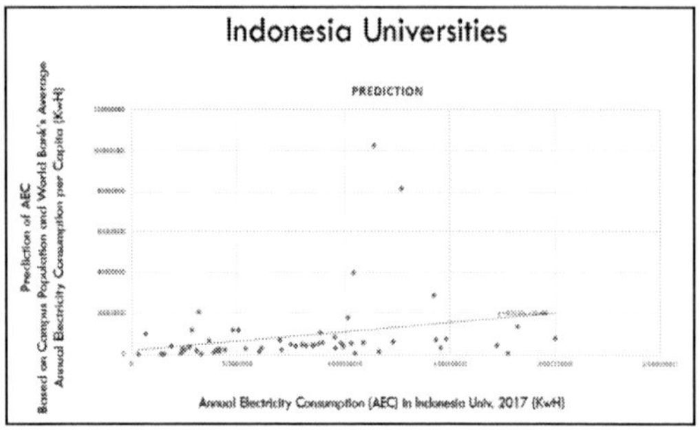


**Fig. 1.** Indonesia universities electricity consumption

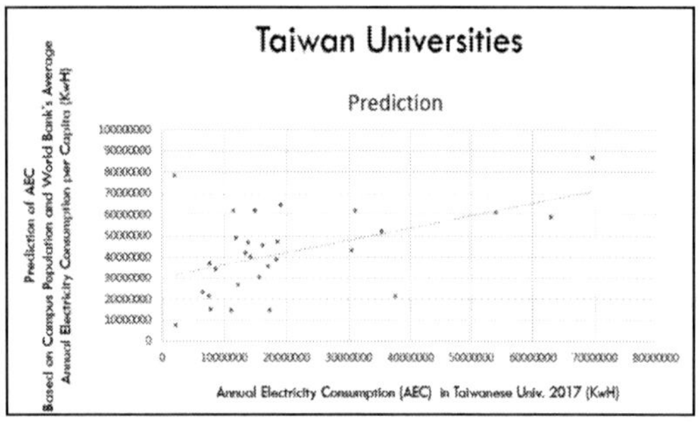


**Fig. 2.** Taiwan universities electricity consumption

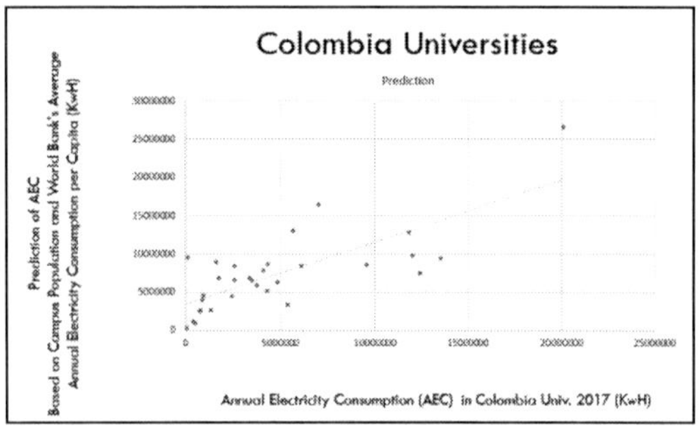


**Fig. 3.** Colombia universities electricity consumption

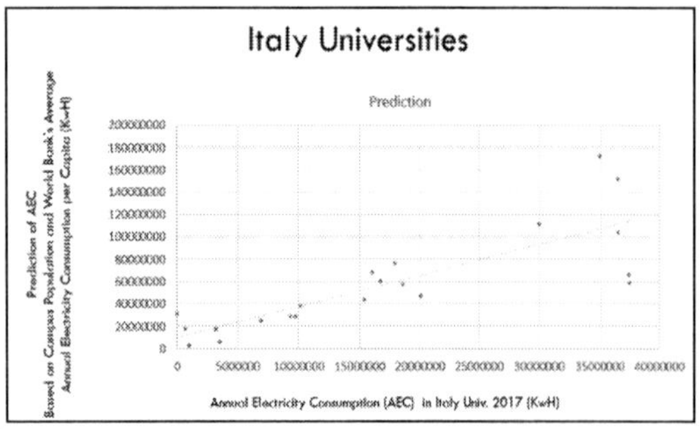


**Fig. 4.** Italy universities electricity consumption

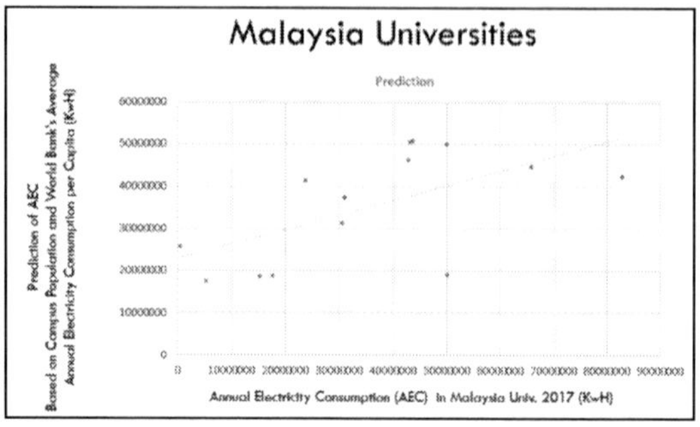


**Fig. 5.** Malaysia universities electricity consumption

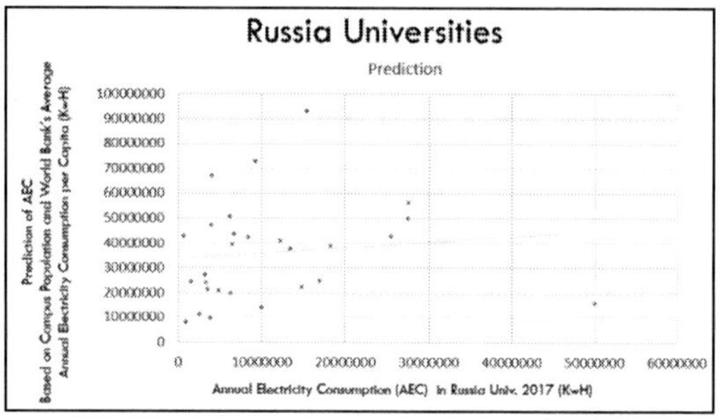


**Fig. 6.** Russia universities electricity consumption

Based on the Figures 1,2,3,4,5, and 6, we can estimate the universities annual electricity consumption. The function of trend-line gradient only work for specific country. As shown in the Figures, different country may have different trend-line obtained by regression method.

The results for Indonesia and Italy, which are shown in Figures 1 and 4, respectively, tend to concertize close with normal line. The results for Taiwan and Russia, which are shown in Figures 2 and 6, respectively, have more scattered pattern. The result for Columbia and Malaysia have almost similar gradient of the trend-line, as shown in Figures 3 and 4, respectively, which is larger than the gradient of the other 4 countries. The quantity of the outlier's data also different.

In these 2D plots, we can observe some outlier sample. Our original hypothesis is that these outliers are generated by wrong data inputs. However, we find that different university may need to power different electronic appliances in their core business. For example, science and engineering related universities may need more electricity than that of social studies. Hence, for UI GreenMetric, these results are useful to decide which universities needs further verification. It will be very ideal if UI GreenMetric can perform rigorous verification to all of the universities. However, this mechanism will reduce complexity and cost of self-assessment. On the following steps, this is also important to maintain the current university network and encourage more universities to join UI GreenMetric and create more green campuses in the world.

### 4.2 Carbon footprint

As explained in the background part. Carbon footprint has correlation with AEC. This is because electricity is one of the source of secondary carbon footprint. In the following, we plotted the annual carbon footprint versus the AEC from universities in Indonesia, Taiwan, Colombia, Italy, Malaysia and Russia.

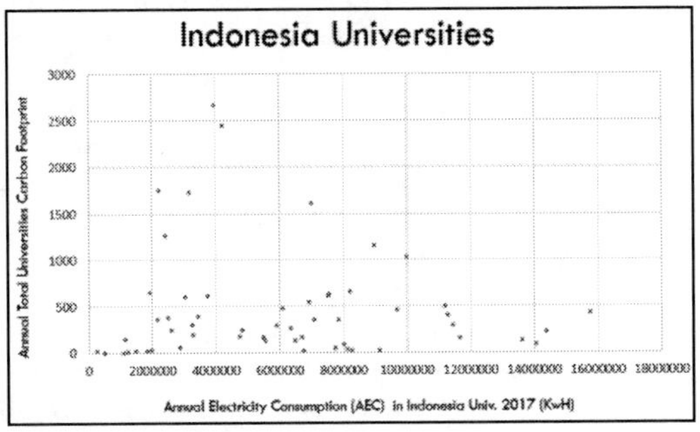


**Fig. 7.** Indonesia universities AEC vs carbon footprint

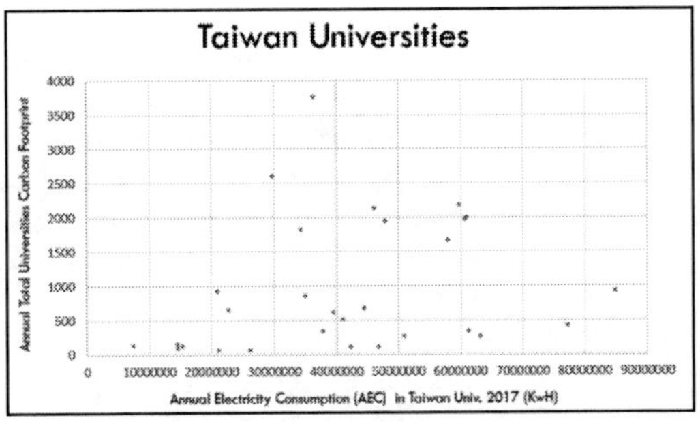


**Fig. 8.** Taiwan universities AEC vs carbon footprint

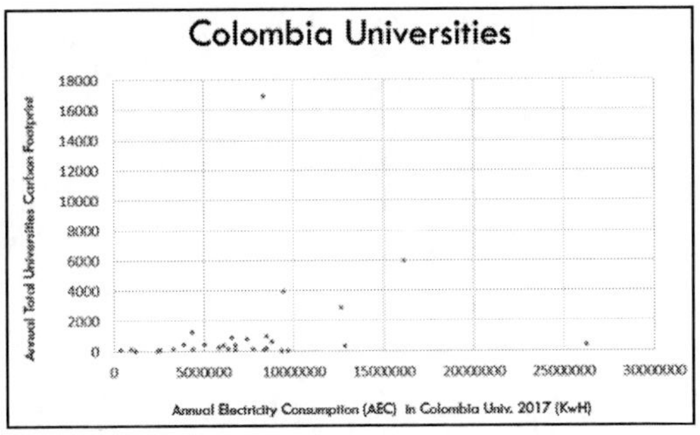


**Fig. 9.** Colombia universities AEC vs carbon footprint

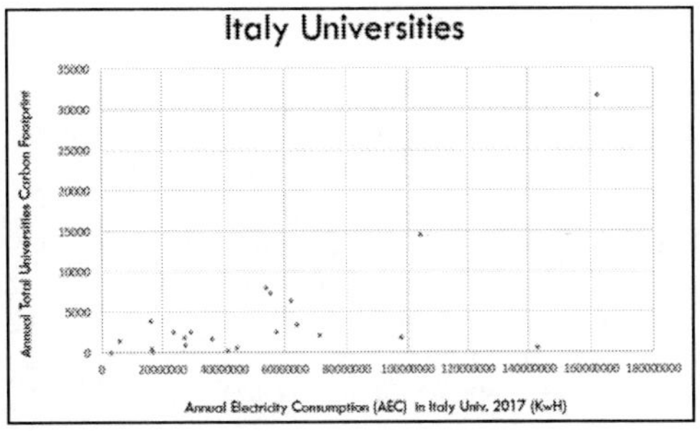


**Fig. 10.** Italy universities AEC vs carbon footprint

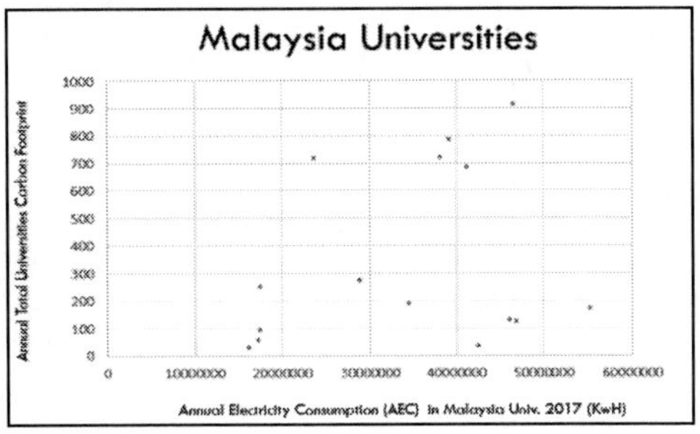


**Fig. 11.** Malaysia universities AEC vs carbon footprint

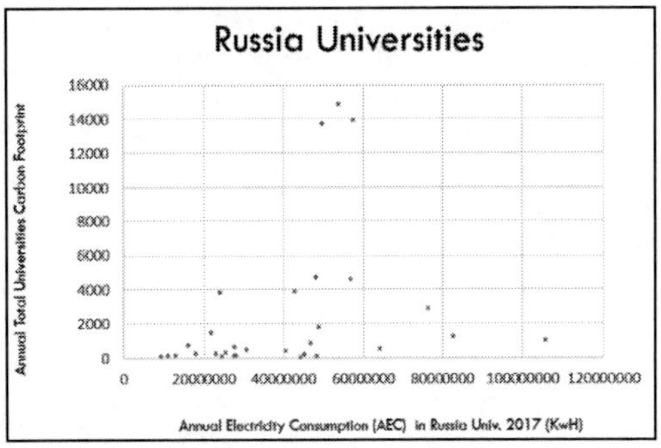


**Fig. 12.** Russia universities AEC vs carbon footprint

Based on these 2D plots on Figures 7,8,9,10,12 and 12, the annual carbon footprint and the AEC have scattered pattern in each country. There is no significant insight can be taken into consideration based on the regression analysis. Our hypothesis of this phenomenon is that the calculation of carbon footprint, as shown in UI GreenMetric guideline, cannot be generalized due to different business process in each university in the same country.

## 5 Conclusion and evaluation

According to the regression test, every country tends to have different pattern. Because of this pattern, method of regression cannot be applied homogenously to all participating universities in UI GreenMetric. Regression method may be relevant and applicable for certain country which has normal pattern around the trend line. Some of countries may have more scattered pattern to make predictive model for AEC less accurate. Overall prediction based on the current pattern is still appropriate for certain countries. For future work more advance method may be applied to predict more accurate AEC from every Universities in different Countries. On the other hand, comparison between AEC and carbon footprint does not have significant result based on regression analysis. This is because the result tends to be scattered. For the future work, we will consider different factors that differentiate university's electricity consumption such as the existence of large laboratory and other highly electricity consumptive equipment. Other method to evaluate the electricity consumption and carbon footprint in universities will be conducted.

## References

1. Braun, M. R., H. Altan, and S. B. M. Beck, *Using regression analysis to predict the future energy consumption of a supermarket in the UK*, Applied Energy 130: 305-313, M. Ben Rabha, M.F. Boujmil, M. Saadoun, B. Bessaïs, Eur. Phys. J. Appl. Phys (2014)
2. Fumo, Nelson, and MA Rafe Biswas. *Regression analysis for prediction of residential energy consumption*. Renewable and Sustainable Energy Reviews 47: 332-343 (2015)
3. Chambers, N; C. Simmons, M. Wackernagel, *Sharing Nature's Interest*. Earthscan Publications Ltd. London, UK (2000)
4. Wiedmann, Thomas, et al. *Multiregional input-output modelling opens new opportunities for the estimation of ecological footprints embedded in international trade*. International Ecological Footprint Conference, Cardiff (2007)
5. Lynas M. 2007, Carbon Counter. Glasgow: HarperCollins Publishers.dalam jurnal Kenny, T and Gray, N. F, 2009, Comparative performance of six carbon footprint models for use in Ireland, Environmental Impact Assessment Review 29 1-6 (2009)
6. Yale University, *Greenhouse Gas Emissions Reduction Progress*, US (2015)
7. Wackernagel, Mathis, and William Rees, *Our ecological footprint: reducing human impact on the earth*, **9**, New Society Publishers (1998)
8. Jeongho Woo, Kyoung-Sik Choi, *Analysis of Potential Reductions of Greenhouse Gas Emissions on the College Campus through the Energy Saving Action Programs*, Republic of Korea (2013)

# Challenges of sustainability efforts of universities regarding the sustainable development goals: a case study in the University of Zanjan, Iran

*Seyed Mohsen* Najafian [1], *Esmail* Karamidehkordi *[2]

[1] President of the University, University of Zanjan, Zanjan, 45371-38791, Iran
[2] Director of International Scientific Cooperation Office, University of Zanjan, Zanjan, 45371-38791, Iran

**Abstract.** The purpose of this presentation is to introduce some activities and programmes of the University of Zanjan in changing its campus environment towards a sustainable campus, emphasising setting and infrastructure, waste management, water management, and education and research. This comprehensive university with over 10000 students and 1000 staff is located in a semiarid area with a campus area of 421 ha and is 6 km away from the Zanjan City, center of Zanjan Province. In over four decades, it has expanded its tree cultivation to over 72 hectares, and its total vegetation area covers over 94 percent of total campus area. The university has increased and improved its investment on sustainability, smart buildings and water management in both buildings and vegetated areas. The waste management programmes have been implemented through using electronic correspondence and document submission in different activities of the university; separating plastics, glasses and papers and waste recycling; toxic waste handling in all labs; composting organic waste; inorganic waste management; and recycling sewage disposal. Though the university has provided free buses and shuttles to both staff and students inside the campus and between the city and campus to reduce private car use, it still needs to encourage bicycle use and improve its facilities to support it. Developing renewable energy for the future is still a challenge for this university and needs both innovation and investment. Students and academic staff have also been encouraged to move their conventional education and research methods and contents to more sustainable approaches, for example in courses syllabuses, student activities, research projects, publications and investments. The GreenMetric World University Ranking Network is expected to enhance its scope to contribute much more on sustainable development goals. A sustainable university should play an important role in innovation and technology research and development in sustainability; enhancing staff and students' sustainability knowledge and social capacities; changing the campus environment to an Eco-friendly and sustainable environment; and enhancing social and human capacities of communities and public and private institutions.

## 1 Introduction

Sustainable development concept and strategies have evolved for over three decades, when the "Our Common Future" report was published during the World Conference on Environment and Development in 1987 (Brundtland Report): "Development that meets the needs of the present without compromising the ability of future generations to meet their own needs" [1]. The world has faced a crucial challenge on how to achieve this goal. Numerous international events and conventions regarding sustainable development have been held to enhance international community' knowledge and actions for this subject, for example the World Commission on Environment and Development (1987 in Norway), the Earth Summit' in Rio de Janeiro (1992 in Brazil), The United Nations Framework Convention on Climate Change (UNFCCC) in 1992, the Millennium Development Goals (MDGs) in 2000, the UN World Summit on Sustainable Development (WSSD), Johannesburg in 2002, the Kyoto Protocol (2005), the UN Climate Conference 'COP15', Copenhagen in 2009, the UN summit, 'Rio+20', Brazil in 2012, the UN Summit, Sustainable Development Goals, 2030 Agenda for Sustainable Development in 2015, and the COP23: UN Climate Change Conference in 2017 [1-3] [4] [5].

Universality, integration and transformation are three main indicators for achieving sustainable development goals (SDGs). The terminology of *universality* means these goals apply to every nation and every sector. Cities, rural communities, businesses, schools, universities, organizations, all are challenged to act. The word *integration* implies that the goals are all inter-connected in a system, so it is not possible to aim to achieve just one goal. Finally, *transformation* indicates that achieving these goals involves making very big, fundamental changes in how we live on Earth [2]. This means every nation and sector; including rural and urban communities, public and private institutions, non-governmental organizations and higher education institutions (HEIs) should work to manage interconnected and complex sustainability challenges to

* Corresponding author: najafian@znu.zc.ir

make fundamental changes in how we live on the Earth. HEIs can contribute to the sustainable development through several mechanisms: 1) innovation and technology research and development, 2) enhancing staff and students' sustainability knowledge and social capacities, 3) changing HEIs environment to an eco-friendly and sustainable environment, and 4) enhancing social and human capacities of rural and urban communities and public and private institutions [6]. The assessments made by the University of Indonesia GreenMetric since 2010 to measure and rank universities around the world is one of the international commitments to encourage the HEIs to take sustainability as a serious key concept into account. Based on UI GreenMetric [7], the University of Zanjan has been ranked as the 47th university among 619 universities in the world and the 1st among Iranian universities.

This paper aims to introduce some activities and programs of the University of Zanjan in changing its campus environment towards a sustainable campus, emphasizing setting and infrastructure, waste management, water management, and education and research.

## 2 Sustainability in the University of Zanjan campus

### 2.1 Setting and infrastructure

According to Higher Education Planning and Research Institution [8], 2500 higher education institutions, including 130 non-medical and 63 medical state-run universities. The University of Zanjan is ranked 20th among all public and Iranian Universities. Base on the Times Higher Education World University Ranking 2018, it is among 1000 universities of the world and 18 Iranian universities [9]. As a public university, this university was established as an agricultural higher education institute in a public granted land in a semiarid rural area 6 km away from the city in 1975, but it was gradually developed as a comprehensive university in five main faculties titled Agriculture, Engineering, Humanities, Science and Art and Architecture in its main campus, comprising 37 educational departments and a research institute with 4 research departments. It has almost 10000 students with 1000 academic and administrative staff. Its main campus is located in an area of 400 hectares, in the south west of the Zanjan City, the center of the Zanjan Province, at the north west of Iran (with a distance of 330 km away from the capital city of Tehran. The second and third campuses are located in the Abhar Township and the Tarom Township of the Zanjan Province.

The total ground floor (first floor) area of buildings is estimated to be 87000 out of 4000000 $m^2$ total university area in its main campus, with 189000 $m^2$ total main campus building area and 55000 smart building areas (in different floors). This means only 2.2 percent of the university area is covered by buildings. Moreover, the average land per students is calculated as 400 m2. The university has significantly increased its woodland area since its establishment, which was originally covered by rangelands and croplands. Woodland and orchard areas inside the campus have increased from 0 ha in 1975 and 30 ha in 2010 to over 71 ha in 2017 (17.75 percent of the university). The first week of March every year, the Iranian celebrate natural resources management and environment week and one of the days is called "Tree Planting Day", in which the university calls its students and staff to participate in this event for plantation. Moreover, 76.34 percent of the campus (304.36 ha) was covered by agricultural lands (rangelands and croplands) and other planted landscape vegetation in 2017, which have been used for research, education, landscape management and production. Water can be absorbed in almost 94.09 percent of the university area, including agricultural lands, woodlands and other planted landscape vegetation. The area of parking space is 23000 ha (in ground floor).

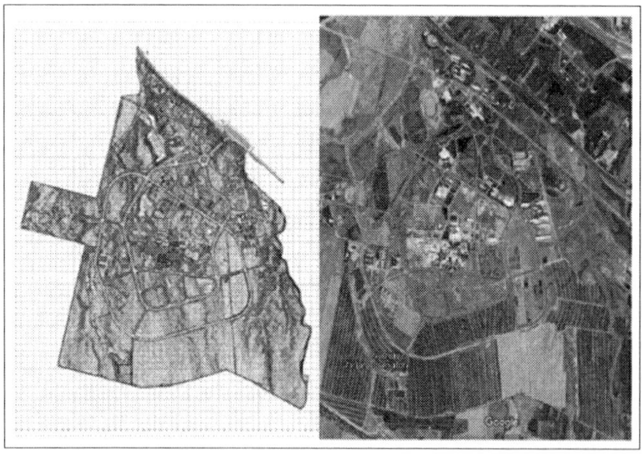

**Fig. 1**. University of Zanjan topography and ariel view

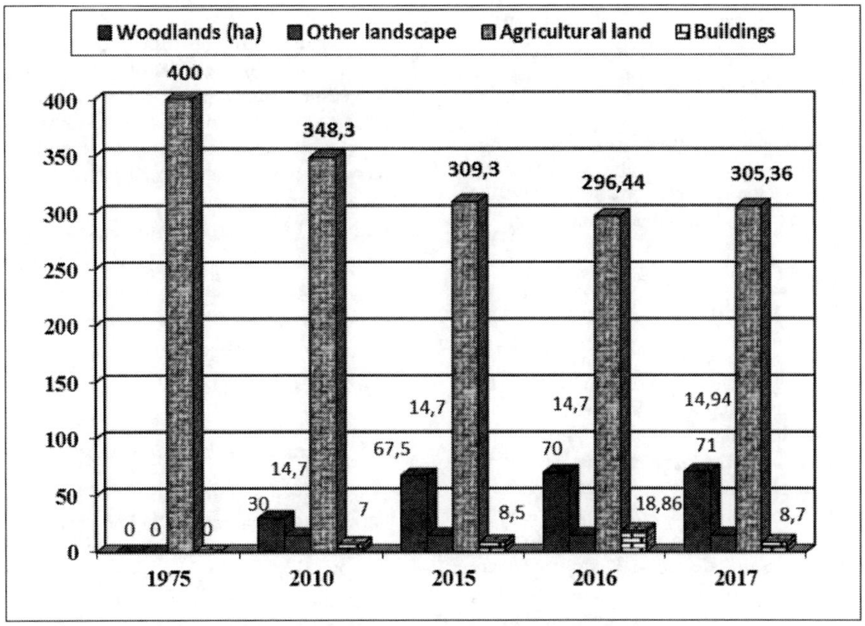


**Fig. 2**. Agricultural lands, buildings and woodland areas in the main campus of the University of Zanjan (ha)

All the woodland areas have been planted by drought resistant trees proper for semiarid areas. Moreover, the irrigation for trees has changed to the dripped irrigation method and other modern irrigation techniques have been utilized in croplands. This has improved the irrigation efficiency and conserved water resources. These conservation activities and other sustainability efforts have demanded an investment of over 25 percent of the total university budget.

## 2.2 Energy and climate change

During the last decade, energy efficient appliances have partially been replacing conventional appliances (40%-60%). The program for smart building is in initial stage, in which 55000 out of 189000 m2 total main campus building area (29%) has been implemented under this program (in different floors) by the end of 2016. Electricity and heating systems have partially been changed through utilizing some more eco-friendly technologies, for example using double glazing windows, smart electricity and heating in some buildings, and using solar energy. Most buildings have natural ventilation and full-day natural lighting. Energy management and climate change policies have been ratified in the long term strategic program of the university, particularly using solar power and biogas technologies, smart use of electricity and heating systems, greenhouse gas emission reduction, green building design through using natural ventilation and lighting, etc. The university has not calculated the campus yearly carbon emission yet, but it has been considered as a mandate for the future. The ten-year strategic program of the university has planned to increase solar energy production and usage in campus from 100 kwh to over 4000 kwh by 2025. Some other practical steps at the university include:

- Recovering the energy efficiency by improving the thermal insulation of existing buildings;
- Considering architectural technologies in the design and construction process of new developments aiming passive energy savings;
- Applying modern HVAC systems for building services, which has improved energy efficiency;
- Adjustments of the internal electricity grid of the university and the main feeder post; and
- Modification of natural gas pressure-reducing and metering station.
- Some planned arrangements for future include:
- Controlling and redesigning main structures of older buildings to make them resistant to earthquakes, according to the new edition of Iranian 2800 standard,
- Double-glazing the openings of previously built faculties, departments and office buildings,
- Repairing and/or installing thermal insulating panels on external walls and roofs,
- Optimising the suspended ceiling systems to improve their functionality, preferably by using recycled parts of the old system,
- Planning for power generation by CCHP power plants by private sector investments,
- Official negotiations for installation of a large network of solar panels and wind turbines to

increase renewable energy generation to %20 of the regular demand in three years. The plan is a part of a larger joint program with the Ministry of Energy, to establish an eco-park with education, research and power generation purposes.

The university is also interested to develop its international collaboration with international institutions for research and development of renewable energy technologies and their implementation in the campus.

## 2.3 Waste management

Vegetation waste is either used by for livestock consumption in the university farm or composted through natural composting or using vermicompost. Inorganic waste such as plastic and papers are collected separately for recycling. The university has a contract with an environmental NGO to give these wastes to that organization for recycling. Waste bins inside and outside all buildings have been detected by different waste signs, including recycle signs. Toxic wastes are also handled by labs and farm through collecting, containing and giving to relevant registered organizations for their disposal. One of the main achievement of the university during the last two years has been "modernising refined water and sewage systems during a long-term plan". Sewage disposal has been managed by using technologies in order that its water is reused and its sludge is treated through reusing and composting for farms. The university strategic plan and policies has put sustainable waste management into its agenda through programs such as using biogas technologies for sewage management, continuing recycling and composting, and toxic waste management. The water produced from sewage treatment will be used in farms and for landscape management.

Another policy for waste management has been paper reduce program for over 15 years. This policy has been implemented through developing the Information Technology in student's halls, catering systems, educational semester registrations, course evaluations, electronic teaching, student assignments, research records, conferences and workshops, journal paper submission and review, student's seminars and theses, using library materials eBook and papers), e-book publication, etc.

## 2.4 Water management

Water crisis is considered as a crucial challenge in all parts of Iran, including the Zanjan Province. The university has considered this issue in its short and long term strategic plans. There are two separate drinkable and non-drinkable water systems in the university. Using new technologies for irrigation has helped the water use efficiency in the campus, despite the increase of vegetation in the university. Drought resistant plant cultivation has also had a contribution in water reduction. The university has improved the irrigation practices in all agricultural lands and woodlands through applying modern irrigation technologies by converting from gravity surface irrigation systems to pressurized irrigation sprinkler systems (such as solid set, hand move, side roll, drip-trickles, center pivot and systems) and gate pipes. The strategic plan of the university has also emphasized on the following aspects:

- Studies for a pilot plan to use recycled/refined water in the irrigation of university green lands for growing agricultural crops and maintaining the landscape. It will also be used in flash tanks at the major water-consuming student accommodations.
- Restoring biodiversity and supporting local wildlife.

## 2.5 Transportation management

Public transportation is provided inside the campus and between the university and the Zanjan City free of charge to both staff and students. The policy and program of the university is to continue public transportation support and to increase bicycle and pedestrian use and to limit the motor vehicles in the campus. It has encouraged the staff to share their cars between the university and the city. The strategic plan for the future has also emphasized on establishing a tram on the campus and establish a bicycle road between the city and the University for 6 km to enhance safety for bicycle riders.

## 2.6 Education and research

On average, 700 papers on sustainability or environment are published in English or Persian and 55 events through international and national conferences, workshops, meetings, campaigns, speeches, etc. are organized. There are 33 students' organizations in the university that participate in facilitating sustainable and environment activities. The University of Zanjan consists of five main faculties including Agriculture, Engineering, Science, Humanities and Art and Architecture. The enrolled students in the academic year of 2017- 2018 have been over 10000 (64% Bachelor, 29% Master and 7% PhD) in 230 undergraduate and postgraduate programs (28% Bachelor, 46% Master and 26% PhD) studying in 37 educational departments. Approximately one fourth of educational programs provide courses and research regarding sustainability and environmental concerns. One third of research projects and budget have directly or indirectly been allocated to sustainable development.

## 2.7 Equity in education

Most Iranian students do not pay any tuition fee for their education, but if they should, they get loans for their education. Almost 55 percent of the students are female. The university facilitates low cost and subsidized accommodation for all students inside the campus through student halls, located in separate areas for male and female students. These student halls have all the facilities required for student life, including IT services, reading rooms, gyms, restaurants, subsidized meals, mosques, and so on. Single foreign students with scholarship can use free-catered accommodation. The university also facilitates accommodation for other married and single foreign students without scholarship, through university private accommodation.

## 3 Conclusion

Education and research for sustainability may need a significant change in not only in curricula and programs change but institutionalizing this concept in these institutions. The HEIs have a profound influence on all other institutions through educating and producing human resources who must care sustainability. The assessments made by the University of Indonesia Green Metric since 2010 to measure and rank universities around the world is one of the international commitments to encourage the HEIs to take sustainability as a serious key concept into account.

The University of Zanjan has increased and improved its vegetation and investment on sustainability, smart buildings and water management in both buildings and vegetated areas. The waste management programs have been implemented through using electronic correspondence and document submission in different activities of the university; separating plastics, glasses and papers and waste recycling; toxic waste handling in all labs; composting organic waste; inorganic waste management; and recycling sewage disposal. Though the university has provided free buses and shuttles to both staff and students inside the campus and between the city and campus to reduce private car use, it still needs to encourage bicycle use and improve its facilities to support it. Developing renewable energy for the future is still a challenge for this university and needs both innovation and investment. Students and academic staff have also been encouraged to move their conventional education and research methods and contents to more sustainable approaches, for example in courses syllabuses, student activities, research projects, publications and investments. The GreenMetric World University Ranking Network is expected to enhance its scope to contribute much more on sustainable development goals. A sustainable university should play an important role in innovation and technology research and development in sustainability; enhancing staff and students' sustainability knowledge and social capacities; changing the campus environment to an Eco-friendly and sustainable environment; and enhancing social and human capacities of communities and public and private institutions.

## References

1. Elliott, J., *An Introduction to Sustainable Development*. Routledge. (2012)
2. United Nation, *Transforming our world: The 2030 agenda for sustainable development*. United Nation: Washington DC. (2015)
3. Kinley, R., *Climate change after Paris: from turning point to transformation*. Climate Policy, **17(1):** p. 9-15. (2017)
4. Hermwille, L., et al., UNFCCC before and after Paris – what's necessary for an effective climate regime? Climate Policy, **17(2):** p. 150-170. (2017)
5. Mal, S., et al., *Introducing Linkages Between Climate Change, Extreme Events, and Disaster Risk Reduction, in Climate Change, Extreme Events and Disaster Risk Reduction: Towards Sustainable Development Goals, S. Mal, R.B. Singh, and C. Huggel, Editors*. Springer International Publishing: Cham. p. 1-14. (2018)
6. Karamidehkordi, E., 2018. *Enhancing the Contribution of Higher Education Institutions in Sustainable Development: A Discussion for the National Workshop on the UI GreenMetric World University Ranking*s, Iran, in 1st International Conference on Green University, February 6-7, 2018: Ferdowsi University of Mashhad, Iran. (2018)
7. UI GreenMetric. 2018. World University Ranking, Overal Ranking 2017. [cited 2018 15 February]; Available from: http://greenmetric.ui.ac.id/overall-ranking-2017/.
8. Higher Education Planning and Research Institution, *Iran's Higher Education Statistics at a Glance*. Higher Education Planning and Research Institution, Ministry of Science, Research and Technology: Tehran. (2018)
9. Times Higher Education, *World University Ranking 2018*. Times Higher Education: London. (2018)

https://doi.org/10.1051/e3sconf/20184804002

# Current practices of waste management at Universitas Diponegoro campus, Indonesia

*Yos Johan* Utama[1], *Ambariyanto,* A[2], *Syafrudin*[3] *and Ganjar* Samudro[3]

[1]The Rector of Diponegoro University, Jl. Prof. Soedarto, SH-Tembalang, Semarang Indonesia
[2]The Vice Rector of Research and Innovation, Diponegoro University, Jl. Prof. Soedarto, SH-Tembalang, Semarang Indonesia
[3]The green metric task force, Diponegoro University, Jl. Prof. Soedarto, SH-Tembalang, Semarang Indonesia.

**Abstract.** The existence of waste is one of the problems faced by the community, including in the college of higher education. Waste is produced not only from personal consumption, but also from other activities including office operations and research activities in laboratories and other personal activities. This condition is also related to increasing the existing activities on the campus of Universitas Diponegoro UNDIP, resulting in increasing waste. Related to this UNDIP gives serious attention to develop integrated waste management system. It is also related to the desire of UNDIP to be a sustainable university, where the principles of greenmetrics are always applied in the management of the environment. This article provides an overview of the curent practices of waste management at UNDIP's campus.

## 1 Introduction

The development of campus environmental management system in universities around the world is currently growing very fast. This is also related to the occurrence of Industrial Revolution 4.0, where technology plays a very important role. Improving services from various aspects becomes an important demand for the campus stakeholders, whether services related to the process of education, research and others [1].

One of the important demands is the service related to the condition of the campus environment. Today's college should be able to provide a good environment in the sense of a healthy and comfortable environment for the learning process in it. Campus is no longer a place to learn, but also serves for recreational areas such as sports and arts activities. Therefore, the management of the campus environment becomes very important in the management of universities in general.

Universitas Diponegoro is one of the largest state universities in Indonesia is located in coastal city Semarang. The university main scientific interest is coastal region eco-development [2]. With the main campus in Tembalang, Semarang (other campus in Pleburan, Semarang and in Teluk awur, Jepara), UNDIP is very concern in regard to environmental management. One of our important policies is to provide a comfortable, safe and healthy campus environment for all academic community who carry out various activities on the campus of UNDIP. This article provides an overview of waste management practices conducted at UNDIP Campus, in Tembalang, Semarang, Central Java province.

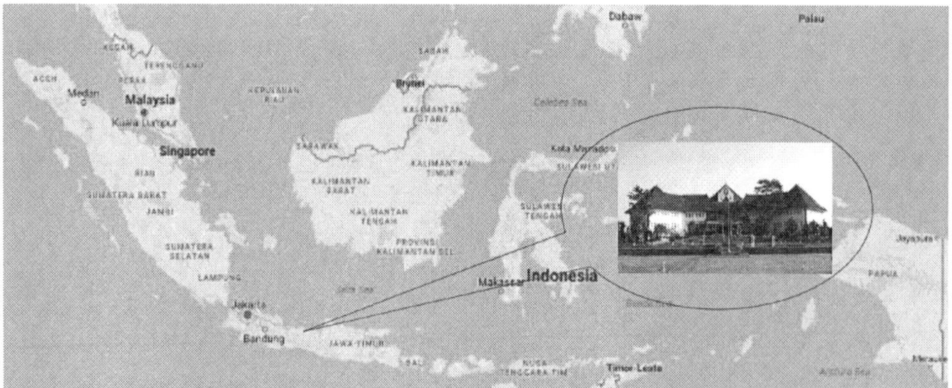


**Fig. 1**. The location of Universitas Diponegoro at Central Java, Indonesia

* Corresponding author: rector@undip.ac.id

## 2 Waste issues at Universitas Diponegoro

As a university with autonomy in 2017, UNDIP is required to develop faster, both in the field of education, research and public services. These developments provide increased programs and activities as well as the involvement of various parties. So there is a significant increase in activity within the campus of UNDIP. The consequences of this increased activity include increased waste.

Problems resulting from increased waste production and pollutants are the main problems associated with environmental issues [3]. Especially for campus UNDIP, then this is closely related to personal activities, office, as well as laboratories or university hospitals. The resulting waste is not only in solid form, but also the liquid that should be treated differently. Especially for waste from laboratories and hospitals that are very hazardous have become toxic pollutants. These kind of wastes require special treatment for handling, because it is not only harmful to the environment but also to humans. While the air pollutants inside the campus, mainly derived from the use of cars and motorcycles on campus by students and university employees. For air waste is also required a special policy for handling. UNDIP's campus.

## 3 Waste management practices

"Understanding waste" is a major aspect to be undertaken before determining a policy for its handling [4]. Understanding means we know where the waste comes from, why waste appears, what are the properties of the waste, what is the effect of waste both on environment and human, and so on. Given the enormous waste generated by humanity and its impact, the environmental management has become an international issue, not only politically and economically, but also educationally related.

UNDIP campus with thousands of people consists of student, lecturer and employee, and other supporting component, has a potency to generate waste from their activities. One of it is solid waste or garbage / rubbish which is a focus of treatment beside other environment fields such as water, drinking / clean water, wastewater, drainage, green space, and energy. For energy usage, UNDIP has the policy to optimized energy availability [5]. In the framework of waste management and environment, UNDIP undertakes various steps and policies in managing the environment, including related to the handling of waste. Some of these programs include:

1. Solid waste management
TPST (Integrated Waste Management Site or integrated solid waste treatment plant) is located inside the campus that plays a role in managing solid waste generated within the campus of UNDIP. This TPST Management Agency is responsible for collecting, managing, and treating solid waste from 14 faculties and other activity buildings as a student, academic and general administration. Before this TPST was formed, the street sweeper in UNDIP always burned the garbage, but this has not happened again.

The quantities of solid waste in UNDIP campus approximately 50 $m^3$/day contributed up to 1% of solid waste generated in Semarang City, whereas the qualities of it are mixed compositions of organic, non-organic and residue. However, given the current capacity, not all solid waste can be handled by TSPT. UNDIP in cooperation with the city of Semarang in handling the remaining waste around 30% in the waste management of the city of Semarang.

There are several kinds of supporting infrastructures such as worker/employee, collector vehicle or it known as "*Roda Tiga*" or three wheels, and waste bin (green as organic waste and yellow as an-organic waste). TPST as a place of solid waste management adopting 3R program of Indonesia's national solid waste management and one of the waste management infrastructure is used as treatment and separation of solid waste compositions and characteristics in UNDIP Campus. TPST received the mixed solid waste or heterogeneous waste, but it is separated at TPST. TPST separated organic, nonorganic, and residue into specific compositions such as organic consist of organic compostable and non-compostable, nonorganic consist of recovery and non-recovery, and residue is stated as non-recovery and inert waste.

2. Organic and Non-Organic Waste
Given that the resulting solid waste is heterogeneous waste, it is first separated between organic and non-organic waste. This separation is an important step in solid waste management [6, 7]. This separation was carried out by the UNDIP waste management team at the faculty and university level. Until now, the composition of waste produced by UNDIP consist of 55% organic, 40% an-organic, and 5% residue. Compostable organic waste is approximately 40% of organic waste consisting of food waste and leaves litter, 20% of nonorganic waste consist of plastic waste, and 20% of nonorganic waste consist of paper, cloth, wood, rubber, glass, metal, and hazardous waste.

3. Hazardous waste management
Various types of waste are produced by laboratories and hospitals have hazardous and toxic properties [8, 9]. Such waste is very dangerous for the environment, humans and other living creatures. Hazardous waste and toxic such as this have the requirements and procedures to manage them [10, 11], therefore UNDIP cooperates with Semarang city government so that this waste can be treated exclusively in accordance with the instructions of this waste processing.

4. Parks and other facilities
In the environmental management process at the UNDIP campus in Tembalang, Semarang, we set up some important support facilities. This facility greatly influences the success of environmental management and is able to create a good environmental atmosphere

for educational process in UNDIP. In addition to TPST, some of the main support facilities are:

a.Rare Plant Corner.

Located near UNDIP integrated laboratory, within this park there are some rare plants that are intentionally planted to maintain its existence. In general, this rare plant originated from Indonesia and is the result of cooperation with the Environment Agency of Central Java Province. More than 30 species of rare plants available in this park.This park is planned to be more developed, so that more rare plants are awake its existence.

b.Campus Forest

The campus forest in UNDIP is a natural forest that has always existed and developed with new plants through tree planting program. This forest is located on the road to the education reservoir. In addition to being useful as a green space in maintaining air quality on campus, this forest is also beneficial for lecturers and students especially in plant identification and forest management.

c.Taman Rusa

Deer Park was established in collaboration with BKSDA (Natural Resources Conservation Center). The type of deer that exists is the protected deer (Timor Deer, *Cervus timorensis*) and is now successfully developed successfully. Deer Park is located in Faculty of Animal Science and Agriculture and has been operated since 2014 started with 7 deer's. To date there are more than 20 Timor Deer in the park.

d.Educational Reservoir

The educational reservoir located on the campus of UNDIP was built in cooperation with the Ministry of Public Works of the Republic of Indonesia. This reservoir plays not only as a reservoir of water from the river that flows in the middle of the UNDIP campus, but also beneficial for lecturers and students in studying the management of reservoirs and their utilization. Even now the location of the reservoir is a place for visits to local attractions, especially on the weekend.

e.Agrotechnopark

Agrotechnopark is located at Faculty of Animal Science and Agriculture and has been operated since in the middle of 2016. In this agrotechnopark other than chicken coop with the Close House system, there are also agricultural crops. Technopark is very useful for lecturers and students, especially in managing various activities related to the field of animal science and the surrounding environment.

f.Parking Lot

UNDIP also provides a parking lot that aims to regulate the car park, so it will further minimize the effect of exhaust gas. Currently under construction of a new parking lot to accommodate cars that do not have a parking space. Based on calculations that we have done, it turns out the carbon dioxide produced by the vehicle UNDIP during peak hours, still much lower than the ability of all the trees in each unit to absorb the carbon dioxide. So far, the air at UNDIP campus is still very clean and not polluted.

5. Supporting activities

a.Tree planting

Tree planting program has been widely conducted worldwide with different opinions [12]. At UNDIP, tree planting is a routine program that continues to be held especially at the university level. Until now all the locations have been planted with trees of various types. One of the university's policies is that every new students must bring one tree, and the planting activities are regulated by UNDIP environmental managers..

b.Environmental related courses

To raise awareness of the importance of environmental management, almost all departments have courses related to the environment. For study programs such as Marine Sciences, Fisheries Science, Animal Husbandry, Agriculture, Biology and so on, are dominated by courses related to the environment. As for other departments also provided courses related to the environment..

c.Greenmetric WUR

UNDIP has been very actively involved in the ranking related to environmental management UIGreenmetrics World University Ranking. We consider this type of ranking is very important to follow, because it will give an overview of the advantages and disadvantages of UNDIP in environmental management. This will make it easier for us to develop new policies to improve the management system.

Waste management on campus UNDIP is closely related to environmental management in general and we are very committed to this. This is done in order to continually improve environmental conditions in UNDIP [1]. We believe that the desire to be sustainable university with a clean, healthy, and convenient campus for all can be achieved and will improve the performance of Diponegoro University.

## References

1. Utama, Y.J., Purwanto, and Ambariyanto. *Developing Environmentally Friendly Campus at Diponegoro University*. Advanced Science Letters, **23(3)**: 2584-2585 (2017)
2. Ambariyanto, A., Utama, Y. J., Darsono, D., Setyiono, B., Zainuri, M., and Noor, R. *Supporting Indonesian Government Maritime Policy, Through Diponegoro University Main Scientific Interest*. Advanced Science Letters, **23(10)**: 10061-10063. (2017)
3. Zurbrugg, C., *Urban solid waste management in low-income countries of Asia: How to cope with the garbage crisis*. Presented for: Scientific Committee on Problems of the Environment (SCOPE) Urban Solid Waste Management Review Session, Durban, South Africa, pp.1-13. (2002)

4. Palmer, J. A. *Environmental thinking in the early years: Understanding and misunderstanding of concepts related to waste management.* Environmental Education Research, **1(1)**, 35-45. (1995)
5. Ambariyanto, A., Utama, Y. J. and Purwanto. *Managing Campus Energy: Compromising between Rapid Needs and Environmental Requirement.* In E3S Web of Conferences. 31: 01003) (2018)
6. Artiningsih, N.K.A., *Peran serta masyarakat dalam pengelolaan sampah rumah tangga (Studi kasus di Sampangan dan Jomblang, Kota Semarang).* Thesis, Program Pascasarjana Universitas Diponegoro.110 pp. (2008)
7. Riswan, R., Sunoko, H.R. and Hadiyarto, A., 2011. *Pengelolaan sampah rumah tangga di Kecamatan Daha Selatan.* Jurnal Ilmu Lingkungan, ***9*(1)**, pp.31-38. (2011)
8. Marinković, N., Vitale, K., Holcer, N. J., Džakula, A., and Pavić, T. Management of hazardous medical waste in Croatia. Waste management, **28**(6): 1049-1056. (2008)
9. Wilson, J. L. Laboratory investigation of residual liquid organics from spills, leaks, and the disposal of hazardous wastes in groundwater. Robert S. Kerr Environmental Research Laboratory, Office of Research and Development, US Environmental Protection Agency. (1990)
10. Paramita, N., Evaluasi Pengelolaan Sampah Rumah Sakit Pusat Angkatan Darat Gatot Soebroto. Jurnal Presipitasi, **2(1)**, pp.51-55. (2007)
11. Harjanto, N.T., Suliyanto, S. and Ismojowati, E.S., Manajemen Bahan Kimia Berbahaya Dan Beracun Sebagai Upaya Keselamatan Dan Kesehatan Kerja Serta Perlindungan Lingkungan. PIN Pengelolaan Instalasi Nuklir, **8 (IV)**: 54-67. (2013)
12. Summit, J., and Sommer, R. Urban tree-planting programs—A model for encouraging environmentally protective behavior. Atmospheric Environment, **32(1)**: 1-5. (1998)

# Methodological proposal for the assessment of environmental aspects in Higher Education Institutions (HEIs)

*Liven Fernando Martínez* Bernal[1], *José Javier Toro* Calderón[2], *José Herney Ramírez* Franco[2*]

[1]Institute of Environmental Studies - Universidad Nacional de Colombia – Bogotá Seat.
[2]Department of Chemical and Environmental Engineering - Universidad Nacional de Colombia – Bogotá Seat*

**Abstract.** This work proposes a methodology specially aimed at Higher Education Institutions (HEIs). The methodology is based on the use of quantitative and qualitative indicators that allow diminishing the assessment's subjectivity and the uncertainty of the results. This proposal was designed within the implementation of Environmental Management Systems (EMS) according to the ISO standard 14.001/2015, so it can be applied by any kind of organization interested in improving its environmental performance by identifying its environmental aspects, as well as assessing and preventing its environmental impacts. The methodological proposal assesses standard criteria like the affected area and the frequency of the activities that generate the environmental aspects. This analysis is complemented by innovative indicators such as the danger of the substances related to the environmental aspect, the time of exposure and the effects on human health. Additionally, some indicators that allow measuring the magnitude of the harm generated by the identified aspects are also included. The proposed approach has two main advantages: first, it allows integrating the environmental analysis with other management systems, such as the Occupational Health and Safety Assessment Series (OHSAS 18001), quality (ISO 9001) and general requirements for the competence of testing and calibration laboratories (ISO 17025), among others; and second, it establishes an indicators system whose frequent assessment allows following up the environmental aspects and evaluate the effectiveness of its management measures. The application of this methodology in the main seat of Universidad Nacional de Colombia revealed that the most significant environmental aspects were: generation of ordinary solid waste, consumption of potable water, consumption of electric power, generation of recyclable solid waste, and wastewater discharge.

## 1 Introduction

The HEIs work as a system where exchanges of matter, energy and information take place, even though their missional purpose is the training and generation of knowledge, rather than the production of goods and services for the market or the business sector.

In the development of their missional functions, the HEIs generate environmental impacts that affect the university communities (students, teachers and administrative staff) and interfere with the learning and teaching processes, as well as with the human health and the natural resources of the campuses.

As a consequence, a high number of HEIs worldwide have turned to the implementation of EMS from a preventive approach as a strategic measure that integrates the university community in general and constitutes a didactic tool for environmental education.

The main basis of the EMS is the determination of those activities or actions developed by the HEIs that may interact with the environment. They are called Environmental Aspects (EA) and their identification and scope is essential for the subsequent design of action plans. The methodologies for their identification and assessment are mostly qualitative, therefore their associated subjectivity and uncertainty may interfere with the objectives of the EMS.

A methodology for the assessment of the EA has been designed taking into account this necessity. This proposal is based on the analysis of objective indicators that provide more credibility and it is framed in the implementation of EMS according to the ISO standard 14.001/2015, so it can be applied by any kind of organization interested in improving its environmental performance by identifying its environmental aspects, as well as assessing and preventing its environmental impacts. The proposed methodology was tested in Universidad Nacional de Colombia (Bogotá Seat) and it allowed obtaining more objective results and criteria that will contribute to a more effective decision-making process in the short, medium and long term. **Table 1.** Setting Word's margins. Use a two-column format, and set the spacing between the columns at 8 mm. Do not add any page numbers.

## 2 Conceptual foundations

According to the ISO 14001 standard, an EA is an element of the organization's activities, products or services that may interact with the environment. Carretero (2007) points out that "an EA is something

* Corresponding author: jhramirezfra@unal.edu.co

generated by an activity [1], product or service that may have incidence on the environment, which is defined as the natural setting that contains those EA as well as other living beings". Two concepts are proposed for the development of this methodology: unit of analysis (UA) and macro unit (MU). The UA refers to a space or a combination of spaces presenting homogeneous physical conditions and where common activities take place, so they have similar EA. Spaces such as warehouses, auditoriums, libraries, cafeterias, sport halls, laboratories, offices, classrooms and green areas can be considered as UA. On the other hand, the MU is the combination of UA that belong to a physical space clearly distinguishable due to its structural elements (e.g., a building). Unlike the UA, several types of activities may take place inside an MU.

## 3 Methodological proposal

This methodology has two main stages: identification of EA and their assessment. The following are the steps for its application.

### 3.1 Staff selection and training

Once the processes developed by the institution have been identified, a group of interviewers must be selected and trained. One MU will be assigned to each interviewer, who must identify all the UA and the person in charge in order to schedule an interview. Thus, will be obtained the information required to assess the EA.

### 3.2 Identification of UA and associated activities

The interviewee must provide all the information related to the activities developed in his/her UA. These activities must be grouped into generic categories, according to their nature. For instance, if the interviewee performs duties such as answering emails and writing documents, these tasks can be grouped into the "office and IT activities".

### 3.3 Identification of EA

Once the generic activities of each UA are defined, then the EA must be identified. In the case of Universidad Nacional de Colombia (Bogotá Seat), 33 EA were determined (see Table 1).

### 3.4 Assessment of EA

The assessment of the EA's significance takes into account the attributes of time, severity and magnitude:

**Table 1.** EA identified at Universidad Nacional de Colombia – Bogotá Seat. Source: based on [1] and [2]

| EA | | | |
|---|---|---|---|
| 1 | Combustion emissions | 19 | Generation of fat, oil and grease waste |
| 2 | Non-combustion emissions | 20 | Disposal of tires |
| 3 | Noise emissions | 21 | Disposal of rechargeable batteries |
| 4 | Emissions of offensive odors | 22 | Disposal of alkaline batteries |
| 5 | Saturation of visual objects | 23 | Disposal of light bulbs |
| 6 | Discharge of domestic wastewater | 24 | Disposal of waste of electrical and electronic equipment (WEEE) |
| 7 | Discharge of water with environmental effects | 25 | Disposal of tonners and cartridges |
| 8 | Discharge of water with sanitary effects | 26 | Disposal of expired medications |
| 9 | Generation of leachates | 27 | Disposal of pesticide containers |
| 10 | Generation of biodegradable waste | 28 | Consumption of potable water |
| 11 | Generation of ordinary solid waste | 29 | Consumption of non-potable water |
| 12 | Generation of inert waste | 30 | Consumption of electric power |
| 13 | Generation of recyclable solid waste | 31 | Consumption of fossil fuels |
| 14 | Generation of bio-sanitary waste | 32 | Consumption of paper |
| 15 | Generation of sharps waste | 33 | Consumption of disposable materials |
| 16 | Generation of human anatomopathological waste | 34 | Consumption of fertilizers |
| 17 | Generation of animal anatomopathological waste | 35 | Consumption of pesticides |
| 18 | Generation of chemical waste | | |

- **Time**: it is assessed according to the frequency of the EA and its duration (in hours). The ranks presented in Table 2 must be used in order to assess this attribute. The logarithm base 5 of the total of hours of EA per year (obtained from the multiplication of the frequency and the duration) was calculated to define the values included in Table 2. Values below 1 were approximated to that number, so that the scale could be limited to the rank from 1 to 5, and thus the significance of the EA is not underestimated (those values are identified in Table 2 with an asterisk).
- **Severity:** it is calculated from the interaction between the coverage (range), danger and effects of the EA on human health. Table 3 includes the description of each component of this attribute.
- **Magnitude:** it refers to the quantification of the EA based on quantitative or qualitative indicators. In order to assess this criterion, the results obtained from the indicators must be classified into 5 categories, each one of them with a particular value: "very high" (10), "high" (8), "medium" (6), "low" (4) and "very low" (2). The use of these categories and their associated values allows to standardize the magnitude of the EA, regardless of the units that were employed for their measurement. Some of the proposed indicators for the assessment of the EA in the case of Universidad

Nacional de Colombia (Bogotá Seat) are presented in Table 4. The defined values and ranges for these indicators are included in Table 5.

**Table 2.** Scale of values for the assessment of the Time attribute (own elaboration).

| Frequency | | Duration (hours) | | | | | | | | | | | | |
|---|---|---|---|---|---|---|---|---|---|---|---|---|---|---|
| | | ≤1 | 2 | 3 | 4 | 5 | 6 | 7 | 8 | 9 | 10 | 11 | 12 | ≥13 |
| Daily | 240 | 3.4 | 3.8 | 4.1 | 4.3 | 4.4 | 4.5 | 4.6 | 4.7 | 4.7 | 4.8 | 4.9 | 5 | 5 |
| Weekly | 54 | 2.5 | 2.9 | 3.2 | 3.3 | 3.5 | 3.6 | 3.7 | 3.8 | 3.8 | 3.9 | 4 | 4 | 4.1 |
| Biweekly | 27 | 2.1 | 2.5 | 2.7 | 2.9 | 3.1 | 3.2 | 3.3 | 3.3 | 3.4 | 3.5 | 3.5 | 3.6 | 3.6 |
| Monthly | 12 | 1.5 | 2 | 2.2 | 2.4 | 2.5 | 2.7 | 2.8 | 2.8 | 2.9 | 3 | 3 | 3.1 | 3.1 |
| Quarterly | 4 | 1* | 1.3 | 1.5 | 1.7 | 1.9 | 2 | 2.1 | 2.1 | 2.2 | 2.3 | 2.4 | 2.4 | 2.5 |
| Biannually | 2 | 1* | 1* | 1.1 | 1.3 | 1.4 | 1.5 | 1.6 | 1.7 | 1.8 | 1.9 | 1.9 | 2 | 2 |
| Annually | 1 | 1* | 1* | 1* | 1* | 1 | 1.1 | 1.2 | 1.3 | 1.4 | 1.4 | 1.5 | 1.5 | 1.6 |

**Table 3.** Values for the assessment of the Severity attribute (own elaboration).

| Criterion | Categories | Value |
|---|---|---|
| **Coverage:** area of influence of the EA | **Widespread**: when the effects of the EA go beyond the UA | 2 |
| | **Local**: when the effects of the EA are located inside the UA | 1 |
| **Danger**: inherent qualities of the substance related to the EA | **Dangerous**: corrosive, reactive, explosive, toxic, flammable, infectious or radioactive substance | 5 |
| | **Not dangerous**: a substance that does not present any kind of hazard | 3 |
| **Effects on human health**: level of damage that the EA or its associated substance may cause on human health | **Extreme**: death or disability | 5 |
| | **Critical**: acute or chronic diseases that generate partial or permanent disability | 4 |
| | **Severe**: diseases that cause temporary disability (*e.g.* hearing loss) | 3 |
| | **Moderate**: temporary diseases that cause any kind of malaise (*e.g.* diarrhea) | 2 |
| | **Minor**: does not have any effect on human health or only generates minor discomfort or irritation (*e.g.* headache) | 1 |

## 3.5 Significance of the EA

After assessing the attributes, the significance of each EA has to be determined by using the following formula:

*Significance = Severity + [Time x Magnitude]* (1)

*Where: Severity = Danger x Effects on human health x Coverage* (2)

*Time = Frequency x Duration* (3)

Depending on the result, the significance can be classified according to the values presented in Table 6.

**Table 4.** Indicators for the calculation of the magnitude (own elaboration).

| EA | Indicator | Formula* |
|---|---|---|
| Combustion emissions | Emission of $CO_2$.eq | $EC = \sum_{\alpha\beta} CC_\alpha \times FE_\beta \times PCG_\beta$<br>EC = Emission of $CO_2$-eq (kg/month)<br>CC = Fuel consumption (TJ/month)<br>FE = Emission factor (kg/TJ)<br>PCG = Global Warming Potential<br>α = type of fuel<br>β = Molecule |
| Generation of recyclable solid waste | Recyclable solid waste per capita | $RSRpc = \frac{\sum RSR}{N}$<br>RSRpc = Recyclable solid waste per capita (kg/per-month)<br>RSR = Recyclable solid waste (kg/month)<br>N= Number of individuals per UA |
| Generation of chemical waste | Solid chemical waste per capita | $RQSpc = \frac{\sum RQS}{N}$<br>RQSpc = Solid chemical waste per capita (kg/per-month)<br>RQS = Solid chemical waste (kg/month)<br>N= Number of individuals per UA |
| Consumption of paper | Consumption of paper | $CP = \sum NRM * 2.26$<br>CP = Consumption of paper (kg/month)<br>NRM = Number of realms of paper per month |

*Note: the abbreviations of the elements integrating the formulae correspond to the initials of the words in Spanish.

Once the significance is obtained, then some measures need to be designed in order to prevent, correct or mitigate the EA. The priority of these measures can be defined according to the category of the EA and the values presented in Table 6.

**Table 5.** Classification categories of some EA (own elaboration). All Unit are in kg/month.

| EA | Indicator | Very low (2) | Low (4) | Medium (6) | High (8) | Very high (10) |
|---|---|---|---|---|---|---|
| Combustion emissions | Emission of $CO_2$.eq | ≤10.21 | >10.21 and ≤102.1 | >102.1 and ≤510.5 | >510.5 and ≤1021 | >1021 |
| Generation of recyclable solid waste | Recyclable solid waste per capita | ≤0.23 | >0.23 and ≤0.26 | >0.26 and ≤0.28 | >0.28 and ≤0.31 | >0.31 |
| Generation of chemical waste | Chemical waste per capita | <0.102 | ≥0.102 and <0.114 | ≥0.114 and <0.126 | ≥0.126 and <0.138 | ≥0.138 |
| Consumption of paper | Consumption of paper | <2,26 | ≥2,26 and <4,52 | ≥4,52 and <9,04 | ≥9,04 and <18,08 | ≥18,08 |

**Table 6.** Scale of values for the interpretation of the EA's significance (own elaboration).

| Category | Values | Significance | Priority for decision-making and actions |
|---|---|---|---|
| Critical | 81-100 | Yes | Immediate |
| Severe | 61-80 | Yes | Short term |
| Moderate | 41-60 | Yes | Medium term |
| Minor | 21-40 | No | Long term |
| Irrelevant | <20 | No | Does not require any action |

### 3.6. Assessment of the EA in the MU

A set of indicators for the MU is proposed in order to complement the analysis. These indicators can be used to assess the environmental management at a larger scale (campus, other land properties of the HEI, or seat) and they can be formulated in absolute terms (e.g. kg) or in relative terms (e.g. kg/per capita or kg/month). The use of absolute indicators allows analysing the performance of the identified EA in a period of time [3]. On the other hand, the use of relative indicators facilitates the comparison of the EA [4] among buildings, other land properties and seats, which allows assessing plans, programs and actions related to the environmental management of the institution. Table 7 presents some of the indicators proposed for the MU of Universidad Nacional de Colombia (Bogotá Seat).

**Table 7.** Indicators for the UM (own elaboration).

| Category | EA | Absolute indicator | Relative indicator |
|---|---|---|---|
| Atmospheric emissions | Atmospheric emissions | Global emission of $CO_2$-eq | Emission of $CO_2$-eq per capita |
| Waste | Generation of non-hazardous waste | Non-hazardous waste generated | Non-hazardous waste generated per capita |
| | Generation of hazardous waste | Hazardous waste generated | Hazardous waste generated per capita |
| Consumption of resources | Consumption of water | Consumption of potable water | Consumption of potable water per capita |
| | Consumption of electric power | Consumption of electric power | Consumption of electric power per capita |
| | Consumption of paper | Consumption of paper | Consumption of paper per capita |

The proposed methodology has two main advantages: first, it allows integrating the environmental analysis with other management systems, such as OHSAS 18001, ISO 900 and ISO 17025, among others; and second, it establishes an indicators system whose frequent assessment allows following up the environmental aspects and evaluate the effectiveness of its management measures. For a detailed description of the methodology please refer to Martínez Bernal et al. (2018) [5] or the working document published in the web page of the Environmental Management Office of Universidad Nacional de Colombia (Bogotá Seat).

## 4 Application of the methodology

Table 8 presents the most important results of the application of the proposed methodology in Universidad Nacional de Colombia (Bogotá Seat). This means, table 8 presents the higher values of EA.

**Table 8.** Results.

| EA | % of UA where the EA was significant |
|---|---|
| Generation of ordinary solid waste | 34,5% |
| Consumption of potable water | 28,7% |
| Consumption of electric power | 19,5% |
| Generation of recyclable solid waste | 13,8% |
| Discharge of domestic | 10,0% |

| EA | % of UA where the EA was significant |
|---|---|
| wastewater | |

Source: own elaboration based on the information provided by the Environmental Management Office of Universidad Nacional de Colombia.

## 5 Conclusions

The application of this methodology in the main seat of Universidad Nacional de Colombia determined that the most significant environmental aspects were: generation of ordinary solid waste, consumption of potable water, consumption of electric power, generation of recyclable solid waste, and wastewater discharge. This information led to the formulation of programs aimed at the prevention, mitigation and correction of those aspects and their related environmental impacts.
The methodology proved to be easy to apply and effective for the identification of the main environmental aspects and impacts. It can be used by any other kind of institutions (not only HEIs), because it integrates the community in general and constitutes a didactic tool for environmental education.

## References

1. A. Carretero Peña. Aspectos ambientales: Identificación y Evaluación. AENOR. Madrid. Spain (2007)
2. L. W. Canter. Manual de evaluación de impacto ambiental. Mc Graw Hill (1998)
3. S. Cecchini. Indicadores sociales en América Latina y el Caribe. CEPAL (2005)
4. D.C. Suárez. Conceptos y formulación de indicadores. Programa de información e indicadores de gestión de riesgos de desastres naturales. BID – Universidad Nacional de Colombia Sede Manizales. (2003)
5. L. F. Martinez Bernal, A. L. Caro, J. C. Duran Dueñas, N. del P. Pacheco and J. J. Toro. Propuesta metodológica para la identificación y evaluación de aspectos ambientales en instituciones de educación superior. Gestion y Ambiente I. (2018).

# Waste management for achieving sustainable management of water and sanitation in Universitas Sebelas Maret Indonesia

*Ravik* Karsidi[1], *Okid* Parama Astirin[2], *Winny* Astuti[3]

[1]Rector Universitas Sebelas Maret, Campus UNS Kentingan, Surakarta, Indonesia
[2]Department of Biology, Faculty of Mathematics and Natural Science UNS, Coordinator of Green Campus Team Universitas Sebelas Maret Indonesia
[3]Study Program of Urban and Regional Planning, Faculty of Engineering UNS, Chairman of Green Campus Team Universitas Sebelas Maret Indonesia

**Abstract.** Universitas Sebelas Maret (UNS) is the biggest university in Surakarta Central Java Indonesia, which was established on March 11th, 1976 through Presidential Decree No. 10 of 1976. Commitmen of UNS toward Green Campus began in 2012 by Rector Decree *7nd of* August 2012. Land size of UNS, whis is only 60 ha becomes constraints in development of Green Campus Initiatives. In 2030 agenda of Sustainable Development Goals gives attention to huge increased of urbanization thus cities and human settlements face the problems of high density, economies of agglomeration link economy, energy, environment, science, technology and social and economic output [2]. The Problem of Water and Sanitation becomes big issues of Sustainable Development Goals (SDGs), which is stated in the Goal 6 out of 17 Goals. Goal 6 not only addresses Clean water sanitation and health but also quality and sustainability of water resources all over the world. Universitas Sebelas Maret is one of Campus in Indonesia, which has awareness toward water and sanitation, especially related to the problem of waste. Liquid waste has become problems for a long time in UNS, when domestic liquid waste has come from residential areas surrounding campus passed through the lake in campus before finally flowing to Bengawan Solo River. This causes euthropication and sedimentation of the lake of UNS, which had impact on extremelly decreased the function of the pond estetically and functionally. By partnership with the Ministry of Public Works anad Spatial Planning in 2017, UNS has built Integrated Waste Water Treatment, which treats domestic liquid waste from residential area surrounding campus and internal campus becomes recycled clean water. In spite of that, UNS also treats organic waste of fall leafs through composting units and manages chemical hazardous waste according to the Indonesian regulation issued by Ministry of Environment. Several efforts of UNS related to Waste Management have contribution to achieving SDGs especially Goal 6: ensure availability and sustainable management of water a sanitation for all [2]. This contributes for improving water quality by reducing pollutions, dumping, minimizing chemical hazardous waste, increasing recycled and safe reuse of water on Campus.

## 1 Introduction

Global environment degradation has increased awareness of people all over the world in all sectors of development including education sector as an agent of change and center for Information Technology. Commitment of Universitas Sebelas Maret (UNS) in improving environment to create more livable and sustanable living on campus, has been expressed through Green Campus Programs. Problem of Waste in campus becomes important issues, which has to be managed for achieving Goal 6 of Sustainable Development Goals: ensure availability and sustainable management of water a sanitation for all [2]. Types of waste in UNS generated campus activities are as follows: liquid domestic waste, toxic liquid waste contains B3 (hazardous toxic waste), organic solid waste especially from falling leaves and domestic solid waste. These wastes need proper waste management, to avoid environmental degradation, decreased of estetic and function

## 2 Waste management in Universitas Sebelas Maret

Several programs related to management of waste are conducted for achieving Green Campus. Those programs are: Recycling Program for University

* Corresponding author: ravikkarsidi@yahoo.com

https://doi.org/10.1051/e3sconf/20184804004

waste; toxic waste recycling; Organiz waste treatment' and inorganic waste treatment.

### 2.1 Recycling program for university waste

Some efforts of UNS related to Recycling Program for University Waste have been expressed in program of recycling domestic liquid waste for reproducing standard quality of water. Therefore, the recyclng reduces consumption of ground water in campus. The domestic liquid waste water can be reused after along process of waste water treatment for lowering concentration of pollutan's parameters, which cause water pollution. Therefore, this process reduces the pollutans absorbed by the environment thus quality of environment in campuswill be improved.

**Fig. 1**. Recycle water pond resulted from the process of Domestic Liquid Waste Water Treatment Instalation Universitas Sebelas Maret, Indonesia

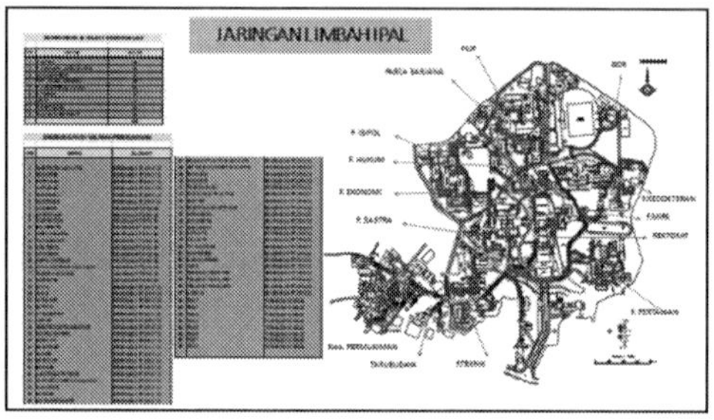


**Fig. 2**. Coverage area of Domestic Liquid Waste Water Treatment Instalation (IPAL)

**Fig. 3**. One of control ponds in Domestic Waste Water Treatment Instalation Universitas Sebelas Maret, Indonesia

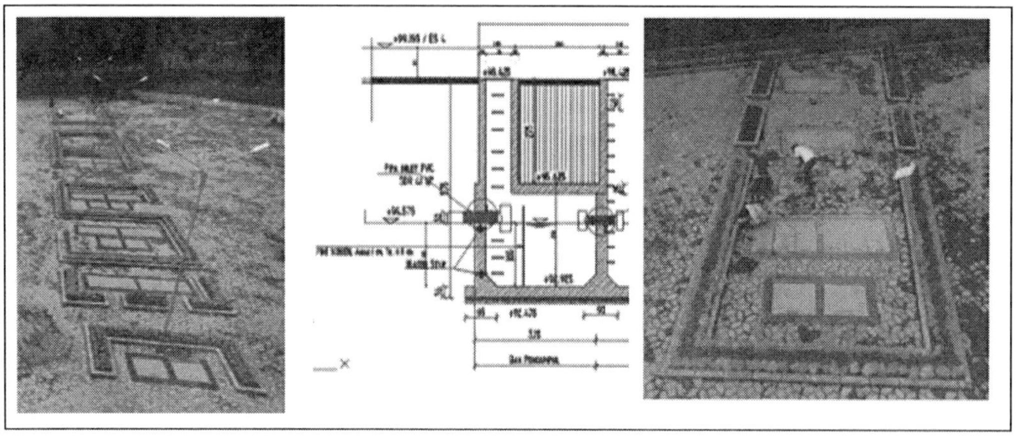

**Fig. 4** . Unit of unaerob domestic liquid waste water treatment UNS.

For a long time UNS has problems of domestic liquid waste water. The sources of domestic liquid waste are not only resulted from activities in campus but also from residential areas in 5 (five) subdistrict (RW) in RW 10, RW 11, RW 12, RW 13, RW 14 in Jebres District, who used to throw their domestic liquid waste to drainage system, which ends in the Pond of UNS. This causes decreasing quality of water of Bengawan Solo River, which flows to the Ponds was below the standard of water quality. In 2017, UNS, funded by Ministry of Public Works, has built Domestic Liquid Waste Water Treatment Instalation (IPAL). In which the installation processes the domestic liquid waste. The capacity of IPAL is 1.293,60 M3/day resulted from student liquid waste of 550,34 M3/day, 306,47 M3/day from lecturers and staff liquid waste and 436,80 M3/day from residential area liquid waste of Jebres District [4].

This instalation operates unaerob method and therefore it is treated in the aerobic recycle water pond (Fig. 1). The liquid waste, which has been processed is analyzed periodically to make sure the Waste Water Treatm3ent Instalation (IPAL) is well performed in processing the waste water. Indicators of well-performed IPAL is the threshold of standard quality of water. The recycle water resulted from IPAL can be used for watering the parks in campus. These prosesses contribute for conservation of water resources in campus as a part of Green Campus Program.

## 2.2 Toxic waste recycling

Management of toxic waste recycling is very important to save environment. In UNS, there are several working units, which dominantly produce toxic waste such as Faculty of Agriculture, Faculty of Engineering, Faculty of Medicine, Faculty of Matematic and natural science, Faculty of Education Science, Units of Center of Laboratory. As Education Laboratory play important roles for improving quality of students education and research of lecturers. In 2009 UNS has built Integrated Toxic Waste Recycle Instalation , which integrated management of toxic waste produced from Faculty of Mathematic and Natural Science (MIPA) and Faculty of Medicine. However, this instalation has not been effectively worked. Right now, the toxic waste is treated by collection of toxic waste into toxic tank for decomposition process in the Center of Laboratory in MIPA Faculty. This process is conformed to the Regulation of Republic of Indonesia no 101 / 2014 related to Management of toxic waste[1].

**Fig. 5** . (a) Toxic waste treatment pond for decomposition process in the Center of Laboratory in MIPA Faculty. (b) toxic waste treatment by collecting waste into tank

## 2.3 Organic waste treatment

Organic Waste, especially falling leaves and papers is one of the bigest waste resulted from activity in campus. Eventhough most of organic waste management is conducted by private, partneship, there are two units in UNS have been operating composting units for Organic Waste Treatment. (Fig. 6) shows composting unit in Faculty of Engineering, which produces compos labelled SEMAR for internal use in campus as well as for income generating activity. Apart from that, a Composting Unit also operates in Faculty of Agriculture, which produces 3 m3/ day in 7 days composting process.

 https://doi.org/10.1051/e3sconf/20184804004

**Fig. 6**. Composting unit in faculty of engineering

The organic waste of falling leaves in UNS is collected into plastic tank. This process is done by outdoor cleaning service by partnersips with outsourching persons, who also maintain campus to be cleaned, maintained and livable for academic sircumstances. There are 178 units of plastic tank of organic waste collection in UNS campus.

### 2.4 Inorganic waste treatment

Management of inorganic waste in UNS is conducted through the concept of 3 R : Reduce, Reuse Recycle. Inorganic waste especially plastic has become a big problem in UNS for a long time. Most of Faculty units in UNS do not manage inorganic waste individually. However, right now 100 % of faculties and units in UNS separates gargabe collection into organic and inorganic (Fig. 7). This is a part of Rector instruction as follows [3]:

1. All units in UNS should manage garbage collection at least into : organic , inorganic waste, and toxic waste.
2. Universitas sebelas Maret by cooperation with private sector operates inorganic waste recycle.

**Fig. 7**. Separation of inorganic and organic garbage collection

Apart from that, in order to reduce production of inorganic waste, especially from mineral water container, UNS by facilitaion from Ministry of Public Works has built Drinking Water Treatment instalation (SPAM) with 150 spots of water dispensers. The civitas academica is encouraged to bring water tumbler and refill their tumbler in the water dispenser for daily activity in campus (Fig. 8).

https://doi.org/10.1051/e3sconf/20184804004

**Fig. 8**. Drinking water treatment instalation in UNS (SPAM)

## 3 Concluding remarks

Commitment of UNS in achieving sustainable living in campus is expressed through management of waste in Campus . Several programs of Green Campus have been conducted such as Building Domestic Liquid Waste Water Treatment Instalation (IPAL) to proceed domestic liquid waste, from residential areas surounding campus integrated with domestic liquid waste generated by campus activities. Another programs are composting unit instalation to proceed organic waste; toxic waste treatment and inorganic waste treatment. These programs contribute to achieve Goal 6 of Sustainable Development Goals (SDGs): ensure availability and sustainable management of water and sanitation for all.

## References

1. Peraturan Pemerintah Republik Indonesia Nomor 101 Tahun tentang Pengelolaan Limbah Bahan Berbahaya Dan Beracun (2014)
2. United Nations, Sustainable Development Goal 6, Ensure availability and sustainable management of water and sanitation for all. Sustainable Development Knowledeg Platform, .https://sustainabledevelopment.un.org (2015)
3. Green Campus UNS, Matery of Green Campus for Socialization of Green Campus in New Student Orientation Program (2017)
4. Ministry of Public Works and Human Settlements. Standart Operational Procedure IPAL Kawasan Jebres Kota Surakarta (2017)

# BASE: a sustainable path for the University of Milano-Bicocca

*Cristina* Messa[1,*], *Giacomo* Magatti[2], *Massimiliano* Rossetti[2] *Matteo* Colleoni[2], *Massimo* Labra[2], and *Marina* Camatini[2]

[1]Rector, Milano-Bicocca University, Piazza dell'Ateneo Nuovo 1, 20126, Milano, Italy
[2]BASE Bicocca Sustainability Office, Piazza della Scienza 1, 20126 Milano,

**Abstract.** Since its foundation, in 1998, the University of Milano-Bicocca has been pursuing the objective to make its structures sustainable from an environmental, social and economic point of view. To this end, in 2015 the University of Milano-Bicocca created BASE (*Bicocca Ambiente Società Economia – Bicocca Environment Society Economy*), an internal office aimed at promoting the interaction between research and training and at stimulating sustainability both within the University and outside. BASE proposes a holistic approach to sustainability including energy efficiency, waste reduction, sustainable mobility, climate change attention and water and food supply. The report will focus on the interventions recently carried out in the various fields, paying particular attention to the issues of waste management and of mobility.

## 1 Introduction

Since its foundation, in 1998, the University of Milano-Bicocca has been actively engaged in order to make its structures sustainable from an environmental, social and economic point of view. The objective has been pursued not only by reducing the environmental impact of its management but also by fostering sustainable behaviour among its employees and students. In 2015 the University of Milano-Bicocca created *Bicocca Ambiente Società Economia – Bicocca Environment Society Economy* (BASE), an internal office aimed at promoting the interaction between research and training and at stimulating sustainability both within the University and outside. Examples are the realization of studies and research on sustainability, training activities on the subject for students and staff and the coordination of the different sectors of the University in the implementation of measures for sustainability. Outside the University, BASE supports the attention towards sustainability by participating in working groups, locally (Bicocca district), nationally (University Network for Sustainable Development) and at international level (International Sustainable Campus Network - ISCN). BASE proposes a holistic approach to sustainability that includes the commitment to energy efficiency, waste reduction, sustainable mobility, climate change attention and water and food supply. After having presented the structure and the operating procedures of BASE, the report will focus on the interventions lately carried out in the various fields. Particular attention will be paid to the issues of waste and mobility management, which have been widely appreciated both at national and international level.

## 2 BASE: a sustainable path for the University of Milano-Bicocca

The universities play a key role in attaining the sustainable development goals set out in the 2030 UN Agenda, firstly because it is the universities, above all, whose task it is to carry out studies and research into the environmental, social, and economic factors associated with sustainable development; secondly because it is in the Universities that students and staff are encouraged to adopt virtuous behavior to protect the environment and its resources; and lastly because the universities can play an important role in promoting collaboration on sustainability issues between public and private institutions and neighborhood associations [3].

That is why ever since it was founded, the University of Milano-Bicocca has been committed to the objective of sustainability. When it was being built, it was dedicated great attention to the environmental quality of the buildings, including the open and closed spaces, this commitment continued by creating BASE (Bicocca Environment, Society, and Economy), an inter-departmental and inter-disciplinary office under the Directorate-General of the University. The BASE Office is coordinated by a president assisted by a Scientific Committee consisting of 9 professors from different university Departments and has the role of providing scientific guidance and supervision for action on sustainability. The Scientific Committee is supported by two technicians who work in collaboration with the university's Infrastructure, Procurement and Communication Area to harmonize the scientific and operational aspects. In fact, it is a particular feature of BASE that it addresses sustainability across the

* Corresponding author: cristina.messa@uimib.it

disciplinary areas of the university and their research, training and management sectors. In that sense, sustainability is both an object of study and research for teachers and researchers, and a management goal for the university administrators. Finally, the external activities of Milano-Bicocca to promote sustainability is shown by its participation in local, national, and international working groups (Bicocca District, the Network of Universities for Sustainable Development, and ISCN, the International Sustainable Campus Network).

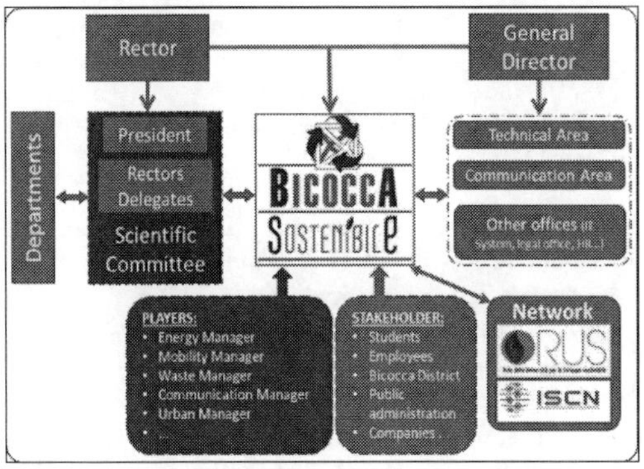


**Fig. 1**. BASE organizational diagram

In operational terms, the academic sustainability policies and interventions of Milano-Bicocca are planned within the framework of the *Three-year University Environmental Sustainability Plan*, which sets out the medium-term objectives and action to be taken to reduce pollutant emissions and energy consumption, promote sustainable mobility, reduce the production of waste, safeguard the quality of water and food resources, and protect the social categories that are most exposed to the negative consequences of the unequal distribution of resources (environmental, economic, and social, see Fig. 2) [4].

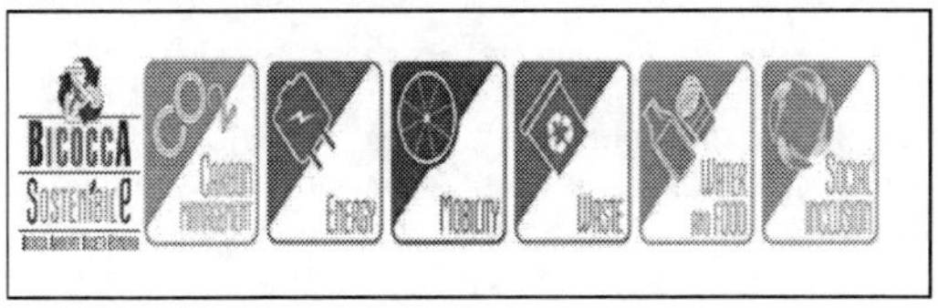


**Fig. 2**. Logos of the Milano-Bicocca University "sustainable path actions"

Within that context, BASE promotes a holistic approach to sustainability as both the goal and the outcome of joint effort by the different disciplinary areas and sectors of the university and the different actors in the surrounding territory, within a plurality of fields for intervention (environmental, economic, and social). The following section focuses on the two areas where Milano-Bicocca has attained its best outcomes and awards, nationally and internationally: waste management and mobility.

## 3 Waste management at the University of Milano-Bicocca

For many years the collection, management, and disposal of city waste within our University was addressed inefficiently and ineffectively and there was no standard method for managing the collection and disposal of waste across all the different buildings. Monitoring carried out during 2015 showed that 27% of that waste was being collected and disposed of differentially. Consequently, given that the amount of undifferentiated waste is an estimated 331 tons per year, the environmental impact of disposing of that waste was very high: 105 tons of $CO_2$e/year (calculated in accordance with the Italian Ministry of the Environment's *Guidelines for the Carbon Management of Universities* on the undifferentiated proportion of waste). In fact, due to the need for it to be transported to the waste-to energy plant and incinerated, every ton of undifferentiated waste was responsible for producing 315 kg of $CO_2$e.

To reduce such a negative effect, a new waste management system was tested in 2016. In a first phase, all bins for undifferentiated waste were removed from all the offices and laboratories in the 28 university buildings and replaced with a paper waste collector only (as most of the waste is paper). Having studied the use of shared spaces in university buildings, approximately 500 islands for differentiated collection were installed at locations with the greatest potential for producing waste. As a consequence, the rate of differentiated waste collection increased from 27% to 70% (of which 50% was paper, 16% plastic and metal, and 4% glass), with a reduction of 45% in greenhouse gas emissions.

To comply with municipal standards, the containers on the islands were colored differently depending on the type of waste they were to contain (yellow for plastic and metal, white for paper, green for glass, and grey for undifferentiated waste, see Fig. 3), and these proved easy to identify and use. A preliminary communication campaign, giving clear information about the purpose of the new waste disposal system and how to use it, also contributed to the success of this initiative.

**Fig.3**. Left: an "Island" for differentiated waste collection at the University of Milano-Bicocca; Right: a BASE water flask

An integral part of this new waste management system was "*PolApp*", an application for smartphones and tablets that enabled users to monitor the islands and report back on the quantity and quality of differentiated waste that was being collected. Each island was identified by a Qr code, making it possible for University

 https://doi.org/10.1051/e3sconf/20184804005

users to detect and send information on how full the containers were, and whether there were any problems, which enabled them to keep the management of the system under control. It also enabled them to compile thematic maps of the quantity and quality of waste being produced in the various buildings of the university.

The second phase of this new waste management system was aimed at reducing the need to collect and dispose of the plastic water bottles dispensed from machines or sold in university cafeterias and bars. For that purpose, 13 water dispensers were installed in the university buildings, and over this past year they have supplied, free of charge, approximately 250,000 liters of natural and sparkling drinking water (enabling a saving of about 18,200 kg of CO2 emissions per year). Finally, the free distribution of steel flasks, marked with the BASE logo, to the university students and staff has achieved the twofold result of limiting the use of plastic bottles whilst at the same time identifying these ecological flasks as the symbol of this new community of environmentally sustainable and responsible consumers.

# 4 Sustainable mobility management at the University of Milano-Bicocca

Since 1998 all public institutions and private companies in Italy with more than 300 employees in a single location (or with more than 800 employees in multiple locations, according to the interministerial decree "*Sustainable mobility in urban areas*" - Legislative Decree 27/03/1998) must appoint a Mobility Manager tasked with:

- encouraging greater use of sustainable modes of transport;
- improving sustainable access for persons and organizations;
- increasing the efficient use of transport infrastructure and types of land use;
- reducing traffic by reducing the number and length of journeys and the use of private vehicles.

As public institutions, all Italian universities have Mobility Managers who are organized under a National Coordination and are committed to making the journeys of university students, academics, and administrative staff more sustainable. Like other Mobility Managers, the university Mobility Managers are required to draw up a home/work/university travel plan, to participate in the governance of urban and metropolitan mobility, and to carry out studies and research on university mobility. In 2017 the National Coordinator of University Mobility Managers, working under the direction of the University of Milano-Bicocca's Mobility Manager, prof. Matteo Colleoni, carried out the first national survey on travel and shared mobility in the Italian universities (based on a sample of 37 universities and approximately 70,000 cases).

This was the first survey carried out in Italy using a common data collection tool, and showed that the different modes of travel were well distributed (with 61% of the respondents making their home/university journeys using public transport) and that there was a positive tendency to share journeys, although the negative aspects included a poor take-up of active mobility (on foot and by bicycle), the time taken by these journeys and their high costs. This last consideration was seen as a cause for concern, since it limits the opportunities for students to access and attend their university and has a negative impact on their entitlement to study. Furthermore, although students and staff make most of their journeys by public transport, these generate greenhouse gas emissions of approximately 31,480 tons of $CO_2e$/year (or 64% of the entire university carbon footprint, the remainder consisting of emissions associated with energy generation, 35%, and waste disposal, 1%).

To address these issues, the mobility management policies of the University of Milano-Bicocca were divided into different sectors of activity aimed at:

- increasing active mobility;
- improving the quality of public transport and the intermodality of the transport system; - increasing the types of shared mobility.

In relation to active mobility, a bike sharing service is currently being completed that will support the on-campus mobility of university staff. The bicycles are fitted with smart locks, are owned by the university and the municipality, and will be supported by a digital platform (APP and web app) that will make the bicycles easier to hire, collect, and return. In relation to pedestrian movement, the university is supporting the Municipality of Milan in making critical road crossings safe and creating cycling/walking routes that encourage active mobility. Since 2016, the University of Milano-Bicocca has been participating in the European *"Bike2Work – smart choice for commuters"* campaign, which will use innovative incentive schemes (based on economic and time rewards and benefits) to shift journeys away from private cars towards cycling.

As it has been mentioned, public transport is the predominant mode of travel by Italian university students, specifically by those attending Milano-Bicocca (with public transport usage rates of 80%). Mobility management policies in that mode are therefore aimed at improving the quality of travel on public transport and increasing the intermodality of the transport system (particularly the use of integrated journeys that combine bicycles and public transport). So far as the quality of public transport is concerned, great attention has been paid to convention agreements with the transport companies that improve public transport and make it fairer and cheaper to use. Employees who purchase tickets for use on public transport (bus, metro and train) receive a financial contribution from the university that covers approximately 30% of the cost. However, since the University cannot afford to pay the subsidy to the very large number of students who live outside the regional capital, those students do not receive that discount. Nevertheless, the University of Milano-Bicocca is currently coordinating negotiations with Trenord, the regional railway company, and the other universities of the region to find an agreement that will also make it cheaper for those students to buy travel tickets. Finally, an agreement between the University of Milano-Bicocca and Trenitalia, the national rail operator, which is already in operation, gives university employees and students a 15% discount on trains. The underlying aim of that agreement, and of the others mentioned, is to provide financial support for mobility in the conviction that making travel easy improves working conditions and, consequently, the productivity of Italian university students in their various disciplines. The railway companies would benefit from the increasing number of clients using public transport with lower levels of pollution.

Lastly, the adoption of shared mobility has been spreading rapidly over the past few years in Italy, particularly in Milan. New car and bicycle sharing and pooling services are expanding the range of modal choices and increasing the proportion of shared mobility, particularly within the urban area. On the campus itself, the University of Milano-Bicocca provides a free low-environmental-impact bus service and gives discounts for bus and van sharing services, car-pooling, car + scooter travel, and cycle hire. Every year there is a *Bicocca Mobility Day,* which provides an opportunity to present the mobility support services to students and staff and to meet with the other local and regional stakeholders and discuss policies for improving sustainable university travel. The underlying objective of these and other initiatives is to transform shared travel from a simple choice of mode to a style of mobility and a way of using the city, and to encourage behaviour that fits well with the principle of sharing resources and being attentive to the well-being of the community and the present and future environment.

To sum up, the University of Milano-Bicocca is aware of the negative effects that the transport sector has on environmental quality and climate change and is devoting great effort and resources to the implementation of sustainable mobility policies. Many years of academic travel management policies have shown that the universities, which were traditionally considered as mere attractors of mobility, can and must actively participate in planning and programming urban and metropolitan policies to achieve a type of mobility that is ever more attentive to the sustainability of the environment [1].

## 5 Summary/ concluding remarks

20,000 teachers and employees and more than 200,000 students work or study at the five universities in the city of Milan, 40% of whom make a daily journey to get there and back. At Milano-Bicocca alone, about 15,000 students and staff need to access and use the university's resources. These high numbers put great pressure on the resources of the university and its environment, requiring the adoption of policies to safeguard its conservation and regeneration. As part of the Italian network for sustainable development, Milano-Bicocca has been implementing these policies by entrusting BASE with the task of addressing the issue in a multi-disciplinary, multi-sectoral way, paying attention to sustainability in its various different aspects. Twenty years since Milano-Bicocca was first set up, it can be said that these policies have been delivering positive outcomes, especially in particular areas such as waste management and mobility. Positive signals are also coming from other sectors such as energy saving, reducing emissions from pollutants, and policies for social inclusion. None of these results would have been possible without collaboration between the scientific and administrative components of the university, and external exchanges with other universities also belonging to the national and international network for sustainable development: demonstrating that as in the natural world, action on sustainability gives the best local results when it is shared and constructed collectively [2].

**Acknowledgements**. The Rector of the University of Milano-Bicocca and the Scientific Committee of BASE wish to thank the University Infrastructure and Procurement Area, particularly its Director Marco Cavallotti and the Engineer Dr. Marco Angelillis, for their support with the implementation of these environmental sustainability policies and the activation of BASE.

## References

1. Colleoni M., Pucci P., *Understanding Mobilities for Designing Contemporary Cities*, Springer (2016)
2. Magatti G., Camatini M., *et al.*, *Energy Consumption Analysis and Carbon Footprint of a Building of the University of Milano-Bicocca: Starting Point for a Sustainability Report*. Energia, Ambiente, Innovazione. ENEA (2013)
3. Ministry of the Environment and Protection of Land and Sea, – G.D. for Sustainable Development, Climate and Energy; University Cà Foscari of Venezia - Ufficio Processi e Progetti Speciali. *Linee Guida in materia di Carbon Management per gli Atenei. Case study: Università Cà Foscari Venezia* (2012)
4. Scientific Committee BASE, *University Sustainable Environmental Plan – 3year-period 2017-2020*, approved by the Administration Committee of the University of Milan-Bicocca, 2017/07/18

# Sustainability in motion at UAO: integrated waste management

*Luis H.* Pérez[1], *Mario A.* Gandini[2]

[1] Rector, Universidad Autónoma de Occidente, Calle 25 No. 115-85, Cali, Colombia
[2] Department of Energy and Mechanics, Faculty of Engineering, Universidad Autónoma de Occidente, Calle 25 No. 115-85, Cali, Colombia

**Abstract.** The Universidad Autónoma de Occidente (UAO), located in Cali (Colombia), launched the Sustainable Campus Program in 2015, in order to implement sustainability actions and strategies within the University campus with the purpose of contributing both to a better environmental performance of the campus operations and to generate knowledge and experience than can be used in the context of cities and regions in their search for reaching sustainable conditions. Integrated waste management is one of the components of the UAO Sustainable Campus Program. Based on the categories and indicators presented in the GreenMetric 2017, the integrated waste management includes the program to reduce the use of paper and plastic in campus, recycling program for university waste, toxic waste handled, organic waste treatment and inorganic waste treatment. **Keywords**:Integrated waste management, sustainable campus.

## 1 Introduction

The Sustainable Campus Program of the Universidad Autonoma de Occidente –UAO- (Cali, Colombia) was launched in 2015. Being a sustainable campus requires performing and promoting local, regional, national and global actions and activities to minimize negative impacts on the environment, economy, society and health, resulting from its teaching, research and projection functions Social. In addition, a sustainable campus considers and promotes environmental, social and health practices and considerations in these three activities and at all levels.

In this order of ideas, a sustainable campus is an environmentally healthy space, which promotes the efficient use of natural resources, the reduction of waste generated in its processes, the recovery of local fauna and flora, animal protection, minimization of the use and disposal of hazardous materials and waste, sustainable and collaborative consumption practices, both in their community and in their local and regional environment. The University actively promotes this type of action in its community and its social and economic environment. The institution actively involves its knowledge, experience and human resources to address and provide solutions to the environmental and social challenges facing current and future society, with a view to establishing a balance between the needs of human beings and those of other beings with which we share the planet, to guarantee its process of evolution and integration in the web of life. For these reasons, it is proposed that the Sustainable Campus project be transformed into a strategic program of the University, taking into account the quantity and complexity of the factors with which it has been decided to work. Although the University has good results in the environmental management of the campus, there is still much to be done in the aspects of each of the aforementioned components. Being coherent and building sustainability at home, require addressing multiple aspects, which merit being taken, each, as projects of a program that must be developed and consolidated in the medium and long term. The Program takes into account both campus operation aspects, such as integrated energy management, integrated water management, air quality and climate change, and integrated waste management. The latter is discussed in this work. Based on the categories and indicators presented in the GreenMetric 2017, the integrated waste management includes the program to reduce the use of paper and plastic in campus, recycling program for university waste, toxic waste handled and organic waste treatment.

## 2 Integrated waste management at UAO

### 2.1 Recycling program for university waste

Currently the University has a policy to reduce the use of pads and plastics on the campus, which includes training and awareness programs to make use of printing exclusively when necessary, as well as encouraging double-sided printing, the use of reusable bags and the use of reusable cups. The recycling program at the University is carried out by the separation at the source approach. All reusable materials are then stored and the deliver to private actor for material recovery.

The University implemented the Integrated Solid Waste Management Plan in 2014, giving continuity to a waste management program established in 2005, with

* Corresponding author: rector@uao.edu.co; magandini@uao.edu.co

the objective of making an adequate disposal of solid waste, managing alternatives for collection, handling, treatment, material recovery and final disposal [1]. To carry out this initiative, the University has arranged various collection points within the university campus, allowing the academic community to separate the different waste fractions at the source. In addition, educational campaigns are carried out every six months to raise awareness among students, teachers and collaborators of the adequate disposal of waste in the bins, identifying them by color:

- Green: not recyclable (dirty napkins, food packaging and sweeping).
- Gray: recyclable (paper, cardboard, kraft, folding, glasses and cardboard plates).
- Blue: recyclable (plastic jars, pet, metal, polypropylene, video and tin).
- Brown: biodegradable waste (food waste of fruits and vegetables, stems and leaves of trees and pruning and gardens).
- Red - dangerous (luminaires, electronic waste, markers, fluorescent tubes, light bulbs, paint jars, oils, chemical residues).

The recyclable waste is stored and classified depending on the type of material, in the warehouse of the Technical Storage Unit to be delivered every 15 days to the company PROGECOL S.A.S, which is responsible for the recovery of recyclable materials in addition to the certification of the proper final disposal of the different material when they are not recovered. Figure 1 shows the generation of recyclable waste during the last 4 years, in which the year 2014 presented the lowest generation with 30,453 kg / year and the year 2016 presented the highest with 37,515 kg / year. The decrease for the year 2017 can be associated with the implemented policy to reduce the use of pads and plastics on the campus, which includes training and awareness programs to make use of printing exclusively when necessary, as well as encouraging double-sided printing, the use of reusable bags and the use of reusable cups [1].

**Fig. 1.** Generation of recyclable waste during the last four years

## 2.2 Toxic waste handled

Toxic waste management is undertaken through high efficiency deactivation treatment for biosanitaries and sharps, incineration for toner cartridge waste - lubricants - markers, chemicals, maintenance and grease trap. Other treatment are employed for luminaires, electronic waste, batteries, vegetable oil from coffee shops and used industrial oil from chillers and the emergency energy plant.

Weekly collection of hazardous waste generated in laboratories, nursing and collection points of the university [2]. In the case of laboratories and nursing, the personnel in charge must label the bags mentioning the elements that it contains and, in this way, prevent and inform the operator that performs the collection. The company that performs the external collection of hazardous waste from the university is RH S.A.S. This company is responsible for collecting them monthly, in addition it gives the university a certificate of final disposal. On average 550 kg / month of hazardous waste and 6,600 kg / year are generated. The second graph shows the generation of hazardous waste during the last 4 years [2]. The high generation of hazardous waste in 2015 is mainly due to Waste Electrical and Electronic Equipment (RAEEs), considering that in June the modernization of the computer equipment located in the study rooms was carried out with a generation of 4,372 kg / month.

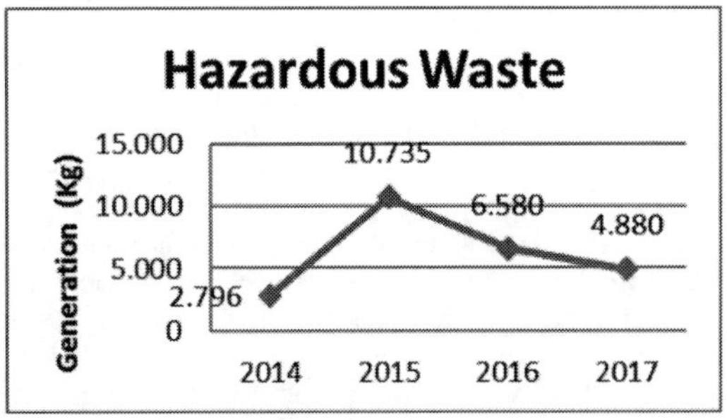


**Fig. 2.** Generation of hazardous waste during the last four years

## 2.3 Organic waste treatment

According to the campus waste composition, organic waste has three significant components: food waste, garden waste and sludge from the wastewater treatment plant. For each of them, a monthly, annual and an average production report is presented. The total organic waste produced by the campus is treated through vermicomposting. It should be noted that the gardens of the University Campus are fed with the compost obtained by this process [3].

The University is currently carrying out two processes for the production of compost. The first is made with plant residues resulting from the pruning of trees, palms and cutting of the meadow. These wastes have an approximate utilization of 90%, considering that the remaining 10% are large residues that cannot be processed by the available chipper, therefore they are delivered to the company PROMOAMBIENTAL, who collects them between two and three times per month, depending on the amount of waste stored. Like the company RH S.A.S, PROMOAMBIENTAL delivers a certificate of final disposition to the University.

The generation of compost from vegetable waste is done in a nursery garden through a composting process.

This process consists of 4 phases: in the first phase a stacking of the vegetable waste is carried out and its size is reduced to facilitate the process, in the second phase the freshly cut material is placed and a natural decomposition process is carried out that lasts approximately 15 days. In the third phase the residue of phase 2 is combined with the one currently in this phase, in addition to adding purge honey and wheat bran to increase the reproduction of enzymes that allows the rapid decomposition of the plant material. In the fourth phase, the process of humus generation ends naturally. This product is strained and mixed with rice husk to be applied directly in the gardens of the university campus and the coarse material is reintegrated into the process. The compost generation inside the university lasts approximately 6 months and generates around 6 and 9 tons per year [4].

The second process to generate organic fertilizer is done through the vermicompost. For the generation of this product it is necessary to collect organic waste from the cafeterias that generate an average of 1,500 kg per month, of which approximately 90% is recollected and treated. The remaining 10% corresponds to material that cannot be treated in this process such as the shell of lemons and oranges.

Initially the cafeterias carry out the separation at the source of the usable organic waste, later they are transported to the nursery garden and deposited in some containers, which allow a natural decomposition to be carried out and then its size is reduced to facilitate the process. The leachate generated in this first phase is applied to the beds of the vermicomposting or in the second phase of composting from vegetable waste. At the moment the University has 14 beds of vermicomposting of which 10 are in operation, and these are fed between one and two times per week.

The method used for the generation of vermicompost is initially composed of a layer of compost generated from phase 3 of plant waste, a California worm base, again a layer of compost and a layer of organic waste that is fed around once or twice a week. The process lasts on average between 6 and 8 months and approximately 800 to 900 kg per bed are generated. The leachates generated in the beds are collected in individual chambers connected to the main collector, which are finally reincorporated in the second phase of the composting process carried out with plant residues.

Additionally, there is a bed specifically to work with sludge from the WWTP. This process starts with 90 kg of sludge, but as the sludge production is variable in quantity and quality, this bed must be fed with organic waste, so the process is continued working under the same conditions as vermicomposting. This bed is fed on average once a week and the process last approximately eight to nine months generating about 500 kg [4].

## 3 Concluding remarks

In 2017 there was a decrease in the generation of recyclable waste, associating with the implementation of waste policy to reuse, recycle and reduce, in addition to the training and awareness programs implemented by the University for the whole community. In this way, it is essential to continue strengthening the policy and the programs implemented to continue reducing this type of waste significantly.

Based on last year's composting production, it is necessary to analyse in detail the delay presented during the process, considering that the generation of 2017 was less than 20% of the historical average.

The production of sludge generated in the WWTP is currently being included in one of the composting lines, however these sludges must be examined to improve their quality and quantity, considering that one of the objectives is the use and generation of organic fertilizer. In the current year a research project dealing with an assessment and optimization of the composting plant has been launched, as well as, a program to reduce the production of food waste in the campus.

## References

1. Universidad Autónoma de Occidente, *Informe general de residuos reciclables durante los últimos 4 años en la Universidad Autónoma de Occidente* (2017)
2. Universidad Autónoma de Occidente, *Políticas para el adecuado manejo y disposición final de los residuos peligrosos, además de la generación durante los últimos 4 años en la Universidad Autónoma de Occidente* (2017)
3. Universidad Autónoma de Occidente, *Manual para la generación de compost en el Campus y cifras históricas (2014 – 2017)* (2017)
4. Universidad Autónoma de Occidente, *Políticas para el adecuado manejo de los residuos ordinarios en la Universidad Autónoma de Occidente y entrega a la empresa prestadora del servicio de recolección* (2017)

# Toward integrated and sustainable waste management system in University of Malaya: UM zero waste campaign

*Sumiani* Yusoff*

Dean, Sustainability Science Research Cluster, Universiti Malaya, 50603, Kuala Lumpur, Malaysia.

**Abstract.** With the escalated increase in municipal solid waste (MSW) generation in Malaysia reaching a shocking 38,000 ton/day in 2017, a sustainable waste management system is much desired. Nationwide, there are 176 landfills but only 8 are sanitary landfill with the rest are open dumpsites. In the campus of University of Malaya, UM Zero Waste Campaign (UM ZWC) was introduced in 2011 to start a long-term campaign to achieve an integrated and sustainable waste management model and ultimately a zero-waste campus. Since year 2015, UM ZWC is fully funded by Sustainability Science Research Cluster of UM (Susci) as one of the living labs of UM as well as by JPPHB under the RMK-11 budget. UM ZWC operating projects including in house composting center, food waste segregation scheme, research composting emission and waste characterization, anaerobic digestion (AD), used clothes collection program, wood waste separate collection, e-waste collection and drop-off recycling collection were initiated under the campaign. Since the inception of the project in 2011 until December 2017, almost over 620 tons of solid waste has been diverted from disposal in landfill with composting, AD, recycling, re-use and energy recovery. A roadmap of UM ZWC was drawn up in 2013, with a goal to achieve 60% landfill diversion by year 2040. In the next 5-10 years, UM ZWC plays a vital role to formalize the recycling collection in UM and further increase the organic waste recycling with green waste shredding and composting. Besides environmental benefits (pollution prevention and carbon emission reduction), UM ZWC brings various benefits such as academic research opportunities for UM, contribute in UM LCCF (Low carbon city framework) target and serve as platform to improve students soft skills and entrepreneur skill. Multi stakeholders participation, support form top management and industrial collaboration are the key factors that are able to drive the development of a sustainable waste management model in UM campus.

## 1 Introduction

Solid waste can be referred as unwanted waste is that derived from the animal and human activities. It is also can be generated from industrial, institution, residential, commercial, construction and demolition activities. Solid waste can be classified based on its contents including materials such as paper, plastics, glass, metal and organic waste. Moreover, it also can be categorized based on hazard potential incorporated with radioactive, flammable, toxic or non-toxic. While solid waste management is defined as discipline associated with control of generation, storage, collection, transportation, processing and disposal of solid waste materials in the best way to deal with the range of public health, conservation, economic and other environmental considerations. The main goal of solid waste management is to minimize and eliminate adverse effects on human health and environment to aid economic development and quality of life.

Generally, solid waste composition in Malaysia largely constitute of municipal solid waste (MSW) 64% with the remaining consisting of industrial waste, commercial waste, and construction waste [12]. In 2007, with a population more than 25 million, Malaysian households produce approximately 18,000 tons of household waste every day. Rapidly growing population, improved quality of life and rising economic growth are the factors that contribute challenges to the management of solid waste [11]. With significant improvement of living standards, it is expected that solid waste generation increases over the years without any transformation in the attitudes and behavior of Malaysians in managing their waste. With the utilization of plastic and paper materials especially in packaging where those materials become easily dispensable to the consumers, solid waste generation increases at uncontrollable rate [13, 2]. The least favored disposal method is landfilling, as waste should be separated and treated (physical, chemical, or biological treatment), but unfortunately these options are costly and time-consuming [14].

Landfill is the most economical and hence most common MSW disposal method in Malaysia. Nationwide, there are 176 operating landfills but only 11 of them are sanitary landfills with seven in Peninsula Malaysia, one in Sabah and three in Sarawak. Besides the operating landfills, there are 114 closed landfills in the country which required post closure treatment and management for at least 30 years. The total capacity of MSW disposed in the current 176 operating landfills is more than 30,000 ton/day and the total size of operating

* Corresponding author: sumiani@um.edu.my

https://doi.org/10.1051/e3sconf/20184804007

landfills is 2,528.2 ha. In total, the size of operating and closed landfills is 3,446.2 ha, which is 0.01% of Malaysia total land area. In Malaysia, approximately 93.5% of municipal solid waste (MSW) is sent without sorting to the landfill or open dumpsites that have no gas recovery, and only 5.5% of MSW is being recycled and 1.0% composted [1]. Table 1 shows the present number of operating and non-operating solid waste disposal sites in Malaysia.

**Table 1.** Number of operating and non-operating solid waste disposal sites in Malaysia.

| State | Operating non-sanitary landfill sites | Operating sanitary landfill sites | Non-operating landfill sites | Total |
|---|---|---|---|---|
| Johor | 12 | 2 | 23 | 37 |
| Kedah | 8 | 1 | 6 | 15 |
| Kelantan | 13 | 0 | 6 | 19 |
| Melaka | 2 | 0 | 5 | 7 |
| Negeri Sembilan | 7 | 0 | 11 | 18 |
| Pahang | 16 | 0 | 16 | 32 |
| Perak | 17 | 0 | 12 | 29 |
| Perlis | 1 | 0 | 1 | 2 |
| Pulau Pinang | 2 | 0 | 1 | 3 |
| Sabah | 19 | 0 | 2 | 21 |
| Sarawak | 46 | 6 | 14 | 66 |
| Selangor | 5 | 4 | 14 | 23 |
| Terengganu | 8 | 0 | 12 | 20 |
| Federal Territory of Kuala Lumpur | 0 | 0 | 7 | 7 |
| Federal Territory of Labuan | 1 | 0 | 0 | 1 |
| TOTAL | 157 | 13 | 130 | 300 |

Source: JPSPN (2015a)

Solid waste has constantly been an issue particularly in the amount of solid waste generated [15]. The increasing affluence leads to uncontrollably high amount of solid waste production despite the potential of source separation and recycling in Malaysia. Lack of public conscientious in today's modern lifestyle has resulted to increasing amount of waste generated and disposed at landfills especially when it comes to packaging, as these materials are dispensable to them [4]. Education and individual upbringing contribute considerably towards environmental awareness, how society perceives the issue, as well as how they decide on their daily behavior, particularly in managing solid waste [3]. Besides the influence of socio-economic factors, perception of infinite resources with no observable environmental consequences to the public led to over-consumption, which produces unnecessary waste ultimately. Without the support and commitment from households, local authorities, private concessionaires, and other stakeholders in organizing solid waste, source separation and recycling practice would be a major challenge.

It was reported that in 2003, the amount of solid waste generation per capita per day ranged between 0.5 up to 0.8 kg but recently it had increase to between 0.5 to 2.5 kg, especially in the major city such as Kuala Lumpur and Petaling Jaya [7]. Table 2 shows that food waste and organic material are found to have highest portion in solid waste generated in Malaysia which ranges between 32% to 68.4%.

**Table 2.** The Material Composition of Municipal Solid Waste Obtained from Various Studies and Site (Chua et al, 2001)

| **Component** | 2001 | 2001 | 2002 | 2003 | 2004 | 2005 | 2005 | 2007 | 2010 |
|---|---|---|---|---|---|---|---|---|---|
| **Food waste & organics** | 68.4 | 32 | 56.3 | 37.4 | 49.3 | 45 | 47.5 | 42 | 43.5 |
| **Mix plastics** | 11.8 | 16 | 13.1 | 18.9 | 17.1 | 24 | NA | 24.7 | 25.2 |
| **Mix paper** | 6.3 | 29.5 | 8.2 | 16.4 | 9.7 | 7 | 18.5 | 12.9 | 22.7 |
| **Textiles** | 1.5 | 3.4 | 1.3 | 3.4 | NA | NA | 2.13 | 2.5 | 0.9 |
| **Rubber & Leather** | 0.5 | 2 | 0.4 | 1.3 | NA | NA | NA | 2.5 | NA |
| **Wood** | 0.7 | 7 | 1.8 | 3.7 | NA | NA | 4.41 | 5.7 | NA |
| **Yard waste** | 4.6 | NA | 6.9 | 3.2 | NA | NA | 2.72 | NA | NA |
| **Ferrous** | 2.7 | 3.7 | 2.1 | 2.7 | 2 | 6 | NA | 5.3 | 2.1 |
| **Glass** | 1.4 | 5.5 | 1.5 | 2.6 | 3.7 | 3 | NA | 1.8 | 2.6 |
| **Pampers** | NA | NA | NA | 5.1 | NA | NA | NA | 3.81 | NA |
| **Other** | 2.1 | 1.9 | 8.4 | 5.3 | 18.2 | 15 | 21.93 | 2.6 | 1.8 |
| **Total** | 100 | 100 | 100 | 100 | 100 | 100 | 100 | 100 | 100 |

Food wastes produce greenhouse gases (GHG) emissions and have an influence on climate change [6]. Generally, these emissions have been identified as an critical environmental concern in the waste sector [8]. In Asian countries, it is estimated that the largest increase of food waste generation could be ranged from 278 to 416 million tones that can contribute to global anthropogenic emissions ranging from 8 to 10% [16]. Apart from the waste treatment, GHG emissions from waste handling, transportation and operation of machinery are also significant especially due to the utilization of fossil-based energy. Indirect GHG saving potential via materials and energy recovery from waste management must be recognized [5].

Food waste composting can be considered as one of the most appropriate approaches for treating biodegradables waste components, also one of the potential waste management elements to divert waste generated to landfill, while simultaneously recycling organic materials by converting them into a beneficial product. The potential of practicing composting are huge as 70% of Malaysian wastes are wet waste, which are not easily recycled as the dry waste. In Malaysia, the average components of MSW are quite similar with the largest categories consisting of food waste (45%), plastic (24%) followed by paper (7%), iron (6%) and lastly 3% for glass and others [9]. Food waste is a main component of MSW which can lead to the emission of odorous compounds and can affect the quality of leachate from landfill and others.

## 2 Literature review

University of Malaya is a public university located in Kuala Lumpur. It is a multidisciplinary Research University that has more than 20,000 students and 2000 academic staff with 14 faculties/academy, 3 academic centres, 11 research institutes and clusters which covers the whole spectrum of learning from the Arts, Sciences and Humanities. In University of Malaya, the Sustainability Science Research Cluster (SuSci) is one of the entity that play a catalytic role to promote research and initiatives in a holistic and comprehensive perspective to resolve the problem that is relevant to global sustainability, social and human life system. SuSci also have its origin in the concept of development as recommended by the World Commission on Environment and Development (WCED) in 1987 and aims to achieve status society and sustainable and balanced life between physical development and maintenance environment. Amongst the many research programs under Susci, the Living Lab projects, promotes translational and problem solving especially in promoting UM eco campus initiatives and environmental conservation and reducing campus environmental impact.

Being a top university and committed toward holistic excellence, UM launched the University of Malaya Eco Campus Blueprint as a sign of commitment toward campus sustainability performance and has targeted 8 thrust areas for improvement where the Zero Waste Campaign (UM ZWC) is one of the university's longest and most consistent endeavors [19]. It is also unique due to the bottom-up and top-down synergy that characterizes its development. It has the following objectives such as to develop policy and innovation systems to divert organic waste (from disposal in landfill) for nutrient (composting) and energy recovery (anaerobic digestion), to streamline recycling activities and strategize efforts to increase recycling rates, to create awareness and inculcate best practices of waste separation at source among campus communities, serve as a long term campaign to achieve an integrated waste management model and ultimately a zero waste campus, initiate projects, research projects and schemes such as the Green Bag Scheme, an in-house composting centre, an anaerobic digestion project, recycling collection system for e-waste, used textiles and wood waste, composting emission study and others.

The project was incepted by final year students in the Environmental Engineering program in 2009 led by Associate Prof. Dr. Sumiani Yusoff who advocated the needs to address the challenges, posed by the inevitable environmental liabilities in waste management and carefully identified the major drawbacks concerning the low environmental performance of MSW management in the country. They initiated a chain of activities to development of recycling management system in the faculty with minimal cost. In July 2009, VeeCYCLE, a student group was formed to run an integrated recycling project in Faculty of Engineering. The project established a recycling management model which has resulted in the development of an organized and effective waste and recyclables collection system in the faculty. 45 sets of an integrated waste and recyclables collection facility called PRO Bin were introduced to replace the existing rubbish receptacles in the faculty. It facilitates the good practice of separation at source.

It was set-up to spearhead the development of a more sustainable waste management model in the UM campus and ultimately achieve the status of a zero-waste campus. This campaign is a daily operation which is seven days a week without interruption that requires observation and a strong commitment to ensuring that all the waste on campus is managed in an orderly manner and in accordance with the establishment of the procedures. This field also requires cooperation from the café operators in UM due to estimated average of 40% of food waste from the overall composition of waste in UM. Hence, organic and inorganic waste are managed by UM ZWC and the university's assets and service department, JPPHB, reducing monthly almost 15-20 tons of waste to landfill, while reducing cost and reducing environmental impacts through reduction in carbon emissions and footprint and leachate contamination avoidance. Hence UM ZWC has promoted the concept of sustainable consumption and production by converting food and green waste into valuable resources such as compost and biogas. Other endeavor includes educational campaign and workshops about waste management, segregation at source and recycling. On a more long-term level, UM ZWC has drawn up a roadmap for the UM Development Unit to achieve 15% landfill diversion by year 2020 (phase 1), 30% by year 2030 (phase 2) and 60% by year 2040 (phase 3) while phase 1 have been achieved by UM ZWC which is 15% landfill diversion by year 2020. The integrated solid waste management system set up through the UM ZWC Living Lab projects has strengthen the green growth agenda toward sustainable development and environmental conservation in UM campus by empowering the campus community through a

https://doi.org/10.1051/e3sconf/20184804007

systematic, concerted, and action-oriented problem solving translational research initiative.

## 3 Methodology

Solid waste generated in UM campus is collected by fixed collection systems. UM communities are supposed to deposit the waste at the locations specified by the Department of Development & Estate Management (JPPHB) every day of the week and will be collected by a specific time. These generated solid wastes are transported by vehicles which can be categorized as collection and haulage vehicles [10]. Collection vehicles collect the waste in where it is generated and then transfer and dispose to the disposal facility which is UM ZWC located near the Damansara Gate. The waste will be segregated by the workers to reduce the volume and pollution potential for landfill sites. Moreover, UM ZWC used a Takakura composting method as a meaningful processing technology for the bulk of the degradable organic fractions. The composting method was eventually evolved into aerated static piles with capacity of 4-5 ton/month (90% food waste and 10% green waste, by weight). In 2013, Cowtec ® anaerobic digestion (AD) 100kg/day unit was installed after research collaboration with CH Green Sdn Bhd. With the AD facility, about 1 ton of food waste is converted to biogas and bio-fertilizer every month. Until end of year 2014, about 120 ton of organic waste had been composted or treated anaerobically by UM ZWC.

2014 is an improvement year for UM ZWC with more collaboration with industries to establish separate collection of various waste streams, collaboration with academic institutions for research, more appearance in environmental conferences, expo and media, and strengthening rapport from UMCARES and JPPHB. The public private partnership (PPP) between UM ZWC and several private entities had resulted in successful separate collection of waste streams for recycling/landfill diversion. At the beginning of 2014, UM ZWC collaborated with Life Line Clothing (LLC) Sdn Bhd to introduce a used clothes collection program which had expanded rapidly in year 2014 that saw the collection of more than 20 ton of used clothes and waste textile. At the end of the year, ZWC formed partnership with TSP Waste Management to kick off a wood waste separate collection system for energy recovery which is implemented successfully with about 5-6 ton/month capacity in the first month.

The support from UM top management, especially DVC (Development) to UM ZWC, is very important to ensure the success of the PPP. For instance, the sites approval to LLC to place the used clothes collection bins and cooperation to collect wood waste separately in a dedicated open top Ro-Ro bin for wood waste recycling. The DVC (Development), Prof. Faisal Rafiq had allocated budget for the upgrading of ZWC facilities in year 2015 such as new ZWC building, green waste shredder, a weighbridge station and composting center. The UM ZWC cabin serve as resource center, site treatment facility and meeting room for visitors to UM ZWC site. Under DVC (Development), JPPHB assists UM ZWC in the provision of several manual workers, waste and recycling data as well as collection receptacles for food waste such as bins and bags. Moreover, this data collection practice is also contributing to UM LCCF (Low Carbon City Framework) project.

On the other hand, various programs were carried out to enable the implementation of the projects, including awareness publicity program for students and staffs, capacity building program for the kitchen staffs and cleaners, discussion and meeting with strategic partners both UM and external bodies as well as several site visits to enhance the students' knowledge in waste management. The programs promote the development of communication, information, negotiation and consultation skills among the students. The projects are poised to further strengthen their roles in realizing IWM model by enabling on site, in-house organic waste treatment operation and expanding the coverage of recycling collection points (PRO Bin).

In 2017, UM ZWC develops an Intelligent Recycle Center (IRC) with Coindex Sdn Bhd to promote behavior and inculcate best practice of recyclables drop-off with this innovative automated recycle center located at lecture hall A&B PASUM. With the new recycling system, UM communities are able to send their source segregated recyclables to the center for conversion into green points which can be used to claim goodies such as compost, USB pendrive, t-shirt and redeemable discounted price in participating cafeteria in UM. This kind of reward-based interactive recycling innovation can bring multiple benefits and contribution to positive social behavioral change (recycling habit) and resource conservation with increase of recycling rate. Furthermore, in near future the IRC can serve as thematic environmental center in UM campus with a galleria of environmental related information to the corridor of lecture hall building. The IRC is anticipated as the cornerstone to develop a formal recycling separate collection in the campus of UM, which is one of the primary concepts of integrated waste management.

## 4 Result and Discussion

Data collection and analysis is very important in development of integrated waste management plan. With the weighbridge station installation in July 2015, UM ZWC is able to capture the waste disposal data. The complete /comprehensive data that UM ZWC fully possesses are food waste collected for composting or anaerobic digestion, green waste collected for composting, wood waste collected for energy recovery, waste textiles collected for reuse/recycle, E-waste collected at UM ZWC site for recycling/recovery, recyclable materials sorted at UM ZWC site and UM transfer station and residual waste disposal data.

UM ZWC coordinated and gathered all the data from different parties. With the data, tonnage per year was calculated as below (Table 3). From the data, it shows the increasing of total waste was diverted for treatment and recycling by UM ZWC from 2012 until 2016. The increasing factor is due to the improvement of the project by collaborating with textile industry to introduced used clothes collection program which had expanded rapidly in year 2014 that saw the collection of more than 20 ton of used clothes and waste textile. At the end of the year, UM ZWC formed partnership with TSP Waste Management to kick off a wood waste separate collection system for energy recovery which is diverted with about 5-6 ton/month from landfill. Used clothes and waste textiles are collected separately with ten (10) units of "drop-off" collection bins while wood waste is collected by JPPHB in separate open top bin for energy recovery in a paper mill.

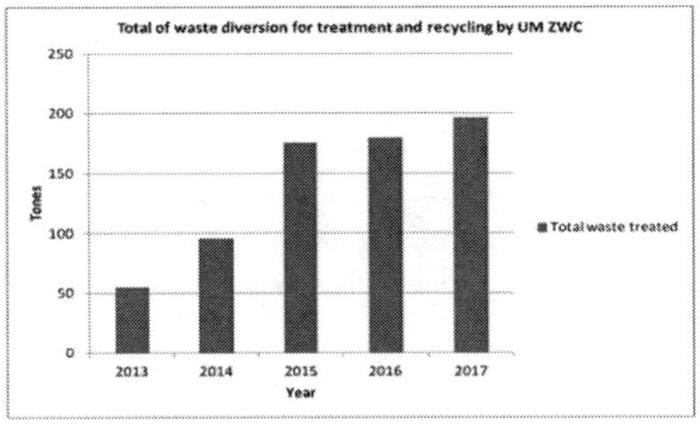


**Fig.1.** Total of waste diversion for treatment and recycling by UM ZWC (2012-2016)

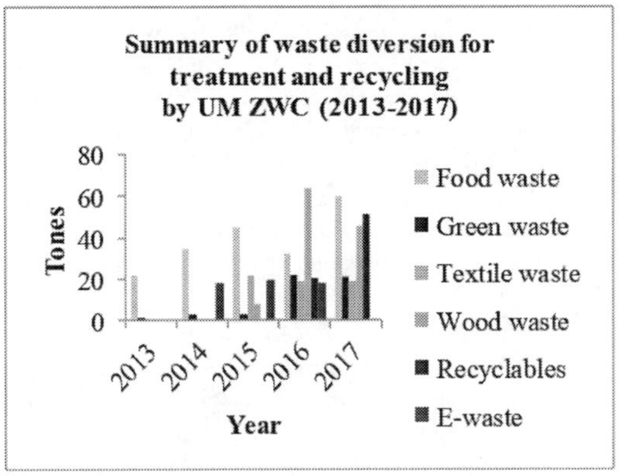


**Fig.2.** Summary of waste diversion treatment and recycling by UM ZWC (2012-2016)

Institutionalization of separate organic waste collection and treatment system in UM is the key to achieve integrated waste management system. Green Bag Scheme is the first program to kick-start the food waste segregation practice in the campus. Organic waste made up almost half of total waste generated in UM campus and thus the recovery and treatment of organic waste is very important to increase recycling rate (landfill diversion rate). In UM, green waste is segregated at source by gardeners and collected separately using a small lorry of JPPHB in a daily basis. About 2 ton of green waste is generated from UM campus everyday and all the green waste is collected and loaded separately in two open top Ro-Ro bins (refer figure 2). It can be shown that the amount of food waste has increased until 2014. For 2015, the amount of treatment and recycling of food waste had decrease due to the closing of dining hall in all residential college of UM campus. UM ZWC had the difficulty to obtain the food wastes that have been segregated. Thus, UM ZWC organized the food waste segregation program to all cafeterias in UM campus with the help of OSH (occupational safety and health unit in UM) by providing the transparent plastics bags that is only to be used for food waste. In 2016, the amount of food waste had risen and was strengthen with the publishing of the Code of Practice Food Waste Segregation at Source guidelines by UM ZWC.

For the recyclables material separation for recycling, it was and is still an up hilling challenge for UM ZWC as the current practice of recycling collection by informal players poses significant hindrance. However, with the persistent efforts by UM ZWC and JPPHB, collection of recycling data is improving from time to time and the introduction of the Intelligent Recycling Center (IRC) this year will hopefully further boost the development of recycling in UM campus. In term of economic, UM ZWC has been saved more than RM 97, 758 from January 2012 until December 2016 (refer Table 3). The monetary saving included the collection fee and landfill gate fee. While for the environmental saving, UM ZWC has been saved almost 726.46 ton $CO_2$-eq. The reduction in economic and environmental aspects due to the waste recovery and composting process that is diverted biodegradable are wastes and composted. [17, 18].

**Table 3**. Economic and environmental cost saving of UM ZWC as of 2011- 2017

| No. | Type of waste | Tonnage | No. of trips reduced | Collection fee saved (RM) | Landfill gate fee saved (RM) | Total monetary saving (RM) | Carbon saving per ton (ton $CO_2$-eq) | Carbon saving (ton $CO_2$-eq) |
|---|---|---|---|---|---|---|---|---|
| 1. | Food waste | 275 | 138 | 48,125 | 15,125 | 63,250 | 2.359 | 648.73 |
| 2. | Green waste | 54.75 | 37 | 12,775 | 3,011 | 15,786 | 1.563 | 85.57 |
| 3. | Wood waste | 162.03 | 81 | 28,355 | 8,912 | 37,267 | 1.552 | 251.47 |
| 4. | Textile waste | 74.86 | 50 | 17,467 | 4,117 | 21,585 | 0.616 | 46.11 |
| 5. | Recyclable materials | 121.42 | 81 | 28,331 | 6,678 | 35,009 | 0.048 | 5.83 |
| | Total | 688.06 | 387 | 135,053 | 37,843 | 172,897 | | 1,037.71 |

## 5 Summary/ Concluding Remarks

The UM ZWC project, comprising of waste segregation and composting biodegradable waste, is a good example of a highly integrated approach accounting for the different elements of solid waste project sustainability. The project also has been made a clear distinction between how to enable improved environment influences performance and outcome of the project, and how the project impacts positively on UM social, economic and ecological environment. Ultimately, by making efforts in implementing food waste management systems, the future perspective of food waste could create opportunities in handling energy demands and moving toward sustainable development. The sustainability of UM ZWC is important for UM's reputation locally and internationally as an example of eco campus which emphasizes on academic excellence and whilst promoting sustainable development. UM ZWC has successfully developed several key projects that serve as milestone to boost recycling rate in the campus. In the next 5-10 years, UM ZWC plays a vital role to formalize the recycling collection in UM and further increase the organic waste recycling with green waste shredding and composting. Besides environmental benefits (pollution prevention and carbon emission reduction), UM ZWC brings various benefits such as academic research opportunities for UM, contribute in UM LCCF (Low carbon city framework) target and serve as platform to improve students soft skills and entrepreneur skill. Multi stakeholder's participation, support form top management and industrial collaboration are the key factors that are able to drive the development of a sustainable waste management model in UM campus.

## References

1. Agamuthu P, Fauziah SH, Kahlil K., *Evolution of solid waste management in Malaysia: impacts and implications of the solid waste bil,*. J Mater Cycles. 11:96–103. (2009).
2. A. Malahkahmad, M.Z.Z. Che Mohd Nasir, S.R. Mohammed Kutty, M. Hasnain Isa. *Solid waste characterization and recycling potential for University Technology PETRONAS* academic buildings. Am. J. Environ. Sci., **6 (5)**, pp. 422–427(2010)
3. Brian, L., Craig, H., James, L., Lisa, K., Richard, W., Tim, S., *Reducing Food Loss and Waste – Installment 2 of 'Creating A Sustainable Food Future'*, World Resource Institute, Washington, DC. (2013)
4. D. Asmawati, A.K. Nor Ba'yah, Y. Fatimah. A study on the knowledge, attitudes, awareness status and behaviour concerning solid waste management. Procedia Soc. Behav. Sci., **18**, pp. 643–648 (2011)
5. Gentil, E., Christenen, T.H., Aoutin, E., *Greenhouse Gas Accounting and Waste Management*, Waste Management & Research, **27**, 696-706 (2009)
6. IPCC, In: Metz, B., Davidson, O.R., Bosch, P.R., Dave, R., Meyer, L. A., (Eds), Contribution of Working Group III To the Forth Assessment Report of The Intergovernmental Panel on Climate Change. Cambridge University Press, Cambridge, United Kingdom and New York, USA. (2007)
7. Kathirvale, S., Muhd Yunus, M. N., Sopian, K., Samsuddin, A. H., *Energy potential From Municipal Solid Waste in Malaysia.* Renewable Energy, **29**, 559–567 (2003)
8. Liamsanguan, C., Cheewalas.H., *The Holistic Impact of Integrated Solid Waste Management on Greenhouse Gases Emissions in Phuket,* Journal of Cleaner Production, **16**, 865-1871 (2008)
9. McDougall, F., White, P., Franke, M., Hindle, P. *Integrated Solid Waste Management: A Life Cycle Inventory.* (2nd Eds), Blackwell Publishing (2001)
10. Onn Chiu Chuen, Sumiani Yusoff et. al. *Working towards A Sustainable Means of Campus Transportation.* In University of Malaya Living Lab: Transforming Research into Action, **1**, Sustainability Science Research Cluster University of Malaya (2018)
11. Shekdar, A.V., Sustainable *Solid Waste Management: An Integrated Approach for Asian Countries,* Waste Management 29 (4), 1438-1448. (2009)
12. Sustainable Waste Management Cycle (EA-SWMC), EU-Perak Solid Waste Management Plan (EU-PSWMP). Perak, Malaysia (2009)
13. M. Abdul Jalil, *Sustainable development in Malaysia: a case study on household waste management.* J. Sustain. Dev., **3 (3)** (2010), pp. 91–102 (2010)
14. M.S. Grodzinska-Jurczak, *Management of industrial and municipal solid wastes in Poland.* Resour. Conserv. Recycl., **32 (2)**, pp. 85–103. (2001)
15. X. Gellynck, R. Jacobsen, P. Verhelst, *Identifying the key factors in increasing recycling and reducing residual household waste: a case study of the Flemish region of Belgium*, J. Environ, Manag., **92 (10)**, pp. 2583–2590 (2011)
16. Saer, A., Lansing, S., Davitt, N.H., Graves, R.E., *Life cycle assessment of a food waste composting system: environmental impact hotspots*. J. Cleaner Prod. **52**, 234–244 (2013)
17. Sumiani Yusoff (Ed) *UM Living Lab: Transforming Research into Action.* **1 and 11**. Sustainability Science Research Cluster University of Malaya (2018)
18. Sumiani et. al. *University of Malaya Zero Waste Campaign: Integrated and Sustainable Waste Management.* In University of Malaya Living Lab: Transforming Research into Action **1**. Sustainability Science Research Cluster University of Malaya, pp 71-85 (2018).
19. Sumiani et al. University of Malaya Eco-Campus Blueprint. University of Malaya. Kuala Lumpur, Malaysia (2016)

E3S Web of Conferences 48, 04008 (2018) https://doi.org/10.1051/e3sconf/20184804008
*IWGM 2018*

# The sustainability efforts of Ton Duc Thang University in the South of Vietnam

*Ut V.* Le[1,2,*]

[1]Informetrics Research Group, Ton Duc Thang University, Ho Chi Minh City, Vietnam

[2]Faculty of Social Sciences and Humanities, Ton Duc Thang University, Ho Chi Minh City, Vietnam

**Abstract.** Established in 1997, Ton Duc Thang University (TDTU) is now among the best universities in Vietnam for all aspects, after 20 years of development. With sixteen faculties of multi-disciplines and around 25,000 students, TDTU has made lots of achievements. In 2017, TDTU is ranked 245th in the world (and ranked 1st in Vietnam) on Sustainable Development by UI GreenMetric World University Rankings (UI GreenMetric); ranked 1st among universities in Vietnam for high-quality research outputs by Nature Index; ranked 2nd among universities in Vietnam (and ranked 1st in research category) by Xephangdaihoc, a group of independent experts of Vietnam. Sustainability is of TDTU's focus. This paper shortly discussed TDTU's latest achievements on sustainability, and then focused on issues and innovation in managing waste at Ton Duc Thang University.

## 1 Introduction

In 2017, TDTU was ranked 245$^{th}$ by UI GreenMetric in the world

**Table 1** TDTU score

| Category | Point | Maximum Point | Percentage |
|---|---|---|---|
| Setting and Infrastructure (SI) | 583 | 1500 | 38.87 % |
| Energy and Climate Change (EC) | 1,016 | 2100 | 48.38 % |
| Waste (WS) | 1,176 | 1800 | 65.33 % |
| Water (WR) | 500 | 1000 | 50.00 % |
| Transportation (TR) | 813 | 1800 | 45.17 % |
| Education (ED) | 659 | 1800 | 36.61 % |
| Total Score | 4,747 | 10000 | 47.47 % |

and 1$^{st}$ in Vietnam [3, 4]

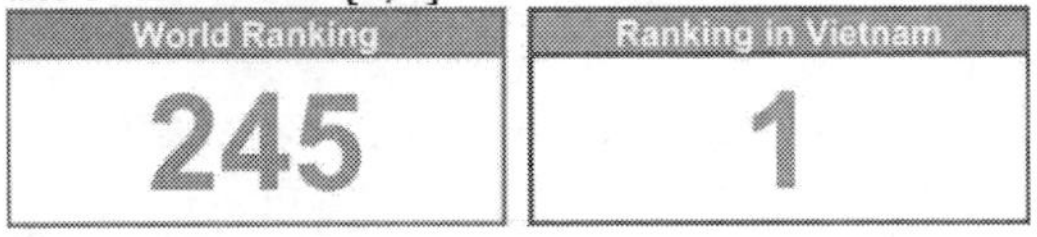


**Fig. 1** TDTU ranking

According to the evaluation from UI GreenMetric, in 2017 TDTU performed under average for most of indicators. For the category of Water, TDTU was at average, exactly 50%. And the performance of Waste was pretty good and also the highest record, 65.33%.

Before joining UI GreenMetric, leaders of TDTU basically thought that they should simply build a clean and well-organized university, beside excellent academic programs and staff. However, after joining this university ranking and some other university rankings, they changed their minds that each indictor of a university can be measured. It is quite lucky that University of Indonesia hosts UI GreenMetric, and sustainability of universities is evaluated every year.

In some countries like Vietnam, universities basically focus on academic programs and their task forces. But sustainability is also an important category that helps to provide staff and students with clean and safe environment. This is essential for improving and maintaining quality of research and education of universities, especially in the current era of much pressure and stress [1, 2].

The 2017 ranking of TDTU by UI GreenMetric was 245$^{th}$. Many indictors of TDTU must be improved in the following years. Thanks to the methodology of the ranking, the University's leaders have been clear what they should pay attention to. Hopefully, TDTU's ranking by UI GreenMetric will be higher in the future.

## 2 Issues and innovation in managing waste at TDTU

By UI GreenMetric, TDTU gave very good performance of Waste. This indicator includes 6 items and TDTU's record is listed as below [3]:

- Program to reduce the use of paper and plastic in campus: Print when necessary
- Recycling program for university waste: Partial (25% - 50% of waste)
- Toxic waste handled: Completely contained, inventoried and handled
- Organic waste treatment: Fully composted, compost used internally and externally
- Inorganic waste treatment: Partially recycled (less than 50%)

* Corresponding author: levanut@tdt.edu.vn

- Sewerage disposal: Centralized treatment before disposal.

Totally, TDTU's record of this indicator is of 65.33%, the highest one of its sustainability.

Basically, waste causes many issues for universities. TDTU is a multi-disciplinary university with over 20,000 students, the waste released is quite huge every day. Thanks to some innovative methods, TDTU can control all issues from waste very well.

### 2.1 Program to reduce the use of paper and plastic in campus

The University issued an official instruction to guide all staff and students [3]:

- Only print out when necessary
- Two-side printing must be taken into account
- Online academic materials are encouraged in using.

In addition, TDTU built a new library equipped with very modern equipment. This is a very luxurious space for our students not only to study but also to entertain. With the multimedia systems in the library, our students have no motivation to print anything out.

**Fig. 2.** TDTU new library

**Fig. 3.** TDTU new library

To make the campuses clean or reduce the use of plastic, beside issuing general principles that all staff and students have to practice, TDTU issued some rules to punish anyone violating such principles. The punishment is quite simple that one has to pay money if violating the rules.

### 2.2 Recycling program for university waste

All wastes are classified thanks to their characteristics. Some of them are recycled for further uses. In addition, the University signed a contract with a waste-treatment company that all waste released in the campus must be treated before being transported out [3].

### 2.3 Treatment program for university chemical waste

To prevent any pollution from chemical wastes, the University has several piped systems connecting all areas where chemicals can be generated. Then they are taken to a treatment station [3].

In addition, many trees and grass are planted in all campuses; and old traffics possibly causing air pollution are not allowed to get in any campus. As a result, there is no traffic pollution at all.

**Fig. 4.** trees and grass planted in TDTU

**Acknowledgements:** Dr Ut V. Le would like to thank all colleagues from UI GreenMetric World Universities Rankings for their cooperation during the evaluation of Ton Duc Thang University and the organizers of the 4th International Workshop on UI GreenMetric "Universities, Impacts and Sustainable Development Goals (SDGs)" in Semarang, Indonesia, for inviting him as an invited speaker and their important supports.

## References

1. Aleixo, Ana Marta; Azeiteiro, Ulisses; Leal, Susana; *The implementation of sustainability practices in Portuguese higher education institutions;* International Journal Of Sustainability In Higher Education, **19**, Issue 1, pp. 146-178 (2018)
2. Allan Lauder, Riri Fitri Sarib, Nyoman Suwarthac, Gunawan Tjahjonod; *Critical review of a global campus sustainability ranking: GreenMetric*; Journal of Cleaner Production, Volume 108, pp. 852-863 (2015).
3. UI Green Metric World University Rankings; *Fact file 2017 for Ton Duc Thang University* (2017)
4. UI Green Metric World University Rankings; *Methodology* (2017). Available online at http://greenmetric.ui.ac.id/methodology

# Water conservation and management: case study in Siam University

*Pornchai* Mongkhonvanit[1,*], *Chanita* Rukspollmuang[2], *Nattiga* Silalai[3] and *Chanida* Phitthayanon[4]

[1]President, Siam University, 38 Petkasem Rd., Bangwa, Phasi Charoen, Bangkok 10160, Thailand
[2]Vice President, Siam University, 38 Petkasem Rd., Bangwa, Phasi Charoen, Bangkok 10160, Thailand
[3]Head, Sustainable Office, Siam University, 38 Petkasem Rd., Bangwa, Phasi Charoen, Bangkok 10160, Thailand
[4]Faculty of Science, Siam University, 38 Petkasem Rd., Bangwa, Phasi Charoen, Bangkok 10160, Thailand

**Abstract.** Siam University (SU) is one of the leading Thai comprehensive private universities of which "Sustainability" is one of the main missions. The mission is guided by the concept of "Sustainable Development (SD)" and the principle of "Sufficiency Economy Philosophy (SEP)" bestowed by the late King Rama IX. SU has endorsed many sustainability policies and projects including those concerned with ecosystem and water management system. Our target groups include SU staff and students as well as members in Phasi Charoen district. SU has encouraged water conservation and management policies and practices including reduction of water usage program for quite some time. Although piped water is still mainly used, we have installed automated hand washing taps counting to more than 75% of total water appliances. Water from the campus natural pond is used and reused for gardening and washing. After this initial stage, we have planned to do more programs concerning water recycling, reuse, and treatment system under the principle of SEP. Considering that our pond and surrounding canals are connected to Phasi Charoen communities, we hope that our water conservation and management system will not only lead to a more sustainable campus but also a more sustainable district.

## 1 Introduction

Siam University (SU) is one of Thailand's leading comprehensive private universities focusing on "Employability", "Diversity" and "Sustainability" as the three main pillars. The university was founded in 1965 as the first private Engineering School. We were later upgraded to "Siam Technical College", "Siam Technical University", and "Siam University" in 1973, 1986 and 1989, respectively.

The university comprises of 13 schools and colleges which offer a broad variety of undergraduate, graduate, and lifelong education programs. The International Program was established in 1995 to offer students the opportunity to earn Bachelor's degrees focusing on International Business Administration and Hotel &Tourism, Service Industry Management. Since the inception, Siam University has produced approximately 80,000 graduates with diversified student body. Our graduates enjoy the same privileges accorded to state university graduates. Our academic collaboration over 100 leading universities and institutions worldwide creates unique possibilities for students who wish to transfer from or to Siam University, or who wish to undertake research projects and graduate studies abroad.

It is our mission to prepare students for their future careers in highly dynamic and globalized environment. We cultivate the sense of ethics and responsibility while respecting different cultures and belief as well as embracing diversity and sustainability. At the same time, we strive to elevate the well-being of our surrounding communities in Phasi Charoen district since SU is the only one university in the area. The university has launched "Sustainability Policy" which is guided by the concept of "Sustainable Development (SD)" and the principle of "Sufficiency Economy Philosophy (SEP)" bestowed by the late His Majesty King Bhumibol Adulyadej or King Rama IX. At present, we have initiated several sustainability policies and projects including those concerned with ecosystem and water management system [1].

## 2 Setting and infrastructure of campus

Siam University is located in Phasi Charoen district within the southwest boundaries of downtown Bangkok, approximately 6 kilometers from Silom, the main financial district of the city. Presently, the campus is about 200 meters from the Bangwa station which is the interchange station of BTS (Bangkok Mass Transit System) and MRT (Metropolitan Rapid Transit). This made it fast and convenient to travel to any part of Bangkok from Bangwa station to Siam Square will take not more than 25 minutes. The BTS is now operating and MRT is due to start in the near future.

Siam University comprises of 21 main buildings with a total floor area of approximately 87,000 square meters. There are two multi-purpose buildings which was graciously given the name by His Majesty King

* Corresponding author: pornchai@siam.edu

Bhumibol Adulyades (King Rama IX) as "Chalerm Phra Keit, Royal Jubilee Building and opened by the HRH Crown Prince Maha Vajiralongkorn (King Rama X) and HRH Princess Maha Chakri Sirindhorn respectively. The total campus area is around 64,000 m2. Within the campus, forest vegetation covers 8,000 (12.5% of total area) and planted vegetation is 4,500 m2 (7.03% of total area). Furthermore, total area on the campus for water sorption besides forest and planted vegetation is approximately 5,500 m2 (8.59% of total area).

## 3 Sustainability policy of Siam University

As mentioned earlier, Siam University embraces the significance of sustainability. We hope to integrate this mission into our teaching, research and community services under our guiding principles of "Sustainable Development (SD)" and "Sufficiency Economy Philosophy (SEP)". We are concern about social, economic, environment as well as energy issues both within the campus and the 54 communities in Phasi Charoen district. It is our belief that becoming "sustainable university" is not meaningful unless we live in a "sustainable district". Hence, our main targets are the 3Ss; namely, students, staff, and surrounding communities (Fig.1).

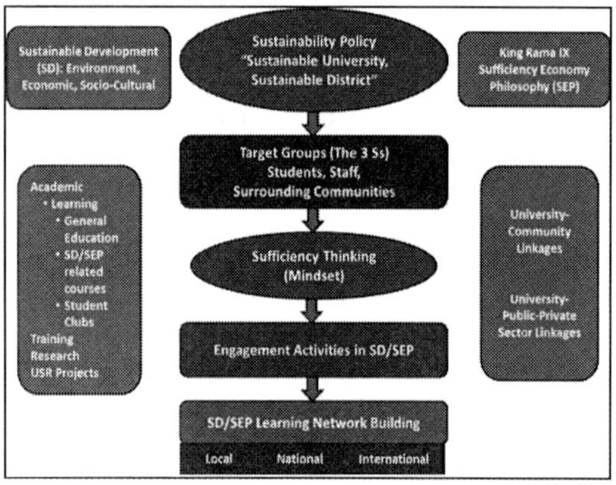


**Fig. 1.** Sustainability policy and implementation of Siam University

SU has followed footsteps of the great King Rama IX by implementing SEP which aims to create a balanced and stable development, at all levels, from the individual, family and community to society at large by developing the ability to cope appropriately with the critical challenges arising from extensive and rapid changes in the economic, social, environmental, and cultural conditions of the world. Sufficiency has three components: moderation, reasonableness, and prudence, with two foundational conditions: appropriate knowledge and virtues [2].

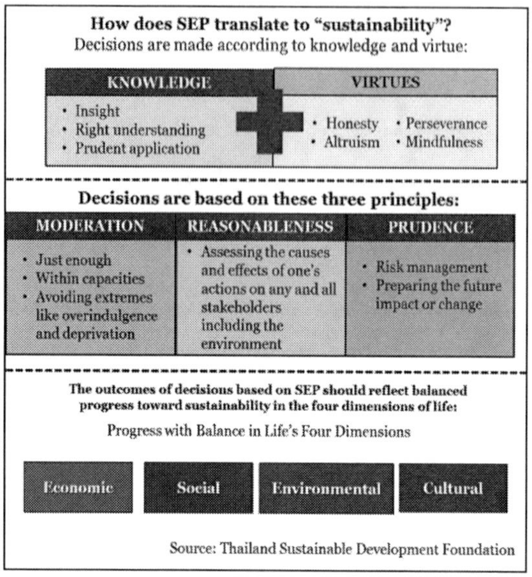


**Fig. 2.** Translation of SEP to "sustainability"

## 4 Ecosystem of Siam University

SU has followed the guideline of SEP in implementing our sustainability projects, especially the principle of moderation and reasonableness. In addition, we also believe in the interconnectedness of resources and living. All-natural resources including water should be preserved, cared, and shared by all in the campus and in the district. In sum, we can divide our ecosystem into one within the campus and that in our surrounding areas.

### 4.1 Ecosystem in Siam University

With emphasis on water initiatives, Siam University has launched policies and programs such as "Green and Clean Campus" policy with the goal to promote safety and healthy space for a better quality of life for all students and staff. We also committed to develop waste management system for both municipal solid waste and hazardous waste in standard practices and meet regulations. Students and staffs shall be encouraged to reduce reuse recycle and proper management on treatment and disposal. It is our policies that (1) solid waste and hazardous waste shall be managed according to the waste hierarchy of reduction, reuse, and recycle, treatment and responsible disposal, (2) products that can be re-used, or recycled shall be encouraged, and (3) efficient use of equipment, materials or chemicals shall be promoted. SU is fortunate to be located near natural canal and pond. The university benefits from these resources as they help reserve natural rain water for our water supply so that we can at least reduce pipe water usage as well as encourage our water conservation program. We have used and reused water from our natural resources and rain water has been used for gardening and car washing (see Fig. 3 as example).

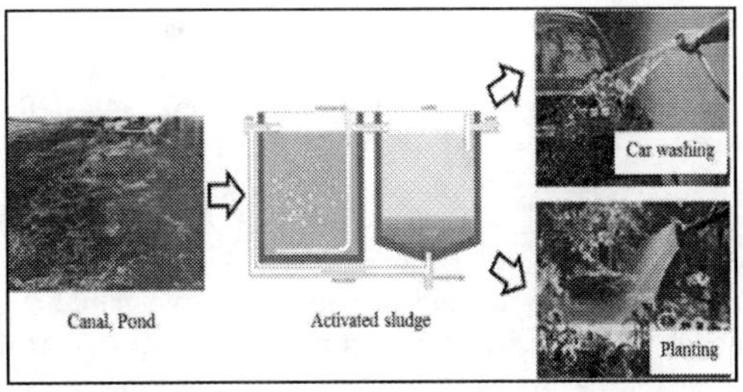


**Fig. 3.** Recycled water for usage in Siam University

Despite our attempts, because we are situated in a rapidly urbanized area, natural water sources are not enough for all public utilities and we have to depend mostly on pipe water in the campus. Nevertheless, Siam University has started some programs in our main green building and others. Water efficient appliances are installed in order to save water usage. Conventional appliances were replaced by automated hand washing taps (734 points). The usage of automated washing taps is more than 75% of total water appliances installed as given in Table 1.

**Table 1.** Water efficient appliance usage.

| Water Efficient Appliances | Number |
|---|---|
| Manual washing tap | 234 |
| Automated hand washing tap* | 734 |
| Total | 968 |

*Automated hand washing tap = (734 X 100 / 968) = 75.8%

In 2017, plans and programs concerning water conservation and management are as follows:

- Design water treatment system of the ponds to reserve rainwater for usage by applying the principle of "Sufficiency Economy Philosophy (SEP)".
- Install more water efficient appliances such as automated hand washing taps to reduce usage of pipe water.
- Recycle water after treatment for gardening and car washing continuously.
- Campaign water conservation program using posters and online media for all staff and students in the campus.
- Encouraging surrounding communities to participate in water conservation and management activities in order to preserve ecosystem.

Siam University has launched some initial water treatment systems for some time; however, we think that there should be more programs and initiatives so that it will be sufficient for other necessary uses. Thus, it is imperatives that we have clear policy to promote more water and energy conservation and management programs. We envision that other water treatment projects emphasizing the application of appropriate technology to improve water quality will be in use. A designate team is studying the possibility to use Chaipattana water turbine, King Rama IX's patented design. We also have a future plan to utilize water for clean electricity using hydropower. Within a few years, the following programs shall be promoted:

- Design high efficient water treatment systems of the ponds to reserve rainwater by applying modern procedures in order to get more clean water for usage.
- Install other water efficient appliances.
- Design systematic data collection on water management system in order to control piped water usage.
- Launch the initial preparation program for waste water treatment from SU laboratories and other resources before running into nearby canals and other ecosystem.
- Encourage students to participate more in environmental conservation activities such as contest of students' projects concerning environmental management.

## 4.2 Ecosystem in the surrounding communities of Siam University

Siam University pledges to take a leading role in promoting green and clean communities around the campus with the participation of the communities in Phasi Charoen district. Five years ago, SU have established a "Research Centre for Community Development (RCFCD)" with the responsibility to study a long-term research project on "Healthy Space Development: A Pilot Project in Phasi Charoen District, Bangkok", a long-term research project funded by the Thai Health Promotion Foundation [3]. This project is well embraced by community leaders and members as well as related public/private organizations. Among all of the programs undertaken, there are many relating to waste and water management. For instance, "Big Cleaning Day" and canal water treatment project which were participated by community members, Siam University staff and students, members of Phasi Charoen District Administration Office, as well as soldiers and local policemen (Fig.4 and Fig. 5).

In addition, activities proposed by SU staff and projects initiated by students who study courses such as Principle of Economics and the Sufficiency Economy Philosophy, Environmental Management, Life and Environment are encouraged. We hope that by doing these activities, environmental conscience will be cultivated for all our target groups: students, staff, and surrounding communities [4].

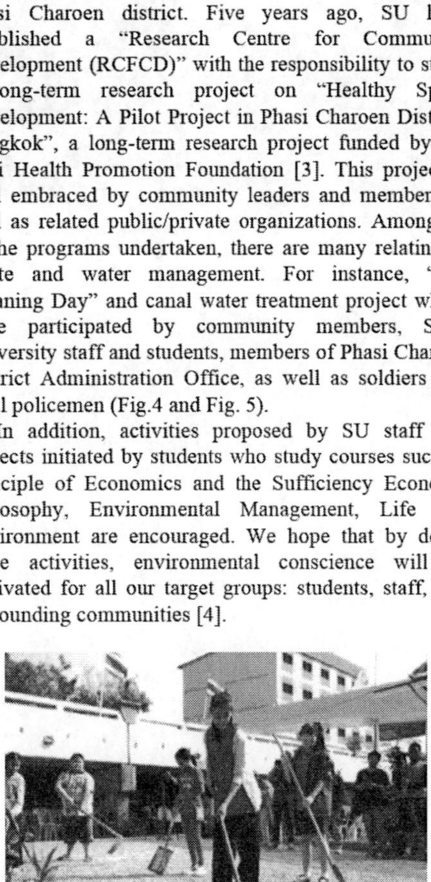

**Fig. 4.** Big cleaning day: under flyover area management in Lertsooksom community nearby Siam University

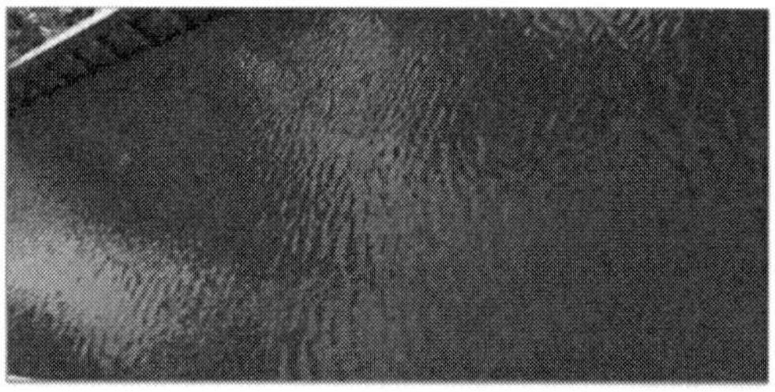

**Fig. 5.** Clean water after treatment in Lertsooksom community's canal nearby Siam University

## 5 Concluding Remarks

Siam University has launched many initiatives to implement our sustainability policies for some time under the guiding principle of "Sustainable Development (SD)" and "Sufficiency Economy Philosophy (SEP)". Our sustainability projects are decided on the principles of moderation, reasonableness, and prudence. With a strong belief in our responsibility to the well-being of everyone in the campus and surrounding areas, the university has endorsed many programs including those concerned with ecosystem and water management system. However, these activities will not be sustained unless all of our target groups (the three Ss) are aware of the necessity and willing to participate. We thus hope that our water conservation and management system will not only lead to a more sustainable campus but also a more sustainable district.

## References

1. Mongkhonvanit, P., Rukspollmuang, C. Sustainable University, *Sustainable District: Siam University as a case study*, presented at the International Sustainable Campus Network (ISCN) 2017 conference hosted by the University of British Columbia and the City of Vancouver, June 26-28, (2017)
2. Thailand Sustainable Development Foundation, n.d., Thailand Sustainable Development Sourcebook. Available online at http://www.thailandsustainabledevelopment.com/the-book/ (Retrieved 5/16/2016)
3. Chancharoen, K. et al., Research Center for Community Development, Siam University. 2013. Research report on "The context study of urban communities of Phasi Charoen District for healthy space development". (Bangkok: Thanthong Printing, (2013)
4. Rukspollmuang, C., et al., *Pathway towards Sustainable University: Policy and Practice of Siam Universit,* Paper presented at the 1st Annual Conference of Sustainable University Network (SUN) Thailand, Mahidol University, November 28-29 (2016)

# Designing catchment area for water resources management in ITS campus

*Joni* Hermana[1]*, *Irhamah*[2], *Dian* Saptarini[3], *Tatas*[4]

[1]Rector of Institut Teknologi Sepuluh Nopember, Surabaya 60111, Indonesia
[2]Department of Statistics, Institut Teknologi Sepuluh Nopember, Surabaya 60111, Indonesia
[3]Department of Biology, Institut Teknologi Sepuluh Nopember, Surabaya 60111, Indonesia
[4]Departement of Civil Engineering, Institut Teknologi Sepuluh Nopember, Surabaya 60111, Indonesia

**Abstract.** Institut Teknologi Sepuluh Nopember (ITS) Campus, with the area of 167.4 Ha, is located within Surabaya coastal region in the eastern part of Java Island. It has initial characteristic with wetlands and swamps ecosystem. As a science and technological university, with the main acitivities in teaching, experimental laboratory works, and student activities, ITS is, currently, using ± 49% of its total vast area as building blocks for supporting academic facilities. Being a campus in a coastal zone, the commonly main problems are high porous soil, brackish surface water, high level of ground water, an obstructed drainage tendency because of delicate slant, and low catchment capability. This paper provides an action program on how ITS manage water resources within campus area in order to suppress environmental damage. Many steps had been taken into account for water catchment role, for instance: maintaining the catchment area on the main ITS master plan, planning catchment pond, surface water stabilization by preventing ground water usage, interrupting drainage water flow as being directly discharged into the city drainage system, rain water harvesting, and also designing floating floor for buildings.

## 1 Introduction

ITS Smart Eco Campus that launched in 2015 is a continuation of the Eco Campus program which began since 2011. ITS Smart Eco Campus is an initiative form from ITS to implement sustainable development program in the campus environment. As an academic society, ITS will utilize technology and science to enable to work in harmony with nature. Therefore, ITS took the initiative to be a pioneer in the development of smart eco campus and utilize every campus element to create environmentally friendly project. Contributions to environmental sustainability programs are the part of the commitments in the Sustainable Development Goals (SDGs).

The implementation for developing ITS Smart Eco Campus is focused in creating an environmentally friendly campus that is embedded with various activities involving students and faculty members. There are five main points of ITS Smart Eco-Campus, namely:

1. Socio Engineering
2. Energy and Water
3. Waste Management
4. Green Area, University Forest and Biodiversity
5. Transportation

ITS is a science and technological university that is located in Surabaya, Indonesia with the effective area of 167.4 Ha. Although ITS had been established as the state university since 1960, however the main ITS buildings campus was built in 1977 and was inaugurated its use in 1982 [1]. In 2017, according to the Ministry of Research, Technology and Higher Education (Kemristekdikti) Indonesia, ITS is ranked as the 5$^{th}$ best university in Indonesia. In addition, ITS also became one of the 10 Indonesia's flagship university version of QS World University Ranking. In 2017, ITS achieved the UI Green Metric World University Ranking [2].

**Fig. 1.** ITS main campus

Currently, ITS has 38 departments that spread across ten Faculties with 970 lecturers and professors, 1135 supporting staffs and with the total number of 20,000 students, starting from vocational, undergraduate, graduate and doctoral study programs. To support and maintain the water needs for that number of faculty members, operation and maintenance laboratory, academic supporting facilities, student dormitories, including for ITS mosque, the average water demand in ITS reaches 32,627 $m^3$/month equal to 1,087 $m^3$/day [2].

ITS uses water from the Municipal Water Supply Enterprise (PDAM) Surabaya to meet its domestic needs, because the unconfined aquifer is unlikely to be consumed. The quality of groundwater in unconfined aquifer within the main ITS campus area and its surroundings is: 4.39 - 6.90 ppt salinity, 1.16 - 6.79 mg/L nitrate ($NO_3$), 0.18 - 2.12 mg/L phosphate ($PO_4$), while for nitrite and ammonia is not detected. Although there is a possibility for groundwater consumption, based on its quality, ITS does not conduct exploration or groundwater exploitation on confined aquifer. because the domestic water requirement is still sufficient from the PDAM, Surabaya.

The main campus of ITS is in the Coastal Zone adjacent to the Madura Strait, with the average of slope the land surface between 0-3% and average elevation around 3-5 m from the sea level. The nearest distance from the main ITS campus to the coastline area is about 3.6 km with the geological areal condition is an alluvial sediment [4] which has not been densed until loose, a mixture of sand, silt, clay [3]. These rocks are the result of deposition from the Brantas River system where ITS lies at the downstream from the Brantas Watershed. Indonesia is a tropical region with an average annual rainfall over 1500 mm.

Based on soil conditions, distance to shoreline, average surface slope, average elevation, rainfall potential, it causes various challenges in terms of water resources management. These challenges include, that the main campus of ITS is a region that:

1. Very small water catchment potential, due to coastal areas with low elevation and ground water depth of less than 1 meter.
2. The small slope of the soil causes the drainage flow very slow and supported by very shallow groundwater conditions, the potential for very high inundation
3. The groundwater is relatively brackish, so it can be used for aquaculture brackish water fisheries.

## 2 Water conservation program implementation

The form of water conservation program is water storage in ground water tank, elevated reservoir, lake or pond, long storage channel and wet land area conservation. Conservation forms are water recycling used for garden sprinkler system and water efficient appliances (water tap, flush toilet, etc). Beside the water reservoirs, because ITS is a former area of swamp and rice field, the majority of campus building structures are elevated, so that under the structure of the building can still be a recharge area and does not change the coefficient of run-off. Here the pictures of the form of water conservation program implementation.

### 2.1 Elevated structures

**Fig. 2.** Floating floor in Geomatics Engineering Dept, Physics Departement and Chemistry Department

### 2.2 Underground and elevated tank

Manarul Ilmi Mosque is the biggest praying center at ITS which requires supply water about 25 to 30 $m^3$ a day. In Friday it increases up to 45 to 50 $m^3$/day. The water is not only utilized for wudhu or ablution, but also for maintaining the mosque and its garden. In 2013, the mosque installed ground tank as water recycling system from used wudhu waste water. One of these goals is to reduce water consumption from municipal water company. The water recycling system has four chambers that one chamber is used for sedimentation and filtration processes while the others are used as water reservoirs (Fig. 3). Each chamber has capacity of 5,000 Liters with gravity water flow system. Automatic sprinklers from the chambers with pumps are used for daily garden watering specially during the dry season [2]. The wudhu water reuse is not only implemented in the Manarul Ilmi Mosque, but also in Mosque of ITS Manyar Campus.

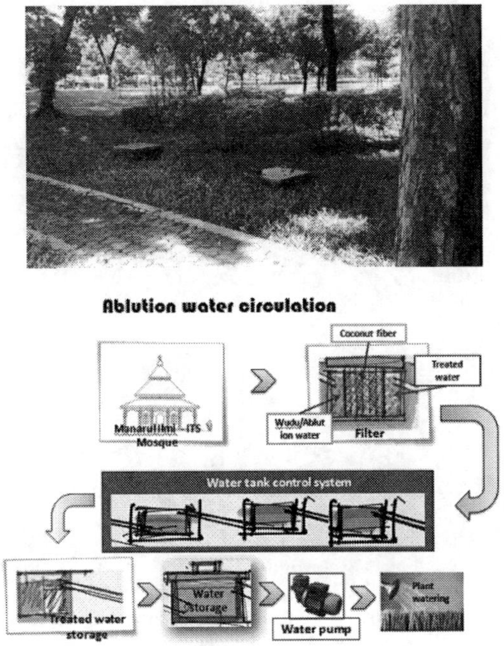


**Fig. 3.** Underground tank as water recycling system in Manarul Ilmi Mosque

One of the activities in the Smart Eco Campus is developing urban farming by planting some organic vegetables with the trademark SayOr ITS (means literally Sayur Organik or Organic Vegetables of ITS). In addition, it is also developed organic livestock and organic fruits. The ITS development effort of organic vegetables finally noticed by PT Takiron by providing assistance in the form of underground tanks to accommodate the rain water that is used to water the plants (Fig. 4).

**Fig. 4.** Urban farming ITS products and underground tanks to harvest rainwater

Besides in the form of underground tank, rain water catchment also uses elevated reservoir. Rainwater is collected through the roof of the building and flowed into the reservoir through pipes.

## 2.3 Lake or pond

The main campus of ITS has several points that become a patch pond. The patch pond serves as a retention pond; there are 10 retention ponds and one detention pond (Table 1). The detention pond is The Graha Sepuluh Nopember Pond, while the others are retention ponds. The capacity of the Graha Sepuluh Nopember pond reaches 18,014.00 $m^3$ equipped with a floodgate that can be flowed to the nearest drainage channel. In total, the artificial pond area in ITS main campus is approximate 49,783.00 $m^3$, with the volume of water at the measurement reaching 31,913.00 $m^3$ or almost 65% of the capacity. Measurement of pond capacity using bathymetry data as measured using USV Boat made by ITS, Himage USV 1 (Fig. 5 - 6) [5].

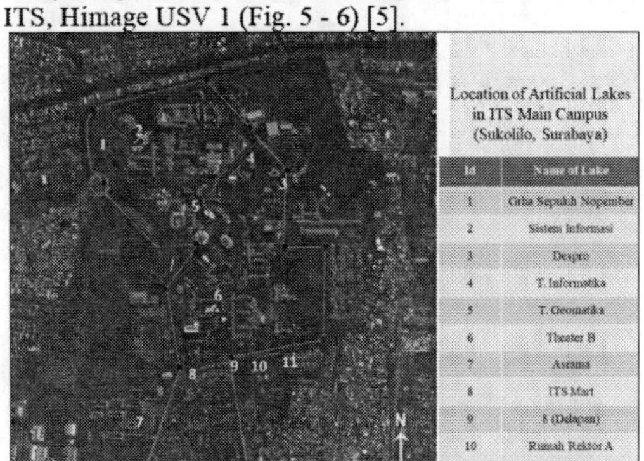


**Fig. 5.** Location of artificial lake in ITS main campus

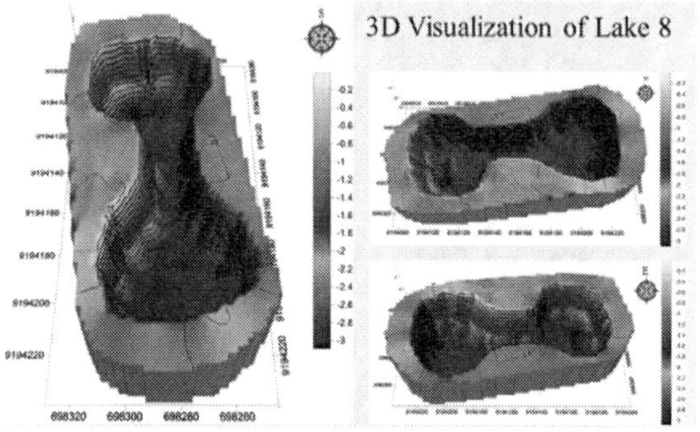

**Fig. 6.** (a) 3D Visualitation of lake 8 with volume 9695 $m^3$ (b) USV boat as bathymeter, Himage USV 1, made in ITS

**Table 1.** Artificial lakes in ITS main campus

| Id | Name of Lake | Area ($m^2$) | Water Volume ($m^3$) | Pond Capacity ($m^3$) |
|---|---|---|---|---|
| 1 | 8 (Delapan) | 4,976.40 | 6,975 | 9,695 |
| 2 | Rumah Rektor A | 435.97 | 223 | 423 |
| 3 | Rumah Rektor B | 797.43 | 239 | 365 |
| 4 | Graha Sepuluh Nopember | 4,990.95 | 11,133 | 18,014 |
| 5 | ITS Mart | 2,168.33 | 745 | 1,145 |
| 6 | Student Dormitory | 2,812.09 | 3,803 | 5,796 |
| 7 | T. Geomatika | 1,065.20 | 1,378 | 2,248 |
| 8 | Information System Dept. | 3,586.15 | 4,217 | 6,168 |
| 9 | Theater B | 1,073.23 | 1,387 | 2,058 |
| 10 | Informatics Eng. Dept. | 1,697.51 | 704 | 2,147 |
| 11 | Design Product Dept. | 2,514.02 | 1,109 | 1,724 |
| Σ | | 26,117.28 | 31,913 | 49,783.00 |

(Source: Geomatics Department ITS, 2017)

In addition to the pond that mention above, there are still some pond in the department environment with a function as retention pond.

Pond near Physic Department for fishery pond 2.5 m depth

Pond in student activities center for garden watering and recreation

The eight-shaped retention pond which has jogging track for floods control, sport activities and recreation

Retention Pond for floods control and recreation (Geomatic Engineering dept.)

Detention pond in Graha Sepuluh Nopember

Retention pond in Material and Metalurgy dept.

**Fig. 7.** Retention and detetion ponds in ITS

## 2.4 Wet land area conservation

ITS also has a wetland area, with a total area around 29.6 ha or about 16% of the total area of Main Campus of ITS. Those wetland areas are utilized by the surrounding community by growing water vegetables, like water spinach, pandanus. In addition, there are also used as agricultural land.

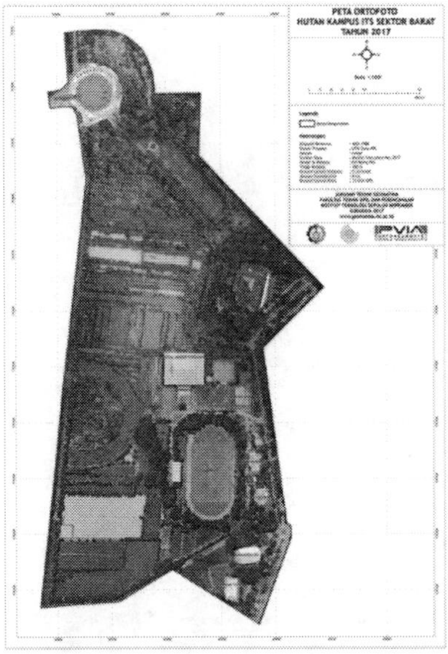

**Fig. 8.** ITS green openspaces with wetland area

### 2.4 Long storage channel

ITS also has a long storage channel, with a channel length reach 470 meters and a channel width around 6 meters.

**Fig. 9.** ITS long storage channel

## 3 Summary

Many programs could be implemented that support sustainable water management within ITS campus, including the stimulation of onsite infiltration of rainfall and runoff, protecting and enhancing surface water quality, promoting ground water recharge, reuse and recycle wastewater, capturing and reuse rain water, and also treating grey water and black water.

## References

1. Utama, W. *The Golden ITS.* Media Informasi dan Komunikasi ITS, ISBN: 978-979-8897-71-9. Surabaya (2010)
2. Hermana, J., Herumurti, W., and Irhamah. Sustainable Water Management in Tropical Region Campus: Study Case of Institut Teknologi Sepuluh Nopember Indonesia. *The 3rd International Workshop on UI GreenMetric. Istanbul* (2017)
3. Wahyudi, Firmansyah, Z., Suntoyo, and Tatas. Study on the Distribution of Seawater Influence to the Coastal Groundwater in Eastern Surabaya Coastal Area. *Prosiding Seminar Teori dan Aplikasi Teknologi Kelautan "SENTA"*, ISSN 1412-2332, Fakultas Teknologi Kelautan, ITS, Surabaya (2012).
4. Padmanagara, S. *Geological of Java and Madura Sheet III East Java. 3rd Ed.* Geological Survey of Indonesia (1969).
5. Hamsa, M. Yuwono, Okta Ferriska, The Study of Capacity/Volume of Artificial Lake in ITS Campus at Sukolilo as a Water Reservoir Areas, *The 7$^{th}$ International Symposium on Earth-hazard and Disaster Mitigation*, Bandung (2017)

# Water conservation and water recycling program implementations at University of Kashan

*Majid* Monemzadeh [*]

Vice-Chancellor for Research of Kashan University

**Abstract.** University of Kashan was founded in 1974 and is the oldest institution of higher education in Kashan. At the time of its foundation, only undergraduate courses in Physics and Mathematics were offered. The current activities of the university are classified into four sections of education, research, development, and side activities. The university is located on a 530 hectares campus, 15 kilometres outside of the town with several affiliated centres located near or at city of Kashan, Ghamsar, Niasar and Tehran. Presently, the university has nearly 300 staff and university lecturers and it has about 8200 students, studying in over 180 courses. Kashan University has awarded and certificated one of the most active universities in the Green Management. Some of the activities of Green Management of Kashan University includes, CCHP Plant, Sewerage Disposal, and Water saving policy, natural ventilation, using efficient applicants. The current paper studies some of the main activities of University of Kashan in Water conservation and water recycling programs.

## 1 Introduction

Kashan is an ancient historical city with an outstanding record in many different aspects of Persian Civilization, culture and development dating back to 7000 years ago. Archaeological discoveries in the Sialk Hillocks which lie 4 km west of Kashan reveal that this region was one of the primary centres of civilization in pre-historic ages. Hence, Kashan dates back to the Elamite period of Iran. The Sialk ziggurat still stands today in the suburbs of Kashan after 7,000 years [1].

Great many scholars, scientists, philosophers and thinkers from this city have contributed to the development of its Islamic history. There are a large number of historical sites in Kashan which includes some of the best and most attractive classical architectures in historical housing of Iran.

Kashan (33° 58' 59" N / 51° 25' 56" E) climate is classified as BWhsa by the Köppen-Geiger system, with hot and dry climate and virtually no rainfall during the year [2], where B is dryness; W is the intensity of dryness where the average annual rainfall (13.5) is lower than the average annual temperature in °C (19.7). h shows that the average annual temperature is above 18 °C, s is the sign of rain in the cold seasons or the winter and a show that the temperature in the warmest month of the year is above 22 °C [1]. According to DeMarttone climatic classification which is based on the dryness coefficient and considers two factors of temperature and rainfall, the climate of Kashan is dry with dryness coefficient of 4.5.

University of Kashan was founded at first as an institution of higher education in 1973. It began its activities in October 1974 by 200 students of mathematics and physics. In 1989 more academic fields were added and the institution continued its activities as Kashan Teacher-Training University under the supervision of the Ministry of Science. In 1994, by offering engineering disciplines, this university was acknowledged as a main university by the Ministry of Science. In 1995 College of Architecture and Arts was added to the university with an emphasis on the local needs, art and industries of the region. Faculty of Chemistry, as the 5th college after Science, Humanities, Engineering, Architecture and Arts, was founded in 2008. Currently, University of Kashan, within the boundary of 530 hectares and with 80000 square meters of infrastructure (building), in addition to educational campuses, sports facilities, dormitory complexes, restaurant complex, amphitheatres, green spaces, mosque, study halls and health-care centre provides leisure, services and comforts to the students. Besides the five mentioned colleges, there are three active research institutes including Nano Technology, Natural Essential Oils, and Energy and also three Research Centres including Kashan studies, Carpet, and Astronomy that contribute to the fact that University of Kashan in recent years has been among the top ten universities in Iran. Accommodating more than 7500 students in 50 disciplines of undergraduate, graduate and postgraduate levels and having about 300 full-time faculty members.

## 2 Water crisis in Iran and city of Kashan

Iran is located at the southwest of Asia and in the northern hemisphere desert belts and this situation has caused a large area of Iran to be affected by dry and

[*] Corresponding author: monemzadeh@kashanu.ac.ir

semi-arid climate, facing Iran with water shortage. Iran faced a growing water crisis [3]. Especially, with the annual rain fall of 50 mm in the central plains of Lut, living conditions in the marginal areas of desert is arduous and intolerable. The southern and central regions of Iran including Kashan are having a severe shortage of rain and, day by day, water resources, especially groundwater, are reducing. The average annual rainfall in Kashan is 116 mm and evaporation are high too; on the other hand, about 25 million hectares of Iranian lands are classified as pasture lands of which region of Kashan is a part. Therefore, soil salinity, dry climate of Kashan, and immethodical deep and semi-deep well drilling has put sources of fresh water subject to salinization.

## 3 Water supply at University of Kashan

University of Kashan absorbs water from the wells drilled at the university. In recent years, however, due to the drop in underground water levels, the depth of the wells has grown. Currently, it takes at least 250 meters to reach water.

It is estimated that, in the coming years, deeper wells should be drilled to receive water.

Due to the shortage of water it faces, University of Kashan is implementing and planning various strategies to store, recycle, and treat water.

## 4 University of Kashan's activities about water

- Conferences and workshop for professors, students, and staff on optimal consumption and water saving
- Drop irrigation for an area of over 2,000 meters at University's campus
- Cultivation of plants appropriate to the climate, plants that require lower amount of water
- Using optimal showers in dormitories. Major water consumption at University are related to bathrooms.
- Construction of 120 m3 wastewater treatment plant applicable for 200 people by charity investments
- Purchasing new washing machines with an energy labels for installation in dormitories. (These replacements cause water and energy savings)
- Construction of artificial turf football field
- Water storage at the Essential Oils research centre, University of Kashan

### 4.1 Example of courses/modules related to water (University of Kashan, Kashan)

**Table 1** Courses/modules related to water

| Courses/modules related to water | |
|---|---|
| **courses/modules** | **Faculty/ Department** |
| • Wastewater Treatment and Recycling<br>• Water Resources Management<br>• Wastewater treatment<br>• Economic and social issues of watersheds<br>• water resource management<br>• water quality<br>• Engineering structures for water and soil protection<br>• Management and exploitation of unconventional waters<br>• Simulation in Watershed | Natural Resource Faculty |
| • Water treatment and sewage treatment<br>• Water and wastewater treatment of food industry | The Department of Chemistry Engineering |
| • Water treatment and sewage treatment | The Department of Biology and Civil Engineering |

### 4.2 Water recycling program implementation

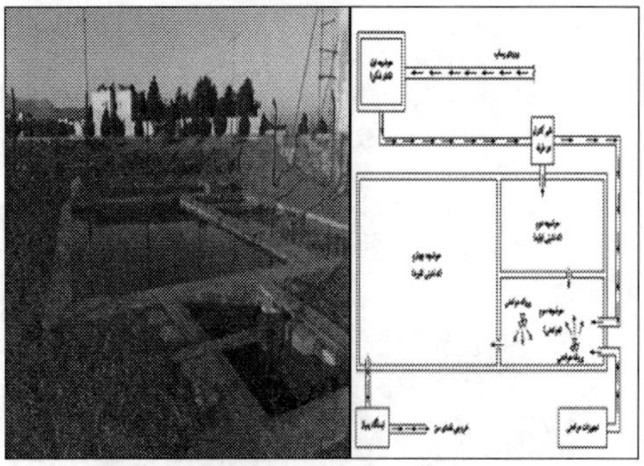

**Fig. 1.** Utilization of the first water recycling plant in Iranian Universities

**Fig. 2.** Utilization of the first wastewater treatment plant in Iranian Universities

https://doi.org/10.1051/e3sconf/20184805003

**Fig. 3.** 2 hectares of Rose Mohammadi (rose flower) irrigated with refined waste water and sewerage disposal at Kashan University

## 5 Conclusion in Our University

In, a reused wastewater treatment system was designed and implemented that was able to recover gray wastewater with a flow rate of 30 cubic meters per day. The purpose of this study was to investigate the effect of aeration system on the removal of pollutants from wastewater for use in agricultural applications. In the first step, the study was conducted with the establishment of a research pilot, quality raw sewage and refined wastewater of a sports complex. Then, according to the results obtained from the wastewater volume and the severity of the input of wastewater pollution, design of the main purification units was carried out. The results showed that using this method, the amount of oxygen required for chemical and biochemical processes, as well as the total soluble solids, have reached a standard for use in the field of green space. The treated wastewater was used for agricultural purposes and water supply needed for the new farm of 2 hectares of rose Mohammadi (Rose Flower).

**Fig. 4.** Especial flowers and green space of Kashan

**Fig. 5.** Conference about green space with optimum water usage

**Fig. 6.** Workshop about water management in Kashan University by Professor Reza Mansouri

**Fig. 7.** Program for planting trees by university staff and the president of the university

These are Examples of Events Related to Environment and Sustainability (University of Kashan, Iran)

**Fig. 8.** Installation of automatic dish washing machine in the Restaurants of University of Kashan

**Fig. 9.** Using the campus drinking water with straw in Kashan University to reduce plastic glass and saving water

Example of the use of water efficient appliances (toilet flush) in Kashan University, Kashan:

- Optimizing all coolers and chillers to saving waste water.
- University of Kashan has used the innovation in reducing use of plastic glass and saving water policy, with this Campus

Drinking Water, there is more than 80% saving in Water and reducing in plastic.

## References

1. Jomehpour, M., *Qanat irrigation systems as important and ingenious agricultural heritage: case study of the qanats of Kashan, Iran*, International Journal of Environmental Studies, **66(3),** 297-315. (2009)
2. Korsavi, S. S., Zomorodian, Z. S., & Tahsildoost, M, *Energy and economic performance of rooftop PV panels in the hot and dry climate of Iran*. Journal of Cleaner Production, **174**, 1204-1214. (2018)
3. Michel, D, 9 *Iran's Impending Water Crisis*. Water, Security and US Foreign Policy, 168 (2017)

# Universitas Padjadjaran concern for sustainable water resource from West Java to national and to the world

*Cipta* Endyana[1,2,3], *Teguh* Husodo[1,2], Hendarmawan[1,2,3], *Tri* Hanggono Achmad[1]

[1]Universitas Pdjadjaran, Kampus Jatinangor, Sumedang 45363, Indonesia
[2]Postgraduate School of Universitas Padjadjaran, Kampus Bandung, Bandung 40132, Indonesia
[3]Faculty of Geology Universitas Padjadjaran, Kampus Jatinangor, Sumedang 45363, Indonesia

**Abstract.** The implementation of Sustainable Development Goals (SDGs) which have been set according to the Peraturan Presiden No. 59 tahun 2017 about The Implementation of SDGs to achieve the purpose of SDGs in Indonesia. Through the implementation of universal, integrated, and inclusive principles of 2030 Agenda for SDGs, a new dawn of development that changes the direction of itself, that enveloping human rights and equality in social, economy, and environment, could be accomplished. UNPAD understand, as a seat of learning and to give real contribution, both in the community and national development, by adopting transformative education paradigm, which seek more activities in university level for better community through the idea of sustainability and sustainable development, and its written as Principle Scientific Pattern of Universitas Padjadjaran: *bina mulia hukum dan lingkungan hidup dalam pembangunan nasional.* Our research was started to perform through multi and transdiscipline-based with all stakeholder involved such as local government, industry side and community. The study areas were focused on most West Java and Banten where region related with industry. Researches were divided into 4 stages, start from study until monitoring stage. As appreciation, Universitas Padjadjaran reached Indonesia Green Award (2013, 2014) and Green World Awards (2017). What's Unpad researcher has done is a step in engagement to ensure the availability and management of sustainable water and sanitation in accordance once of SDGs in the environmental sector life.

## 1 Introduction

In the context of global, the strengthening focus of university role on the sustainable development agenda is in line with the global agreement of Sustainable Development Goals (SDGs), which is an Agenda aiming for better world by 2030 through 17 aims and 169 targets. In line with the agreement of SDGs, on the university level, there is a Higher Education Sustainability Initiative platform to support SDGs. This platform is adopted by 272 Universities from 47 countries, and for Universitas Padjadjaran its imprinted on the *Pola Ilmiah Pokok* (PIP) or Principal Scientific Pattern: "*bina mulia hukum dan lingkungan hidup dalam pembangunan nasional*" (building a noble development of law and environment in national development). The attempt of fulfilling SDGs' goals by Universitas Padjadjaran, is started by looking at the needs of Cities and Districts within West Java through the program of "*Unpad Nyaah ka Jabar*", which is expected to have a national wide impact and becoming a regional effort as "from West Java to the world through sustainable development".

"*Bina mulia hukum dan lingkungan hidup dalam pembangunan nasional*" is the cornerstones of a lofty philosophy, that will always have remembered and implemented during the implementation of *tridharma* in UNPAD. Based on this idea, the role and attention of Universitas Padjadjaran on environmental field, is being formed into study program that specifically studied environmental issues since late 1990s and changed Pascasarjana Program into School of Pascasarjana by 2015, which focused on the environmental sustainable issues, both in education and research, and encouraging the accomplishment of Sustainable Development Goals (SDGs) through inter- and trans-discipline approach. Strengthen existing SDGs study centers and establishing new study center which relates to the efforts of achieving SDGs by Mei 2016. With an approach that is also inter- and trans- disciplinary, those study centers expected to be sharpen UNPAD's role on the accomplishment of sustainable development goals, in addition to the roles that have been done by academic community of UNPAD, directly or indirectly.

PIP that had formulated to fulfilling the vision of Universitas Padjadjaran, is implemented on the *tridharma* UNPAD. Among them, then formed into a study program that specifically studied environmental issues since late 1990s and changed Pascasarjana Program into School of Pascasarjana by 2015, which focused on the environmental sustainable issues, both in education and research, and encouraging the accomplishment of Sustainable Development Goals (SDGs) through inter- and trans-discipline approach. Strengthen existing SDGs study centers and establishing new study center which relates to the efforts of achieving

* Corresponding author: cipta.endyana@unpad.ac.id

SDGs through Mei 2016, and on February 2018, a center for *inter- and trans- disciplinary research on the environment and sustainability science* was inaugurated.

The UNPAD's strategic program of *Rencana Induk Riset* (RIR) inserted into the implementation of leading research that determined rationally by paying attention to the human resources, available facilities, and critical and strategic issues, whether its local, regional, national, or worldwide. The leading research of UNPAD are planned as a semi top down and determined based on the research goals written in RIR UNPAD 2012-2016. Leading research is set by paying attention to the competence, advantages, and research roadmap of each faculty; national, regional, and university research policy; and also, the strategic and critical national, regional, and worldwide problems/issues. All of the UNPAD's leading research are an interdisciplinary-oriented to give a real contribution to help solving the problem exist in regional/national/West Java or even world wide. UNPAD's leading research could be in form of a basic research, applied and action research that oriented to the economic independence, competitiveness, and poverty reduction for better community. During 2012-2016, UNPAD has 5 fields of leading research: 1. Food, 2. Environment, 3. Health, 4. Energy, 5. Policy, Culture, and Information.

## 2 Research in Universitas Padjadjaran: overview

Based on the RIR Universitas Padjadjaran 2016-2016 which implemented in the implementation of *tridharma* on 5 research fields [17], Eco Campus cluster through water resources field focused, raised the importance of UNPAD to conduct management and conservation of water resources. In the concept of Hydrology, ground water is one of the components within the hydro cycle that highly related to the underground water availability, precipitation, infiltration, percolation, evapotranspiration, and surface run-off. Conceptually, the condition of water recharge and discharge area, could be identify by studying the water spring spread available. For this matter, the knowledge of condition of potential water catchment area and the ground water behavior is highly important, particularly from the water recharge area to the discharge area.

As resources, water is a essential resource for human being, animals, and plants. Therefore, it is substantial of knowing and understanding water sources, its characteristic, also the good management and utilization. Jatinangor and Sumedang are one of the example of volcanic area that have good potention of water resources. Its landscape that varies from valley to hills, is an area well suited for the availability of recharge area and discharge zone of ground water. However, the geological characteristic of volcanic deposits which always change in short period of time and complex geological structure, quite influential on the ground water flow system in the area. The outlet of ground water to surface could be caused by the intersection of ground water level as the result of the contact of permeable rocks with the impermeable, and the present of a fault.

Universitas Padjadjaran is an education institution with a large number of academic community, reaching up more than 30,000 people, requires vast area of land and also a large quantity of ground water. Ground water scarcity will occur if the management done by UNPAD is not suitable with the aqifer system characteristic and the existing ground water system on the area. Therefore, UNPAD understand that there is a importance of conducting research about the characteristic of groundwater and aquifer system existing within the area as a consideration to determine an integrated work program in term of conserving ground water. The research strategy used, is based on the research roadmap on environmental field, so the main achievement of environmental friendly campus could be fulfilled. Fulfilling the condition of environmental friendly campus consist of several stages of eco-building, eco-place, eco-behavior, eco-relationship, and eco-services. (Fig. 1) [17].

To ensure the suitableness of this water resources research as a leading environmental research includes: 1) Framing the ground water conservation master plan as a guidance of recommendation in campus construction and infrastructure development based on the ground water conservation. 2) Determine a critical priority zone of ground water use from the upstream (water catchment and recharge area), managing the available potential within the campus and its conservation, and increase the capability if water recharge area within campus for downstream. 3) Socialization and education as a form of delivering research result and masterplan to the stakeholders, in order to supporting the recommended working programs. 4) Implemented the working programs based on the recommendation of masterplan study for groundwater conservation within designated location and timeframe.

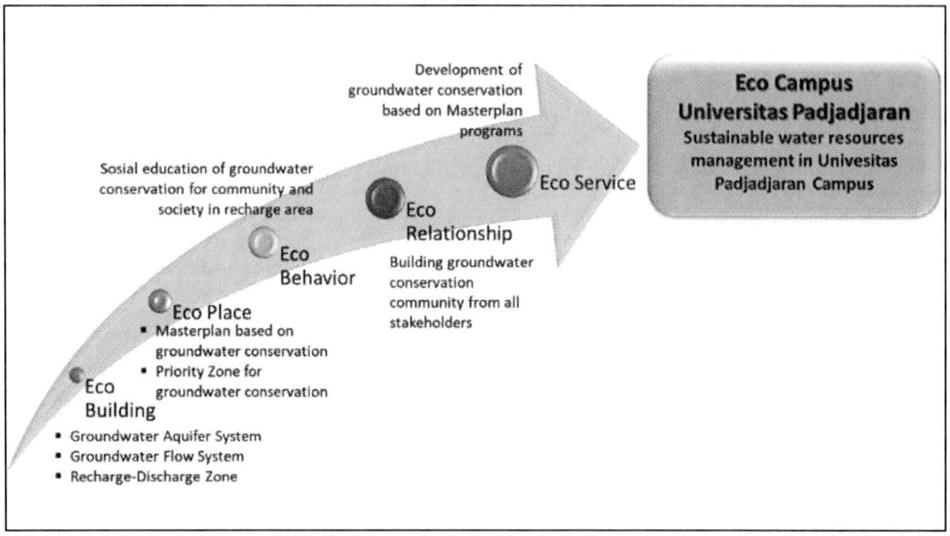


**Fig. 1.** Leading research on environmental of Universitas Padjadjaran.

## 3 Integrated and Collaboration in Environmental Research

Universitas Padjadjaran, on its research development of Environmental field, is trying to contribute the research result for the whole community, particularly people of West Java. Integrated research model is developed by Universitas Padjadjaran as Research University in its participation on development, have an up-to-date research design and could give a correlated, efficient, and effective solution for the environmental issues for both government and industrial, as the stakeholder on development.

The needs of government and industrial for mentoring by the university is indispensable, where both sector requires research to resolve up rising problems caused by development process and for the practical solution as a best practice based on the theory from the university. Cooperation and collaboration, up till today on the level of West Java Province, among others is being conducted on Kabupaten Bogor, Sukabumi, Cianjur, and Sumedang. While on national level, the collaboration is being conducted on some places in Sumatra, Borneo, and Sulawesi. Research collaboration with industrial sector until today is done with regard to exploring ground water resource and monitoring the quantity and quality of ground water, among others is collaborating with private company (PT. Danone, PT. Biofarma, PT. Chevron), and others company.

Integrated Research Design Collaboration developed, is not only being focused on the government and industrial sector only, but it is also developed for the community that merging as environmental care community. The university has *Tri Dharma Perguruan Tinggi* which consists of Education, Research, and Community Services. Therefore, in addition to research and education, Universitas Padjadjaran has the responsibility to increase community Capacity Building in informal knowledge. On its relation to environmental management, in line with improvement of the community quality that grouped into environmental communities, they can participate and have an active role on the development as supervisors and volunteers in the handling of environment.

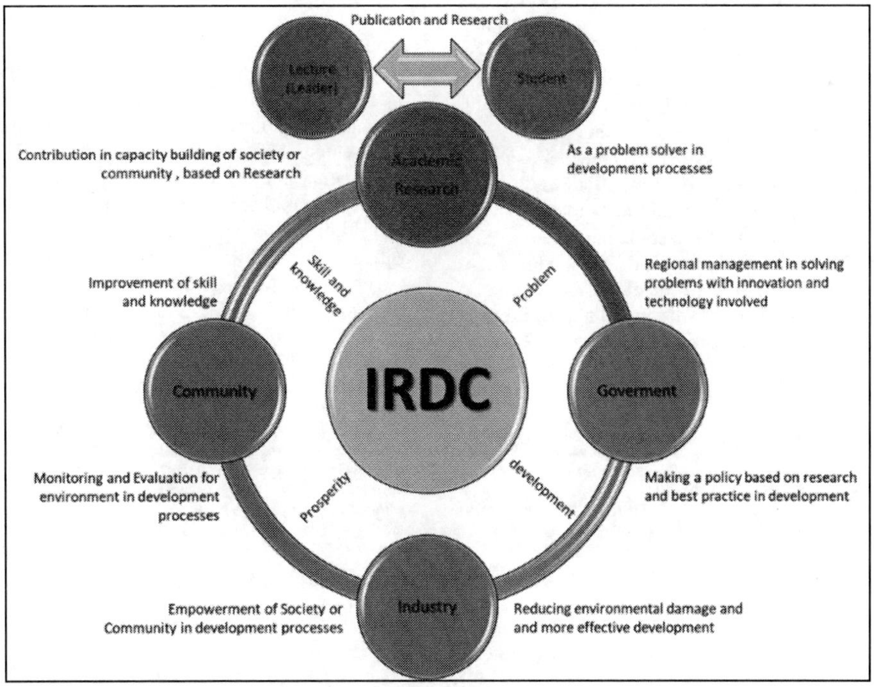


**Fig. 2.** Integrated research design collaboration

## 4 Sustainability water resources management

According to the hydrology cycle concept, not all infiltration water (ground water) flows into river or other water reservoirs, but there are some infiltration water remains in the top soil to be evaporated back in to the atmosphere through soil evaporation and transpiration. The evaporated water comes from soil through the mechanism of plant physiology, so the evaporation is called transpiration. In other word, intersection occurs during and immediately after the rain. While the transpiration process takes place when there is no rain. Based on the short description above, it is known that ground water conservation research is a continuous research, requires a long implementation stages from framing steps, planning integrated master plans and implementing. Some of the steps framed until today are:

### 4.1 Aquifer characteristic studies stage

Research on this early period is a basic research to understand and determine surface and undersurface lithology. The research was conducted for three years. In the same time, ground water research on site was also being conducted to ensure source and ground water slow, water movement duration, and measuring its infiltration and run off ratio [15,16]. Further updates on this research mention that a ground water isotope study is being conducted by Environmental Team of Universitas Padjadjaran for a year in a vast area (Hendarmawan, 2014) [1,2,3].

In short, the result of this early research is a direction of ground water conservation strategies that could be done in short amount of period with an artificial model [4], such pond, infiltration-well, and bio pore. And also, in addition to considering technical factors, supporting non-technical factor is also worth mentions as a recommendation, such as social involvement of the community for the soil conservation [5]. The result also indicates a recommendation for an infiltration zone or area that requires being well-managed [6,7], like the infiltration and preservation program of Desa Sindangsari and Desa Cileunyi Wetan. And for the water movement duration, this research indicated that two drilled-well on UNPAD showing that the groundwater in it has a lifespan in range of 1090-2120 years before now (Before Present). And for the research result regarding upstream and downstream, the area of UNPAD indicates as a infiltration area or zone of shallow ground water of downstream [13], and this result encourages UNPAD to restoring the ground water immediately through artificial infiltration like bio pore, infiltration well and injection well [8,14].

### 4.2 Planning and implementing stage

Research on the conservation planning is a follow through research and this stage was conducted for 2 years. Combining the result data of ground water basin study, land use data, demography, recharge and watershed map using spatial analysis, statistical and couple of water balance simulation scenario. As for the result of the master plan of groundwater conservation research, priority ground water conservation zone was determined, based in three main indicators: hydrology, agriculture, and social-economy [9,10,11]. The direction of recommendation generated among other are: implementing concept of eco building, developing water pond and injection well [12,13]. On this stage, some recommendations is already being implemented, such as 1) making of bio pore on a specific site like the Faculty of Nursery, Medical School, Faculty of Dentistry, and at the south part of campus, 2) completing the injection well near Rectorate building and Medical School building with the minimum depth of 75 meter, and 3) involving people in building water pond and infiltration well in Desa Sindangsari Kecamatan Sukasari Jatinangor.

### 4.3 Monitoring stage

Monitoring stage is aiming to keep the quantity of ground water within the campus of UNPAD Jatinangor available for 4 years. This stage is already being implemented, however not all area of the campus is represented. The important result from this stage is the implementation of the solution for the ground water conservation is no longer requires researcher on data collection, but ground water conservation communities which comes from the local community and local government, while the researcher of Universitas Padjadjaran would acting as supervisors.

In short, Universitas Padjadjaran's step on conducting the management of water resources is shown in this roadmap below (Fig. 3)

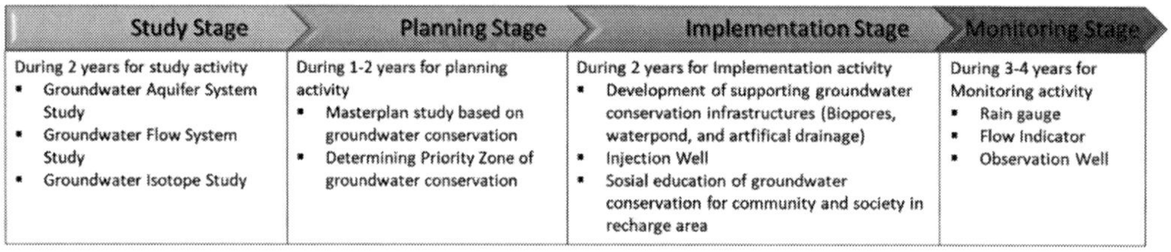


**Fig. 3**. Roadmap of water resources conservation management system.

## 5 Achievement

The integrated research conducted in span of 10 years by Universitas Padjadjaran has giving an active role in the environmental competition, both on national and international scale. And for it, the achievement acquires are:

- Indonesia Green Award 2013
- Indonesia Green Campus Award 2014
- Green World Award 2017, an award from Green Organization which offers Universitas Padjadjaran as the Ambassador for Green Organization on Education and Best Practice Sector for environment to worldwide.

## 6 Closing

Sustainability water resources management research is one of the effort of Universitas Padjadjaran to improve the quality of Life on the campus environment and nearby community based on the Sustainable Development Goals (SDGs) [18], particularly the SDGs target of (6.4): significantly improving the efficiency of water use in all sector, and ensuring the sustainability of utilization and availableness of freshwater to counter water scarcity; (6.5) implementing integrated water resource management in all level including on point cross sector cooperation; (6.6) protecting and restoring ecosystem relates to the water resources, particularly on mountains, forest, wet lands, river, ground water, and lakes; and (6.7) supporting and strengthened local people participation in improving water management.

**Acknowledgement.** This Research of Sustainable Water Resources is conducted with the long terms period and systematic stages. The collaboration between academic, government, industries, and communities become important to make this reseacrh accomplished. This research has been supported by many researchers at Universitas Padjadjaran, Ministry of Research, Technology and Education of Indonesia, Ministry of Environmental and Forestry of Indonesia, Government of West Java Province and its belowed Sumedang, Sukabumi, Cianjur, Bogor, Bandung District, and several industries contrubution those are PT. Danone, PT. Bioframa, PT. Chevron, and others.

## References

1. Clark, I., Fritz, P., *Environmental Isotopes in Hydrogeology*, Lewis Publisher (1997)
2. Payne, B., and Y. Yutsever, "Environmental Isotopes as a Hydrological Tool in Nicaragua", Isotope Technique in Groundwater Hydrogeology **1** (1974)
3. IAEA – Vienna, *Stable Isotope Hydrology, Deuterium and Oxygen-18 in the WaterCycle.*
4. Domenico, P.A. and Schwartz, W.F., Physical and Chemical Hydrogeology. John Wiley and Sons, Inc., Canada, 824p (1990)
5. Brooks, K.N., P.F.F. Folliott, R. Touchan, *Sustaining the Benefits of Integrated Watershed Management and Coping with Climatic Variability.* Paper for discussion, CIFANS, University of Minnesota, USA (2010)

6. Chiras, Daniel D., *Environmental Science*, 8th Edition, Sudbury, Massachusetts: Jones and Bartlett Publisher (2009)
7. Chao-Hsien Liaw and Yao-Lung Tsai, *Optimum Storage Volume of Rooftop Rain Water Harvesting System for Domestic Use*, Journal of the American Water Resources Association; 2004; **40**, 4; Proquest Agriculture Journals pg. 901 (2004)
8. Sharpe, William E. and Swistock, Bryan., *Household Water Conservation*, College of Agricultural Sciences, Agricultural Research and Cooperative Extension College of Agricultural Sciences, The Pennsylvania State University (2008)
9. Thomas, Terry, *Rainwater Harvesting: Practical Action*, School of Engineering, University of Warwick, Coventry CV4 7AL, UK (tanpa tahun)
10. To'th, J., A theoretical analysis of groundwater flow in small drainage basin. Journal of Geophysics Research, 68: 4795-4812 (1963)
11. Worm, Janette & Hattum, Tim van., *Rainwater Harvesting For Domestic Use*, Agrodok 43, Agromisa Foundation and CTA, Wageningen
12. Mau, D.P., Winter, T.C., Estimating ground-water recharge from streamflow hydrographs for a small mountain watershed in a humid climate, New Hampshire, USA. Ground Water, 35: 291-304 (1997)
13. To'th, J., A theoretical analysis of groundwater flow in small drainage basin. Journal of Geophysics Research, 68: 4795-4812 (1963)
14. Fetter, Jr. C.W., Applied hydrogeology. Bell and Howell Company, Colombus, Ohio, 488p (1980)
15. Freeze R.A., Cherry, J.A., Groundwater. Prentice-Hall, Englewood Cliffs, NJ, 604p (1979)
16. Hooper, R.P., Christophersen, N., Peters, N.E., Modeling of Streamwater Chemistry as A Mixture of Soilwater End membersan Application To The Panola Mountain Catchment, Georgia, USA. Journal of Hydrology, 116: 321-343 (1990).
17. Universitas Padjajaran, Rencana Strategis 2015-2019, Available online at http://www.unpad.ac.id/wp-content/uploads/2012/07/Renstra-Unpad-2015-2019.pdf (2015)
18. Kementerian PPN/Bappenas, Available online at http://sdgs.bappenas.go.id/

# Dublin City University – toward a sustainable campus: water

*Samantha* Fahy*

Dublin City University, Glasnevin, Dublin 9, D09 W6Y4, Ireland

**Abstract.** Dublin City University, originally established in 1989, is located approximately 5km to the north of Dublin City Centre and is the most significant and comprehensive provider of university education on the rapidly growing and economically important Irish Eastern corridor. DCU is committed to embedding an ethos of sustainability across the entire institution. Sustainable operations are a core element of the DCU Sustainability Plan. Baseline ecological footprint metrics have been agreed and ambitious targets are set on an annual basis to reduce these with the eventual target of delivering a carbon neutral campus. Such baselines have been agreed for all campuses for water consumption. Several projects have been undertaken both on an operational and research footing to address consumption levels. Such projects include full water surveys of all campuses leading to significant savings due to fixing of leaks. In addition to operational projects, DCU focuses on educating and information its internal and external communities raising awareness of the importance of this essential resource and seeking behavioural change to reduce consumption and increase the implementation of water efficient principles and practices.

## 1 Introduction

Dublin City University (DCU), originally established in 1989 by the University statue, is located to the north of Dublin City, Ireland and is the most significant and comprehensive provider of university education on the rapidly growing and economically important Irish Eastern corridor.

There have been some significant changes in DCU over the past couple of years. September 2016 saw the completion of the process by which St Patrick's College Drumcondra (SPD), Mater Dei Institute (MDI) and Church of Ireland College of Education (CICE) were fully incorporated with DCU. From this date, all students and staff of St Patrick's College, Mater Dei Institute and the Church of Ireland College of Education are students and staff of DCU.

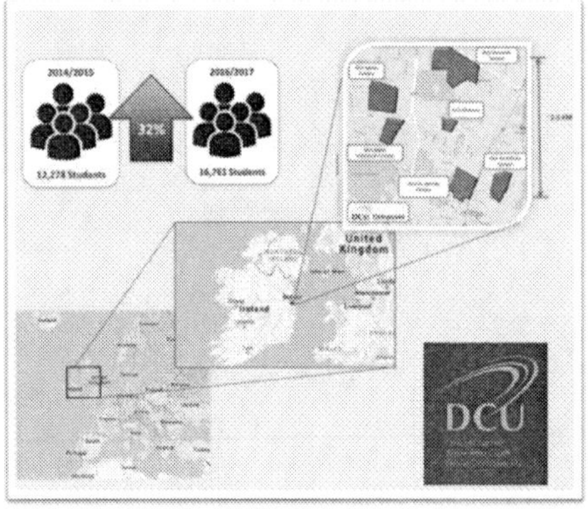


**Fig. 1.** DCU location 2018

This newly expanded DCU hosts over 17,000 students across five faculties Humanities & Social Sciences, DCU Institute of Education, Science & Health, Engineering & Computing and DCU Business School with a continuing focus on personal, academic and professional life-skill development. Fundamental to the DCU ethos is open access to education for all. The DCU Access programme was established more than two decades ago and identifies and supports students from groups currently under-represented in higher education. In 2017, DCU supported nearly 1,300 access students, making it the largest such programme by far in any Irish university. In addition, DCU is the first Irish "University of Sanctuary", awarded by the City of Sanctuary, in recognition of a range of initiatives demonstrating commitment to welcoming asylum seekers and refugees into the university community and to fostering a culture of inclusion for all. The Age-Friendly University (AFU) initiative at DCU is committed to widening access to higher education, and this includes enhancing the lives of older members of the community through innovative educational programmes, research agenda, curriculum development, online education, health and wellness activities, arts and culture programme and civic engagement opportunities.

## 2. University strategy 2017-2022

2017 seen the launch of the new **DCU Strategic Plan: TALENT, DISCOVERY, AND TRANSFORMATION (2017 – 2022)** [1]. In presenting this plan the DCU President Prof. Brian MacCraith said

* Corresponding author: samantha.fahy@dcu.ie

*"Our purpose must surely be to address the major challenges facing the world today and to develop the talent and knowledge that society needs."*

Continuing its mission statement DCU aims to transform lives and societies through education, research, innovation and engagement.

DCUs vision is to be a globally significant University of Transformation and Enterprise that is renowned for among other things its commitment to sustainability. This is clearly articulated in one of the nine strategic goals for the University over the next five years – Strategic Goal 8: Place sustainability at the core of the university.

At the core of the Strategic Plan 2017 -2022, DCU will seek to further develop and implementation the DCU Sustainability Plan created by a cross-institutional Sustainability Council. This council will also play a key role in monitoring the implementation of the plan.

Sustainability will play a key role in academic and research activities, with a view to enhancing the insight and commitment of students, the principles of sustainability and sustainability literacy will be integrated into all the undergraduate curricula. Specific degree programme(s) will be developed at postgraduate level. Research focusing on sustainability will continue to be focused on and fostered.

## 3. Management and operations

There is strong evidence that Higher Education Institutions (HEIs) are in a prime position to demonstrate strong leadership to promote and support the embedding of sustainable practices not just within their own institution but as a broader example to the industry and society [2-4]. University graduates can be empowered with an understanding of sustainable and the role they must play, research departments can develop the innovation solutions (technical and non-technical), campuses, working together with industry can act of living labs to test and demonstrate these and other innovative solutions, all on a campus that utilising sustainable practices in its daily operations demonstrating to society the realisable possibility of practicing what one preaches.

Water is perceived as an infinite resource; however it is easily impacted by human activity and needs to be conserved for the benefit of health and life on the planet.

According to current estimates [5] each person in Ireland consumes ~150 litres of water per day (54,750 litres of water per person each year) most of which is flushed away every day. Annual consumption rate on the DCU campuses are between 180k m3 – 220k m3 per annum. Major stakeholder have been identified and for the past number of years DCU has commissioned expert *reports on* water consumption on the DCU Campuses. Large number of major and minor leaks were identified and fixed leading to significant (+30%) consumption reductions.

In November 2017 DCU was awarded the ISO50001 management standard for both Energy and water management on campus. This award was achieved not only through the reduction in leaks but also in the implementation of several infrastructural measures over the past 3-5 years including waterless urinals, reduced flow water taps and shower heads along with several measures to engage the university community and increase awareness of consumption habits and how to reduce these.

## 4. Green campus at DCU

Engaging the internal and external communities at DCU has been a focus of Sustainability strategy for some time. In 2012 DCU signed up to the An Taisce Green Campus programme which "encourages practical environmental education and the empowerment of campuses to become exemplar in environmental stewardship" and which has been in operation in Ireland since 2007 [6]. This programme raises the profile of sustainability actions across campus and creates the platform to extend and translate the learning from Glasnevin Campus to the new incorporated campuses. In early 2018 following a rigorous assessment DCU St. Patricks Campus was also awarded Green Campus Status.

A primary focus of the Green Campus programme is educating and informing all members of the DCU Community, all staff (academic, research, administrative and technical), all students (undergraduate and postgraduate) along with societal groups and enterprise partners who are linked to DCU. The operational projects have of course a very significant impact on the water consumption levels within the university however to translate this impact to a global scale behavioural change is needed across the board.

The Green Campus programme at DCU had undertaken many engaging and compelling projects working with a broad range of stakeholders that have focused on developing a understanding of the limited resources available and identification of sustainable solutions.

### 4.1 Water consumption in the DCU sports complex

A student driven project was undertaken in 2015/2016 to investigate water consumption in the sports complex. This project demonstrated the importance of knowing where your water is consumed – results showed that the pool consumed 3,000-5,000$M^3$ per annum while in comparison the shower facilities were a factor of greater at ~46,000 $M^3$ per annum.

### 4.2 Rain water harvesting at DCU

Over the years there have been several projects undertaken on RWH at DCU including the installation of a RWH system on the 1.6-acre DCU Community Garden. Total volume of capture water is ~ 3,000 liters however this proved not to be sufficient for summer

https://doi.org/10.1051/e3sconf/20184805005

conditions and in addition a 5-6k liters holding pond was also installed in the garden to ensure sufficient water in the growing season. In addition to providing a reserve of water this pond has also enhanced the biodiversity within the garden attracting frogs, newt and other water dependant plants and animals into the garden.

**Fig. 2.** Installed RWH system and filtration system at DCU community garden

Supporting the RWH system in the community garden a student undertook their final year project on the designed a relatively cheap and robust sand filtration system that removed a large portion of contaminant in the water so that it was in essence potable. The model for this filtration system was disseminated to other community gardens in the greater Dublin area.

A feasibility study was undertaken to assess the potential for RWH as a recourse on the Glasnevin Campus indicating that there was significant potential for the reuse of water on campus however prohibitive regulation has limited the ability to implement this solution

### 4.3 Promoting water awareness

On an annual basis the DCU Green Committee hosts either green weeks or green days to promote awareness of sustainability issues. This includes a regular intergeneration water rocket-launching competition where primary (6-12yrs), secondary (12-18yrs) and university (18-99yrs) students design, create and launch their own rockets using recycled plastics bottles and harvested water from the DCU Garden.

**Fig. 3.** Participants in world water day at Dublin City University Glasnevin Campus.

Also on world water day DCU media students run awareness campaigns using social media such as facebook/snapchat/ instagram. Many are surprised on what they have discovered like how much of water you have to use to flush a toilet and how much water you can save when you shorten your shower time.

**Fig. 4.** Water awareness campaign with Students on social media

### 4.4 Hot water in public sanitary facilities

DCU, along with nine other partner Universities, are participating in the Global Consortium for Sustainability Outcomes (GCSO) project on trialling the removal of hot water from public sanitary facilities. DCU plan to remove hot water from the majority of sanitary facilities in one designated building for the period of one year and closely monitor the impact and perspectives of staff, students and other building users. Hot water is not required for hygiene purposed and due to legionnaire's **disease,** all public hot water system must hold hot water at circa 60 degC to prevent spread of the disease. The provision of hot and cold water is however required under national building and health & safety regulations. Significant water and energy saving and associated greenhouse gas emissions reduction for the building have been estimated over the period. Working closely with Dublin City Council and the Health and Safety Authority this project aims to provide the necessary data to amend legislation and regulation for all future buildings in Ireland.

### 4.5 Water education and research at DCU

The DCU Water Institute is developing and applying innovative approaches to the specific challenges of water quality, delivery and management. The DCU Water Institute will be a state of the art research facility, situated on the DCU campus, and focused on solving national and global water-related problems through excellence in research, education and innovation. Through the work of its 30 Principal Investigators, the Water Institute aims to achieve safe, secure drinking water, healthy aquatic ecosystems and reliable, energy efficient, quality water supplies for a sustainable economy.

https://doi.org/10.1051/e3sconf/20184805005

### 4.6 Community engagement and social enterprises

DCU does not reside in an ivory tower, it is very much engaged with it local communities and enterprises and support many initiatives to promote awareness and understanding of sustainability and the sustainable development goals.

**Fig. 5.** Geodesic dome using aquaponics

One such initiative is a partnership with a local social enterprise that has been established in the DCU Community Garden. The Grow Dome Project uses harvested rain water to grow salads and herb for commercial use.

## 5 Conclusions

Water, in particular potable water, is one of the greatest challenges facing global societies today. Academic Institutions are ideally positioned to educate and communicate the significance of the challenge faced and to implement and demonstrate the measures necessary to protect this essential natural resource. DCU has taken a holistic approach to this challenge, using its campuses as living labs to demonstrate new technologies, empowering and enabling staff and students to engage in the process. Operational excellence is key to developing and implementing solutions. DCU has demonstrated significant savings in water usage over the past number of year through developing and deploying best practice in the identification and elimination of leaks within the DCU water network. Furthermore

DCU has tested and implemented several waters saving technologies to reduce water consumption. However, without the engagement and contribution of the members of the DCU community, including all staff, students, researchers and visitors the impact of technological solutions will be limited.

DCU has focused significant efforts at educating and informing its community through compelling interactive initiatives. Embedding behavioural change is fundamental to scaling sustainable solutions to the broader community.

DCU seeks to demonstration leadership in addressing the grand challenge of our time by placing sustainability at the core of the institution as an exemplar for industry and society. Using the campuses as living laboratories, DCU is supporting and enabling the conception, development and implementation of viable solutions.

## References

1. Dublin City University, Available online at http://www.dcu.ie/sites/default/files/ marketing/ digitalmedia/presidents-office/strategic-plan/ index.html
2. W. E. Kirwan, Innovative Higher Education, 35, 101–111, (2010)
3. M. Ralph, W. Stubbs, Higher Education, 67, 71- 90. (2014)
4. L. Sharp, International Journal of Sustainability in Higher Education, 3(2), 128–145 (2002)
5. The independent survey was carried out by Ignite Research among 2011 people aged 16+ in ROI
6. Green-Campus, An Taisce Environmental Education Unit, Available online at http://www.greencampusireland.org/about/

https://doi.org/10.1051/e3sconf/20184801001

# Tra Vinh University and strategies heading to green campus

*Nguyen* Thi Mai Khanh*
Department for International Collaboration and Project Promotion, Tra Vinh University, Vietnam

**Abstract.** Tra Vinh University (TVU), the only university in the province for developing labour force in Tra Vinh. This article is to share the strategies that Tra Vinh University used to make the university become green. Being aware that environtment and climate change are issues should be concern as its impact to human life, Tra Vinh University identidied green campus is one of the mission that have to be done along with the development of the university. Different strategies have been applied and bring Tra Vinh University significant affect in terms of energy saving and green activities through the university curriculum, green projects and environment tracking in different aspesct. One of the environment issues that has been carried out frequently is waste water from daily activities and laboratories. The waste water is check physically and via database from the labs in different time of the year. The positive result from the test gives good signals to TVU's green target.

## 1 Introduction

Tra Vinh University (TVU) is located in the Mekong Delta and is the only university in Tra Vinh Province.Tra Vinh University began as Tra Vinh Community College in 2001 under the framework of the Vietnam-Canada Community College Project (VCCCP), which was funded by the Canadian International Development Agency (CIDA). The project implemented successfully the Canadian community college model in a Vietnamese context under the leadership of Dr. Pham Tiet Khanh, Rector of Tra Vinh University, and President of the Vietnam Association of Community Colleges (VACC).

TVU has the mission to provide with multi-level, multi-discipline and articulation training programs including certificate granting courses, short and long-term courses, undergraduate and postgraduate programs to all learners at all ages with special care for women, ethnic minority and people with special need. TVU responds flexibly to the long-life learning demand for learners; closely link to enterprises and employers in order to ensure skill-driven training and create the best employment opportunity for learners. TVU aims to become a typical university in professional educational system in Vietnam and success in international integration. Its motto is to "bring about quality learning opportunity for community. The vision of TVU is to become one of the best community-driven universities in not only the region, but also all of Vietnam.

As other university in the region, Tra Vinh University targets to develop the University into a Green University. Rector of Tra Vinh University has signed Talloires Declaration in 2013 as a commitment to implement the Green Project within the Campuses.

## 2 Strategies heading to green campus at Tra Vinh University

### 2.1 Significant green activities

With the target of developing a green university, Tra Vinh University has applied different strategies and organized many activities of which three main activities bring significant results since the signature of Talloires Declaration in 2013.

First of all, the green project for sustainable development in order to raise awareness of green ideas among students and staff. It was started by a project co-organized with a university in Australia, Swinburn University of Technology. The students from the two universities have been involved into variety of green activities under the framwork of the project. Also, four proposals on green campus were developed by the students with very good ideas including creating waste water treatment system.

**Fig. 1.** The students from Tra Vinh University and Swinburn University of Technology

* Corresponding author: ntmaikhanh@tvu.edu.vn

Secondly, water consumed reduction and energy saving are applied using automatic systems such as automatic taps and solar energy system for heating water at the residence. In fact, these systems help the university reduce the usage of water and electricity.

Finally, many measures applied to prevent the factors that cause damage to environment as dust and smoke, normal solid waste substance, dangerous solid waste substance, waste water and noise. The solid waste substances and waste water are collected and treated in a safe way to the environment. Especially, waste water treatment was conducted so that it does not harm the nature when it goes into the ground. The tested results of waste water can prove that the waste water on campus is safe. It is also tested by the tree planting in this source of water.

## 2.2 Waste water management at Tra Vinh University

The average volume of water using at Tra Vinh University is about 9.900 – 10.000m3/month mainly for daily use of the students, staff and for gardening.

The volume of waste water from daily activities is about 6500 – 7000 m3/ month. This waste water comes from washrooms of administration buildings, classrooms, medical clinic, labs and student residence. The waste water is collected and treated through two section septic tanks before going to the general drainage system on campus.

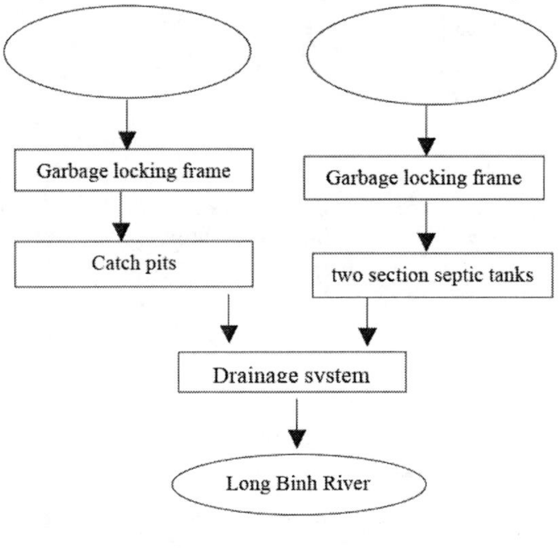


**Fig. 2.** Water collection and treatment on campus

Tra Vinh University has set up domestic waste water treatment system with the capacity of 560m3/day so that waste water from campus 1 and the student residence can be collected and completely treated.

In order to control the impact of waste water on the environment, the samples of waste water from different location on Campus are collected and test every three months especially the waste water from the labs. The test result was compare with the national technical regulation to check the level of effect and pollution so that the solution of control can be implemented immediately to reduce the contamination. National technical regulation on domestic waste water QCVN 14:2008/BTNMT, National Technical Regulation on Industrial waste water QCVN 14:2011/BTNMT, National Technical Regulation on Industrial Emission of Inorganic substances and dust QCVN 19:2009/BTNMT, National Technical Regulation on Industrial Emission of Organic Substance 20:2009/BTNMT, National Technical Regulation on Noise QCVN 05:2013/BTNMT are used as a guideline for keep the campus safe from water contamination.

## 2.3 Report on waste water controlled on campus:

The tested results from the laboratory are reported base on comparison with national technical regulation as in Table 1. (2017) and Table 2. (2016). This will help the university to check if the contamination limitation of waste water on Campus.

**Table 1.** Laboratory result of waste water from labs in 2017

| No | Tested items | Unit | Results | Allowed limit value |
|---|---|---|---|---|
| 1 | pH | - | 7.6 | 9.0 |
| 2 | Biochemical Oxygen Demand ($BOD_5$) | mg/L | 15 | 60 |
| 3 | Chemical Oxygen Demand (COD) | mg/L | 24 | 180 |
| 4 | Turbidity & suspendid solids (TSS) | mg/L | 38 | 120 |
| 5 | Nitrogen (N) | mg/L | <2.3 | 48 |
| 6 | Phosphorus (P) | mg/L | 0.6 | 7.2 |
| 7 | Cadimi (Cd) | mg/L | <0.002 | 0.12 |
| 8 | Chromium (III) ($Cr^{3+}$) | mg/L | <0.03 | 1.2 |
| 9 | Chromium (VI) ($Cr^{6+}$) | mg/L | <0.03 | 0.12 |
| 10 | Manganese (Mn) | mg/L | <0.025 | 1.2 |
| 11 | Iron (Fe) | mg/L | 0.5 | 6.0 |
| 12 | Zinc (Zn) | mg/L | 0.14 | 3.6 |

 https://doi.org/10.1051/e3sconf/20184805006

| | | | | |
|---|---|---|---|---|
| 13 | Hg | mg/L | <0.0005 | 0.012 |
| 14 | Pb | mg/L | <0.005 | 0.6 |
| 15 | Ni | mg/L | <0.1 | 0.6 |
| 16 | Cu | mg/L | <0.05 | 2.4 |

**Table 2**. Laboratory result of waste water from labs in 2016

| No | Tested items | Unit | Results | Allowed limit value |
|---|---|---|---|---|
| 1 | pH | - | 7.72 | 9.0 |
| 2 | Biochemical Oxygen Demand (BOD5) | mg/L | 23.9 | 60 |
| 3 | Chemical Oxygen Demand (COD) | mg/L | 103 | 180 |
| 4 | Turbidity & suspendid solids (TSS) | mg/L | 25 | 120 |
| 5 | Nitrogen (N) | mg/L | 7.01 | 48 |
| 6 | Phosphorus (P) | mg/L | 0.14 | 7.2 |
| 7 | Cadimi (Cd) | mg/L | <0.002 | 0.12 |
| 8 | Chromium (III) (Cr3+) | mg/L | <0.03 | 1.2 |
| 9 | Chromium (VI) (Cr6+) | mg/L | <0.03 | 0.12 |
| 10 | Manganese (Mn) | mg/L | <0.025 | 1.2 |
| 11 | Iron (Fe) | mg/L | 0.2 | 6.0 |
| 12 | Zinc (Zn) | mg/L | 0.14 | 3.6 |
| 13 | Hg | mg/L | <0.0002 | 0.012 |
| 14 | Pb | mg/L | <0.12 | 0.6 |
| 15 | Ni | mg/L | <0.12 | 0.6 |
| 16 | Cu | mg/L | <0.06 | 2.4 |

## 3 Summary

The results from the two tables are for the waste water from the labs where the level of pollution is at the highest rate [1, 2]. It is clearly that half of the tested items has significant reduction of contamination and under controlled from year to year. The others can be seen increased but not much and also under the allowed limitation of the national technical regulation. This is a good signal for waste water treatment at Tra Vinh University in developing into a green university.

In the coming years, more international sustainable development programs for students should be implemented in order to raise awareness of environment to the young people. Also, more research should be done to find out more sources of renewable energy to reduce pollutants.

## References

1. Tra Vinh University, *Report on Environment Observation of Tra Vinh University 2016* (2016)
2. Tra Vinh University, *Report on Environment Observation of Tra Vinh University 2017* (2017)

# How universities can work together with local communities to create a green, sustainable future

*Yuhlong* Oliver Su[1*], *Ku-Fan* Chen[2,3], *Yung-Pin* Tsai[2,3], and *Hui-I* Su[3]

[1]Office of President, National Chi Nan University, 1 University Road, Puli, Nantou 54561, Taiwan, ROC
[2]Department of Civil Engineering, National Chi Nan University, 1 University Road, Puli, Nantou 54561, Taiwan, ROC
[3]Environmental Protection and Occupational Safety and Health Center, National Chi Nan University, 1 University Road, Puli, Nantou 54561, Taiwan, ROC

**Abstract.** Over the past few years, National Chi Nan University (NCNU) has made significant achievements in its development as a "green university." Besides continuing to implement sustainable development of its campus, NCNU has also been implementing the "Citizenship Cultivation Plan," "University without Boundary Project" and "University Social Responsibility (USR) Plan," using the local community as a test-bed, and working actively to integrate NCNU's curriculum with the community, so as to enhance awareness of local environmental issues among students, faculty members and local residents, and build consensus of sustainable environmental development. NCNU has implemented a range of action plans and practical courses of environmental education to work together with local residents for environmental issues, such as water resources, air quality, waste, etc. Through these activities, NCNU's students and faculty members and local residents are able to develop a better understanding of the environment, and how they can make a bigger contribution towards environmental protection. Regarding the international voluntary service activities, NCNU has endeavored to fulfil its social responsibility as a university. Under the dynamic leadership of President Dr. Yuh-Long Su, College of Science and Technology and College of Management have joined forces with "Chunghwa International Sunshine Charitable Association" established by Taiwanese businesspeople in Southeast Asia, using NCNU's knowhow and technology to build the water purification systems in Cambodia, providing clean and safe drinking water for people in remote and disadvantaged areas. NCNU has strongly rooted in the local community with the content of the United Nations' Sustainable Development Goals (SDGs) to build capabilities in four key areas – educational development, local action, industry-academia collaboration, and global integration – and to expand the scope of its activities from the local to the international. In the future, NCNU will continue devoting itself to sustainability and environmental issues, working wholeheartedly to meet its social responsibility goals as a university.

## 1 A Vision of Transforming National Chi Nan University into a Green University

To effectively implement the vision of being a green university, National Chi Nan University (NCNU) has adopted a campus plan which incorporates ecological principles and environmental protection measures into the university's operations management, thus fostering an ecologically diverse campus and a more sustainable living environment [1] [2]. Academic and administrative units work together to encourage relevant research, club-based activities, and promotional activities and combine these efforts with three major programs (civic development, university without boundary, and university social responsibility (USR)) to enhance the scope and comprehensiveness of courses that involve sustainable development education. In addition to encouraging faculty and students to get involved in the community, the concepts of sustainable development and environmental protection are integrated into the university's core educational mission to construct a learning environment which integrates pedagogical knowledge and community life and provides opportunities for practical, hands-on participation [3] [4]. By focusing on potential problems and needs within the community and bringing attention to production, lifestyle, and ecological issues that impact local residents, we can more readily understand the tenets of sustainable development. To date, NCNU students and faculty have established a long-term presence in six communities to carry out local engagement and develop action plans for issues that impact the local community, such as eco-towns, social welfare, green economy, cultural revitalization, and civic engagement. Students also team up with local residents to find and implement solutions to sustainability issues, gathering new momentum for community development. NCNU's vision for achieving sustainable development also combines local characteristics with the sustainable development goals (SDGs) formulated by the United Nations [5]. This vision also considers social welfare service providers, friendly environment participants, sustainable

* Corresponding author: yosu@ncnu.edu.tw

 https://doi.org/10.1051/e3sconf/20184806001

development educators, community development leaders, and corporate social responsibility achievers to be the main pillars of sustainable development planning and implementation, as illustrated in (Fig; 1. NCNU has developed a four-pronged approach consisting of educational development, local action, industry-academia cooperation, and global integration (Fig 2) and is continuing to make a concerted effort on issues relating to environmental sustainability in order to help prepare students to become modern citizens who have the wherewithal to care for society and gainfully participate in public life.

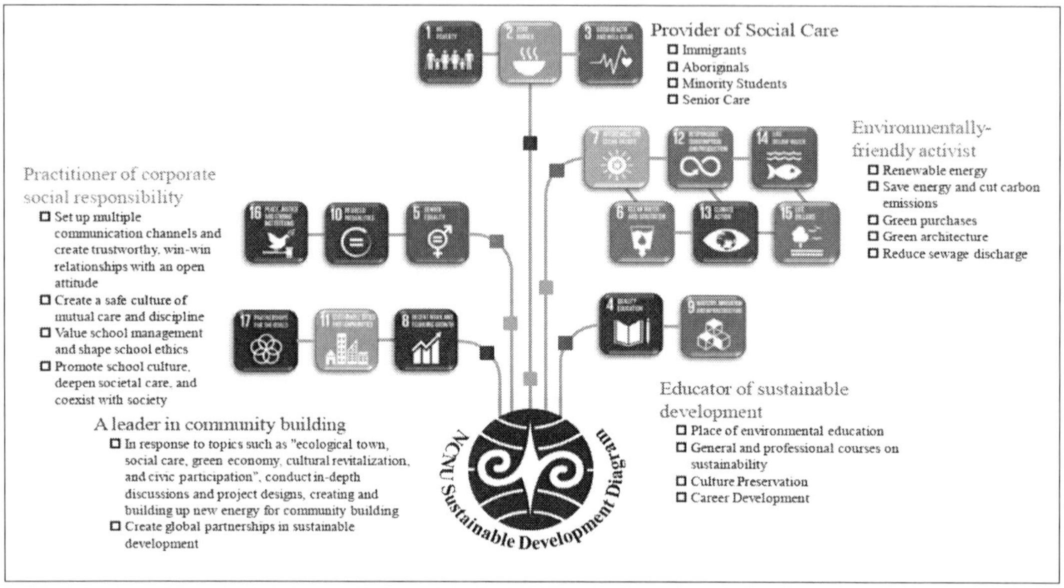


**Fig. 1.** Diagram showcasing NCNU's vision for sustainable development.

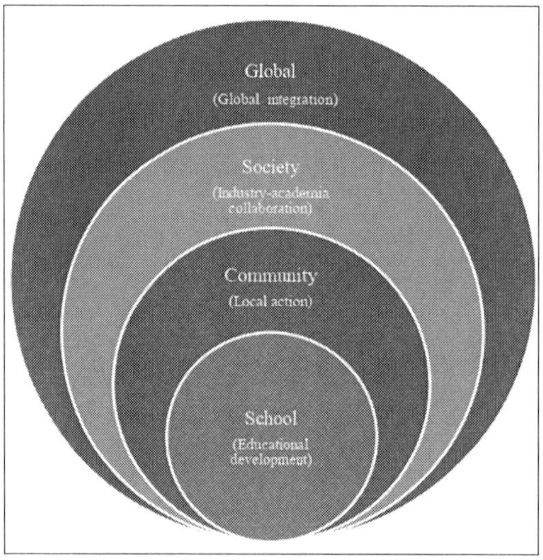


**Fig. 2.** Blueprint of NCNU's sustainable development from local to global implementation.

## 2 Cultivating civic awareness through an open and engaging university

NCNU strives to instill civic consciousness in students by enabling the university's faculty and students to partner up with local residents. NCNU applies the university's knowledge and specialized competencies in domains ranging from ecology and production to life and culture, lending assistance on local governance and facilitating development which is mutually beneficial to the university and local community, as well as taking time to reflect on progress made during the implementation process to devise more diverse learning pathways for the future, as illustrated in (Fig. 3).

https://doi.org/10.1051/e3sconf/20184806001

Methods for implementing the civic development program are contingent on the following: 1. Collaborative learning and sharing: Aims to form a collaborative learning environment and atmosphere for students, faculty, and local residents in a way that facilitates sharing and experience exchange among participants, thereby promoting action and continuous advancement. 2. Community outreach: The university teams up with the community to raise questions, discuss opinions, and work together to conceptualize objectives and content of action plans, integrate resources, and delegate tasks. 3. Curriculum innovation: University faculty are encouraged to utilize diverse educational methods and place an emphasis on the learning content and learning quality of students by making reforms and adjustments to the curriculum. Experience sharing through teacher networks enables educators to learn from each other and motivate one another. 4. Student support system: Students are accompanied and encouraged to form interdisciplinary teams and examine real issues in life, propose solutions, hold practical consultations, and exchange opinions and feedback. 5. Living labs: Students, faculty, and people in the local community are inspired to have greater awareness of real issues and implement five strategies, including experimentation, which apply social innovation perspectives in the real world.

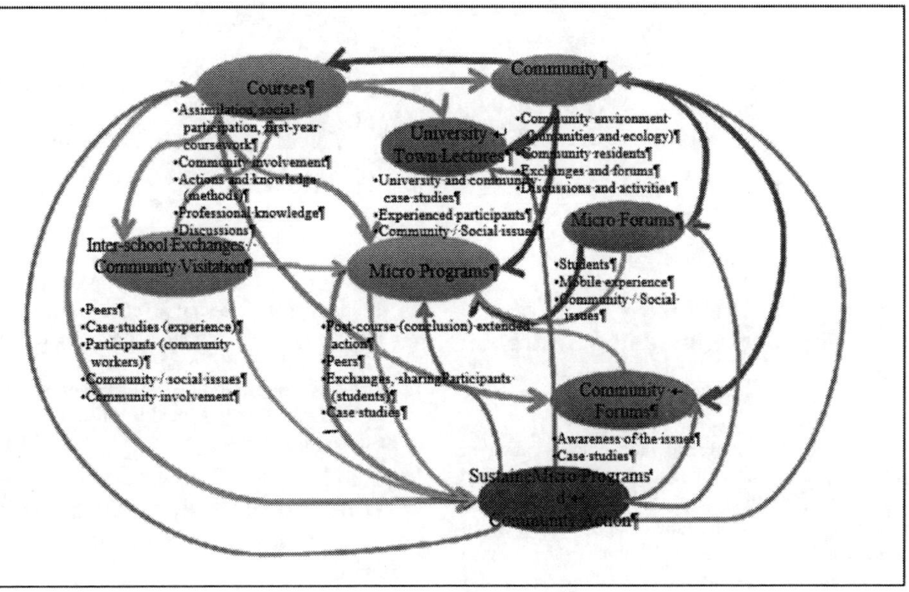


**Fig. 3**. Participation-based learning environment avenues of diverse learning.

## 3 Developing new talent by breaking down barriers to learning

NCNU utilizes four approaches, namely collaborative learning, collaborative education, local symbiosis, and collaborative creation, and achieves implementation through various programs. Various action plans allow these four approaches to be mutually supportive of each other, as illustrated in (Fig. 4). Various learning resources from within and without the university are linked together, including diverse types of educational labs, workshops, learning sites, and regional collaborative locations both inside and outside of the university. Meanwhile, integrate relevant resources within the university relating to social participation, expand and elevate organizational competencies, and jointly develop and cultivate young talent with practical abilities who can remain or return home.

The university without boundary program aims to achieve four goals: 1. Help educated youths cultivate the ability to return home and transform learning methods in a way that enables the university to become an incubator for developing talent who pay attention to urban and rural issues and strive for innovative development. 2. Foster a collaborative learning and working environment for NCNU and the local community, provide systematic support to methodically expand and develop the learning field, allow the university to forge closer ties with the local community, and achieve synergy between university and local knowledge in a way that is mutually useful and supportive. 3. Design a "learning without barriers" model for NCNU and the local community which includes promotional activities, invention showcases, interactive platforms, and other approaches to break away from formulaic curriculum barriers relating to space, time, subjects, and relationships and overhaul existing learning methods. 4. Facilitate change in NCNU's learning ecosystem via cross-unit and cross-level integration to establish an institutional support framework that provides an impetus for changing interdisciplinary collaboration, thereby progressively transforming the university's learning ecosystem.

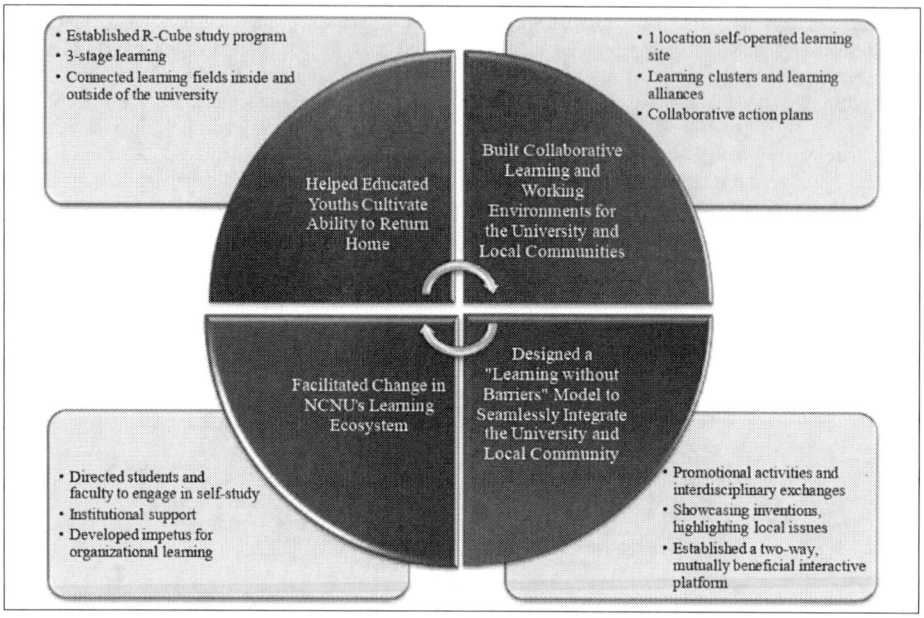


**Fig. 4.** Transforming NCNU's learning ecosystem.

## 4 Fulfilling university social responsibility and increasing outreach experience

NCNU fulfills university social responsibility [6], allows faculty and students to embrace their vitality, aids in regional development, solve difficult local problems, and utilize technology to effectively promote the welfare of the general public. One of the most eagerly talked about achievements is the "$PM_{2.5}$ Team", which is comprised of NCNU students and faculty who developed a "$PM_{2.5}$ micro detector". The device is capable of transmitting monitoring data to the university's cloud platform via Wi-Fi and Bluetooth. The records, calculations, and analysis derived from the data allow changes in air quality in Puli Township to be ascertained with high precision. Data displayed by the monitoring device can generate greater alertness and gradually change public habits, thus resulting in self-perpetuating education. Currently, NCNU students and faculty are also providing assistance to Taumi Eco-village, located nearby the university, to construct artificial wetlands for treating wastewater discharged by the site's guest houses as well as to install smart utility meters which will help meet the energy-saving needs of the guest houses.

Over the years, NCNU has made substantial accomplishments owing to its participation in community development. The university has established an interdisciplinary teacher network for holding practical discussions which is staffed by 47 local practice faculty members (accounting for approximately 1/6 of the university's total faculty), and each academic year 1,222 students (accounting for about 1/5 of the total student body) participate in social practice courses, including the development of 44 research topics pertaining to social practice and collaboration with 7 government departments, and a total of 39 student and faculty clubs have also taken action locally, the effects of which are illustrated in Fig. 5.

## 5 Thinking globally and acting locally to connect local communities with the world

As the only national public university in Nantou County, NCNU is playing an important role in providing assistance on regional development in Nantou County, thereby fulfilling its function as a local storehouse of knowledge and promoting development through "university-led local collaborative governance" and "locally-led university ecosystem innovation". NCNU endeavors to fulfill its university social responsibility by providing services on that impact other parts of the world. Under the leadership of NCNU President Dr. Yuh-Long Su, a global volunteer team created by the university's College of Science and Technology and the College of Management has teamed up with Taiwanese businesses in Cambodia to provide knowledge and technology to build water treatment systems in rural communities that provide a safe source of drinking water. In the future, NCNU will also engage in global crowdfunding efforts to build more water treatment facilities that benefit rural residents in Cambodia, highlighting that Taiwan is committed to doing its part as a global citizen to help others.

 https://doi.org/10.1051/e3sconf/20184806001

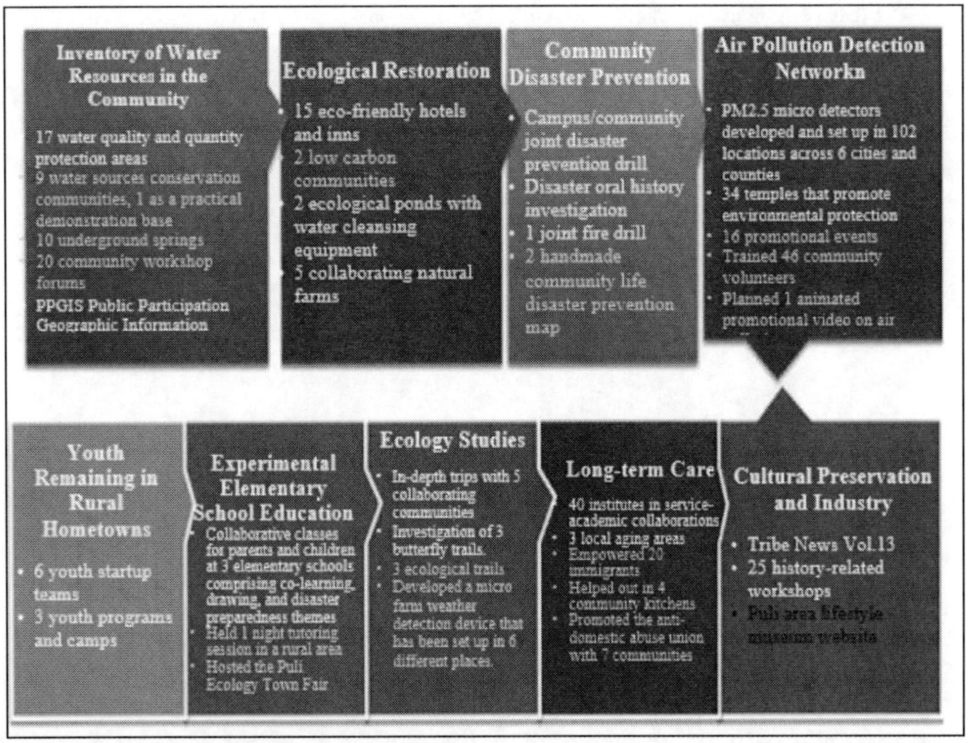


**Fig. 5.** Illustration showing involvement of National Chi Nan University in social practice projects and their effects from 2013 to the present.

**Fig. 6.** By lending assistance to people in Cambodia, NCNU is fulfilling its university social responsibility (USR).

## 6 Conclusion

NCNU is committed to fulfilling its university social responsibility and thoroughly incorporating the concept of environmental sustainability into coursework and campus life. NCNU will furthermore continue to embrace a vision of making positive impacts on local elementary and junior high schools, local communities, and society at large while becoming increasingly connected with the world. As an academic institution, NCNU also plays a part in ensuring the stable operation of society, the economy, and protecting the environment of our communities, and thus has an obligation to lead society toward sustainable development. With respect to environmental issues, humanity is at a crossroads, and environmental education is paramount; this is why NCNU developed a four-pronged approach — educational development, local action, industry-academia cooperation, and global integration — based on the school's local characteristics and the UN's sustainable development goals (SDGs). Going forward, NCNU will continue to think globally and act locally, make a concerted effort on issues relating to environmental sustainability, and strive to fulfill its duties as a socially responsible university.

## References

1. Su, Y.O., Liu, I.C., Chen, K.F., Chang, Y.T., *Green NCNU in The Heart of Taiwan*. The 3rd International Workshop on UI GreenMetric, April 10-11, Istanbul, Turkey (2017)
2. Lauder, A., Sari, R.F., Suwartha, N., Tjahjono, G., 2015. Critical Review of a Global Campus Sustainability Ranking: GreenMetric. *Journal of Cleaner Production*, **108**, Part A, pp. 852-863.
3. Suwartha, N., Sari, R.F., Evaluating UI GreenMetric as a Tool to Support Green Universities Development: Assessment of the Year 2011 Ranking, **61**, pp. 46-53 (2013)
4. Delanty, G., *Challenging Knowledge: The University in the Knowledge Society*. Bukingham: Open University Press., (2001)

5. UNESCO (United Nations Educational, Scientific and Cultural Organization), *Education for Sustainable Development Goals: Learning Objectives*. ISBN 978-92-3-100209-0, Paris, France (2017)
6. López, S.G., Benítez, J.L.S., Sánchez, J.M.A., *Social Knowledge Management from the Social Responsibility of the University for the Promotion of Sustainable Development*. Procedia- Social and Behavioral Sciences, **191**, pp. 2112-2116 (2015)

# Accelerating the transformation to a green university: University of Bahrain experience

*Riyad* Y. Hamzah, *Naser* W. Alnaser, *Waheeb* E. Alnaser

University of Bahrain, P.O. Box 32038, Isa Town, Kingdom of Bahrain

**Abstract.** Many universities are striving to have an environmental impact on the society as they are considered as small communities aiming to be eco-friendly and having low $CO_2$ emission. This concept has been emerging after the worldwide concern on the ozone depletion issue and global warming. As a result, many titles have appeared like "Green Universities", "eco- friendly Universities," "Environmental sustainable Universities" and "Environmentally responsible universities," etc. This paper proposes a mechanism that allows universities to go green or become environmentally sustainable higher education bodies in a short span of time. It simply advises the universities to best practice the Environmental Sustainable Development Goals (ESDG's) Components incorporated in the seventeen United Nation Sustainable Developments Goals (SDG's) that were announced on the 25 th September 2015 by all leaders of the countries which were aimed to end poverty, protect the planet and ensure prosperity for all as part of a new sustainable development agenda where each goal has specific targets to be achieved over the next 15 years. These ESDG's listed within the SDG's are the following: 1) Good Health and Well-being. 2) Quality Education. 3) Clean Water and Sanitation. 4) Affordable and Clean Energy. 5) Industry, Innovation and Infrastructure. 6) Sustainable Cities and Communities. 7) Climate Action. 8) Life below Water. 9) Life on Land. Therefore, incorporating such 9 Goals in the strategic planning of each worldwide university that has aligned its goals with the Country National Strategy - which by default includes these SDG's - will accelerate and boost each university to transform to Green and Environmentally Sustainable campus. The paper also sheds light on the experience of University of Bahrain in this respect.

## 1 Introduction

Countries, including their governments, institutions, industries (private and local), universities and even its schools have become very much concerned with the sustainable developments in general and environmental sustainable developments after they had all come to believe that striving towards sustainability will assure the continuation of living in peace and environmentally hazardous environments. Therefore, pinpointing these "Sustainable Development Global" or "Global Goals" and committing to adopting them by each country will help in protecting the planet and assure enjoying peace and prosperity by its inhabitants. The SDGs came into effect in January 2016 on the successes of the Millennium Development Goals (MDG), which started a global effort in 2000 to mainly reduce income poverty, provide much needed access to water and sanitation, drive down child mortality and drastically improve maternal health. The SDG's were born at the United Nations Conference on Sustainable Development in Rio de Janeiro, Brazil, in 2012. The objective was to produce a set of universal goals that meet the urgent environmental, political and economic challenges facing our world.

According to UN [1] these 17 Goals build the SDGs work in the spirit of partnership and pragmatism to make the right choices to improve life, in a sustainable way, for future the generations of any nation). The broad goals are interrelated though each and has its own targets to achieve with total of 169 targets. SDGs cover a broad range of social, environmental and economic development issues and because they are very vital they were known as "Transforming our World: the 2030 Agenda for Sustainable Development" or 2030 Agenda. SDG's provide clear guidelines and targets for each country to adopt in accordance with their own priorities and the environmental challenges of the world at large, i.e. setting these SDG's in each national strategic plan. UNDP is uniquely placed to help such countries to implement these Goals through its work in some 170 countries and territories. These 17 Goals are the followings:

1) No Poverty. 2) Zero Hunger. 3) Good Health and Well-Being for people. 4) Quality Education. 5) Gender Equality. 6) Clean Water and Sanitation. 7) Affordable and Clean Energy. 8) Decent Work and Economic Growth. 9) Industry, Innovation and Infrastructure. 10) Reduced Inequalities. 11) Sustainable Cities and Communities. 12) Responsible Consumption and Production. 13) Climate Action. 14) Life Below Water, 15) Life on Land. 16) Peace, Justice and Strong Institutions. 17) Partnerships for the Goals.

## 2 Discussion

Universities and industries are more concerned to the Environmental Sustainable Developments Goals (ESDG) Components incorporated in these 17 Sustainable Developments Goals (SDG's), these are:

1) Good Health and Well-being (Goal 3).
2) Quality Education (Goal 4).
3) Clean Water and Sanitation (Goal 6).
4) Affordable and Clean Energy (Goal 7).
5) Industry, Innovation and Infrastructure (Goal 9).
6) Sustainable Cities and Communities (Goal 11).
7) Climate Action (Goal 13).
8) Life below Water (Goal 14).
9) Life on Land (Goal 15).

Ranking bodies for Environmental Sustainable Universities or Green Universities should include in their criterion these listed 9 goals, as much as possible in a measurable scale or indicators. These will make them faster to be achieved since the mother of these universities are committed to adopt the SDG's and accordingly the ESDG's- which already set in their Strategic Plan (or Government Working Action Plan) which is already budgeted and will receive favorable support.

According to our modest analysis, the UI Greenmetrics for Green Universities ranking which is established by University of Indonesia in 2010 [2] looks like that it has benefited a lot from such ESDG, which includes the followings main criterion:

1. Setting and Infrastructure (SI), which is related to Goal 5 and 6 in our ESDS (or Goal 9 and 11 in SDG's).
2. Energy and Climate Change (EC), which is related Goal 4 and Goal 7 in our ESDS (or Goal 7 and Goal 13 in SDG's).
3. Waste (WS), which is related Goal 1 and 8 in our ESDS (or Goal 3 and Goal 14 in GSD's).
4. Water (WR), which is related Goal 3 in our ESDS (or Goal 6 in SDG's).
5. Transportation (TR), which is related Goal 7 and 9 in our ESDS (or Goal 13 and 15 in GSD's).
6. Education (ED), which is related Goal 2 in our ESDS (or Goal 4 in SDG's).

To clarify and support such conclusion on the matching of UI Greenmetric and ESDG one can refer to components of such Environmental Sustainable Development Goals [3]. This is explained as follows:

**I. Criteria number 1** in UI Greenmetric (Setting and Infrastructure), which is classified to match with Goal 5 and 6 in ESDG (or Goals 9 and 11 in SDG's), is explained as follows:

A. Goal 5 in ESDG (or Goal 9 in SDG's) is concerned with building resilient infrastructure, promote inclusive and sustainable industrialization and foster innovation. Industry plays a critical role in innovation and research, which are crucial for job creation, poverty eradication, gender equality, labour standards, and greater access to education and health care. Together, it will lead to promote inclusive and sustainable industrialization (higher education campus) and technology development.

B. Goal 6 in ESDG (or Goal 11 in SDG's) is concerned with making cities (University Campus) and human settlements inclusive (students and staff), safe, resilient and sustainable. This is because rapid urbanization puts pressure on supplies of fresh water, sewage systems, the living environment and public health. Embracing in the technological and social benefits of cities (universities) by making sure they are safe for everyone (students and staff) will lead to a sustainable campus valid for decades to come.

**II. Criteria number 2** in UI Greenmetric (Energy and Climate Change), which is classified to match with Goal 4 and Goal 7 in ESDG (or Goal 7 and Goal 13 in SDG's), is explained as follows:

A. Goal 4 in ESDG (or Goal 7 in SDG's) is concerned with ensuring access to affordable, reliable, sustainable and modern energy for all. As is explained [3] clean, sustainable energy is not just about the environment. In fact, there are lots of death cases resulting from gas emissions during energy production and operation.

B. Goal 7 in ESDG (or Goal 13 in SDG's) is concerned with taking urgent action to combat climate change and its impacts. Climate change has the potential to derail other efforts toward sustainable development by altering weather patterns that threaten the food production and increasing sea levels which will displace coastal communities (universities). This compels universities to increase their awareness and convey urgency to the public that they are combating climate change.

**III.Criteria number 3** in UI Greenmetric (Waste), which is classified to match with Goal 1 and Goal 8 in ESDG (or Goal 3 and Goal 14 in SDG's), is explained as follows:

A. Goal 1 in our ESDG (or Goal 3 in SDG's) is concerned with ensuring healthy lives and promoting well-being for all at all ages (particularly young students as they are the power of the future), preventable disease (from pollution or bad interior air quality in lecture room and laboratories or from building materials and furniture), untreated drug & alcohol abuse, preventable birth defects, avoidable traffic and industrial accidents which will lead to long and healthy life.

B. Goal 8 in ESDG (or Goal 14 in SDG's) is concerned with conservation and sustainable use of the oceans, seas, and marine resources for sustainable development. In fact, a lot of waste (liquid or sometimes solid waste) has ended up in the seas or oceans. Oceans and seas are being threatened and destroyed by human (students and staff) activities like marine pollution, overfishing, and destruction of marine habitats.

https://doi.org/10.1051/e3sconf/20184806002

**IV. Criteria number 4** in UI Greenmetric (Water), which is classified to match Goal 3 in our ESDS (or Goal 6 in SDG's) is explained as follows:

A. Goal 3 in our ESDS (or Goal 6 in SDG's) is concerned with ensuring availability and sustainable management of water and sanitation for all. It is vital to have an access to clean water at home (or universities) therefore it must be conserved and treated or reuse safely for irrigation. Diseases caused by contaminated water kill more people every year than all forms of violence, including war. By prioritizing clean water, universities can improve the health and livelihoods of its students.

**V. Criteria number 5** in UI Greenmetric (Transportation), which is classified to match Goal 7 and Goal 9 in our ESDS (or Goal 13 and Goal 15 in SDG's), is explained as follows:

A. Goal 7 in ESDG (or Goal 13 in SDG's) is concerned with taking urgent action to combat climate change and its impacts. One of the most damaging issues to the environment is gas emissions from transportation, especially if students use their own car in frequent going and coming trips more than two or three times a day! The world's industrialised nations have changed the balance of the earth's carbon cycle over the last 150 years by burning large amounts of fossil fuels. Climate change has been proving and has been witnessed to affect the climate by altering weather patterns that threaten our food production and increasing sea levels which will displace coastal communities.

B. Goal 9 in our ESDS (or Goal 15 in SDG's) is concerned with protecting, restoring and promoting sustainable use of terrestrial ecosystems, sustainably manage forests, combat desertification, and halt and reverse land degradation and halt biodiversity loss. In fact, nearly all these elements are vulnerable due emissions from emission of transport gases [3].

**VI. Criteria number 6** in UI Greenmetric (Education), which is classified to match Goal 2 in our ESDS (or Goal 4 in SDG's) is explained as follows:

A. Goal 2 in our ESDS (or Goal 4 in SDG's) is concerned with ensuring inclusive and equitable quality education, especially in teaching courses concerned with the environment and sustainability based on critical thinking, experimentation, research and contractual studies. Strong and quality education must be always being promoted and be provided lifelong to allow efficient learning opportunities for all. As the world's population grows, more resources and policies are needed to make sure that students at the universities (and everywhere) get a good education. According to UN [4], the world needs two million teachers and four million new classrooms to make sure every student can get an education. Full access to quality education (especially in environmental issues) is the first step to achieving sustainable development, poverty eradication, gender equality and women's empowerment. Let's make the sound investment in quality education by ensuring that primary and secondary schools are free for every child by 2030.

The University of Bahrain, for example, has aligned its "Green Initiatives" in its own strategic plan (2015 - 2018) - which has 7 Goals, 30 Initiatives, and 29 Key Performance Indicators. This strategy is aligned (directly and indirectly) with the nine Environmental Sustainable Development Goals (ESD's) which are embedded in the seventeen Sustainable Development Goals (SDG's) where the National Strategy or the Government Working Plan, approved by the National Council (Parliament and Consultative Council) has made the University receive good rank and points , from the first entry in 2016 i.e. ranked 242 out of 516 worldwide universities with a score of 4,475 while in 2017 its rank was 307 out of 619 worldwide universities with a total score of 4,457. The backward move with a lower score in 2017 was attributed by UI Greenmetric management as due to inaccuracy in their scoring in 2016 in the education although in 2017 we had introduced an Internationally unique PhD program in Environmental Sustainable Developments (with partnership with regional UN offices in Bahrain) and had upgraded our 15-year-old MSc in Environmental Sustainable Development (and partnership with UNESCO chairs) had also won about USD 1 million as contractual research in renewable energy for water production and PV panels testing and accordingly improving. Comparison of the performance of these two years in UI Greenmetric ranking is as follow:

1. In 2016, ranking in Education was 180 globally (813 points) and in 2017 the score is 638!
2. In 2016 ranking in Transportation was 281 globally (773 points) and in 2017 the score was 813!
3. In 2016 ranking in water is 287 globally (340 points) and in 2017 the score was 340.
4. In 2016 ranking in Waste is 279 globally (999 points) and in 2017 the score was 1023.
5. In 2016 ranking in Energy and climate is 362 globally (581points) and in 2017 the score was 638!
6. In 2016 ranking in Setting and Infrastructure was 362 globally (969 points) and in 2017 the score was 978.

This comparison clearly supports the suggestion of taking advantage of the existing UN ESDG's and SDG's in accelerating the Green practice in the universities.

Much research has been conducted seeking mechanisms that enhance the dissemination and the practice of Green or Environmental Sustainable Campus [5-14]. The reason is universities can be considered as "small cities" which may have heavy impacts on the environment due to their activities, and movement of goods and persons inside campuses [16]. Universities can be seen as complex buildings, in terms of waste generation, transportation, water and materials consumption, energy and electricity consumption, given the scientific, social and educational activities that take place within their boundaries [15]. Therefore, it is strongly believed that taking the ESD's embedded in SDG's set by United Nations will assist universities to be "Greener" and "Environmentally Sustainable" universities.

## 3 Conclusion

Aligning the Green Initiatives in certain universities in their own strategic plans which are aligned with the nine Environmental Sustainable Development Goals (ESD's) embedded in the seventeen Sustainable Development Goals (SDG's) will boost and accelerate such universities to be Green or Environmentally Sustainable because such SDG's are already mandated, approved and budgeted in the Government Strategic Work Plan which all legislation parties like the National Council (Parliament and Consultative Council) or Senate have already approved and embraced.

## References

1. United Nations, available online at http://www.undp.org/content/undp/en/home/sustainable-development-goals.html
2. UI GreenMetric World University Rankings, available online at http://greenmetric.ui.ac.id/wp-content/uploads/2015/07/UI-Greenmetric-Guideline-2016.pdf
3. Trombulak, S.C. and Christopher, A. F (2000) Review of Ecological Effects of Roads on Terrestrial and Aquatic Communities, *Conservation Biology*, **14**, issue 1, pp18-30.
4. https://www.globalgiving.org/sdg/?rf=ggad_15&gclid=CjwKCAiA24PVBRBvEiwAyBxf-T-GAFIhsznA2nCOAEbqAcNztXxd2-bKxstZLMsP1yr6jxETOXtjVhoCjXwQAvD_BwE
5. Hajrasouliha, A. (2017) Campus score: Measuring university campus qualities, *Landscape and Unrban Planning*, **158**, pp166-176.
6. Geng, Y, Liu, K., Xue, B., Fujiita, T. (2013) Creating a "green university" in China: a case of Shenyang University, Journal of Cleaner Production, **61**, pp13-19.
7. Ragazzi, M and Ghidini, F. (2017) Environmental sustainability of universities: critical analysis of a green ranking, *Energy Procedia*, **119**, pp 111–120.
8. Azar, E and Al Ansari, H. (2017) Framework to investigate energy conservation motivation and actions of building occupants: The case of a green campus in Abu Dhabi, UAE, *Applied Energy*, **16**, pp 563-573.
9. Tiyarattanachai, R. and Hollmann, N. M. (2016) Green Campus initiative and its impacts on quality of life of stakeholders in Green and Non-Green Campus universities, *SpringerPlus*, **5**:84, DOI 10.1186/s40064-016-1697-4.
10. Kingston, K., Cassil, B. and Wilson, B. (2017) Inspiring a sustainable culture at a school or university, *American School & University*, Asumag.com, pp 24-26.
11. Chalfoun, N. (2014) Greening University Campus Buildings to Reduce Consumption and Emission While Fostering Hands-on Inquiry-Based Education, *Procedia Environmental Sciences*, Vol 2, pp 288-297.
12. Subramaniam, D., Sulaiman, H., Saleh, Al Omar, W., Salim, M (2016) Institutionalize waste minimization governance towards campus sustainability: A case study of Green Office initiatives in Universiti Teknologi Malaysia, *Journal of Cleaner Production*, **135**, pp 1407-1422.
13. Patel, B. and Patel, P. (2012) Sustainable campus of Claris life sciences through green initiatives, *Renewable and Sustainable Energy Reviews*, **16**, pp 4901-4907.
14. Massimo, D.E., Fragomeni, C., Malerba, A. and Musolino, M. (2016) Valuation Supports Green University: case Action at Mediterranea Campus in Reggio Calabria, *Procedia- Social and Behavioral Sciences*, **223**, pp 17-24.
15. Alshuwaikhat, H.M. and Abubakar, I. (2008). An integrated approach to achieving campus sustainability: assessment of the current campus environmental management practices. *J Clean Prod*, **16**, pp 1777-1785.

# The internalization of conservation mindset in disruptive era

*Fathur* Rokhman*, *Amin* Retnoningsih and *Hendi* Pratama

Universitas Negeri Semarang, Indonesia

**Abstract.** Environmental conservation is not all about physical preservation of our surroundings but also the development of the human components of the conservation. Only with the right mindset, conservation can be maintained and, in the end, sustainable. Conservation awareness has been a common theme for academics and environmental activists but it is not always the case for the common masses. This paper focuses on illustrating the challenges and strategies of planting the appropriate mindset of conservation to young generation in Indonesia. Most of examples and cases provided in this paper are taken from our experience in managing the university we currently work for.

## 1 Introduction

Humans are changing the face and the composition of the planet but they are rarely aware of that fact. Most of the times, a single human think that what he/she does can only affect his/her local surrounding. In accumulation, human population accrue their 'local' destruction of the environment to the global level. Take farmers and palm industry in Kalimantan as an example. Every farmer thinks that burning a little bit of forest will not harm the planet and a single industry can only take a few acres of lands. However, when the deeds have been conducted en masse and going for some years the following effects have shown up [1].

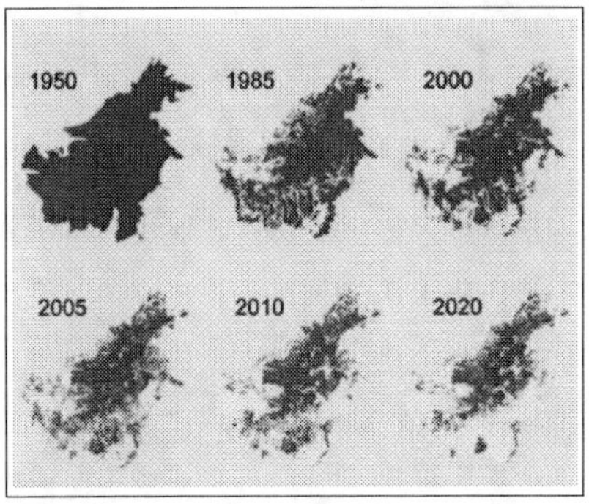


**Fig. 1.** Satellites images of Kalimantan's forests over the year

That is the first challenge among many that we have to tackle as human beings. As a rule if thumbs, humans are not able to behave and act responsibly in global level. People will not burn their fuel tank because of obvious reason that they can predict the immediate effect of their action. However, people will not suddenly stop to buy and drive cars even if they know that their cars contribute to global warming via their exhaust gases. This is because people are not able to fathom effects which have no immediate context with their actions. Kahneman (2011) states that this phenomenon is caused by evolutionary process of our brain [6]. Homo Sapiens' brain has been hardwired effectively to detect danger of predators in the past. Human brain has been trained to decide immediate effect of a response via sensors in their body and create fight or flight reaction. The brain has remained that way for thousand years and it will not change radically in the near future.

The second challenge of conservation the pressure of economic growth for the environment. The main cause of this phenomenon is the fact that liberal approach to happiness has been accepted as a global norm. Everyone has the right to pursue their happiness via possessions and accumulation of wealth. Majority of human population has agreed to the mechanism of 'better economy for better life'. However, economic growth itself is not without shortcomings. For every incremental change of our economy, there are natural resources need to be sacrificed. People cut trees for houses. People dig the earth to extract metals for car making. Every single economical gain shall hurt the environment with different levels of suffering.

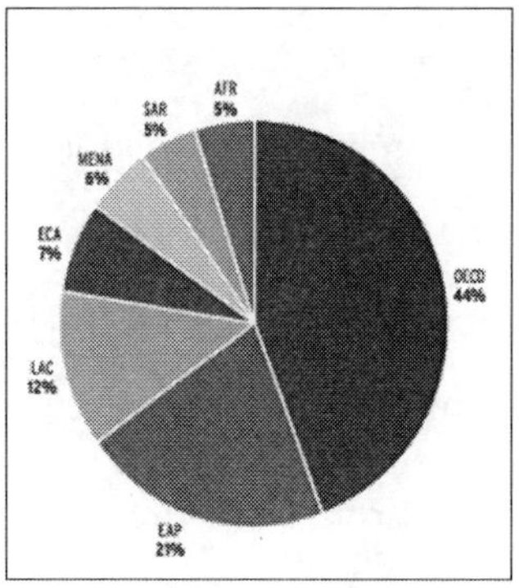


**Fig. 2.** Waste generation based on regions [9].

* Corresponding author: fathurrokhman@mail.unnes.ac.ids

Fig. 2. shows the distribution of waste generation around the world. The strange fact is that the Organization for Economic Cooperation and Development (OECD) whose members are mostly advanced countries has dominated the world's waste generation. We call it strange because OECD countries are the ones who have the technology and upper hand understanding on conservation. However, they have failed to curb their waste production. This evidence supports our case earlier that economic growth has always hurt the environment. OECD countries are among the richest countries in the world consequently hurt the environments the hardest.

The third challenge of conservation in modern era is that we do not have the passion of creating the technology to help our planet. The existing technology we have now is not helping the environment. The green technology we create is only reducing, restoring or sometimes hiding the damages we have created. We do not have the technology to create fresh air, we somehow have technology or initiatives to reduce pollution [4]. We do not have the technology to create water, we only have the technology to reduce toxic and waste in our ocean. Even then, the technology is still not effective as expected [8].

## 2 The 'ideal' conservation

Conservation has many forms but there are guiding principles which bind the forms together. Conservation is the management of human use of the biosphere so that it may yield the greatest sustainable benefit to present generations while maintaining its potential to meet the needs and aspirations of future generations.

There are two types of basic conservations: in situ conservation and ex situ conservation. In situ conservation is the conservation of species in their natural habitat for example natural parks, nature reserves. Ex situ conservation is conserving species in isolation of their natural habitat for example zoos, botanical gardens, seed banks [3].

Modern conservation needs more than just *in situ* and *ex situ* conservation because conservation is not always about biodiversity. Conservation is about the humans too. It is also about cultural conservation [5]. At its very core, cultural conservation has some values:

(a) Aesthetic Value: healthy environment is manifested in beauty.

(b) Recreational Value: preserved nature is a soul entertainment for humans.

(c) Educational and Scientific Value: preserving nature shall be based on education and science

(d) Theurapeutic and Character-Building Value: preserved nature served as therapy and character building for humans.

## 3 Conservation at our place

At our university, conservation has become a crucial part of our vision. The vision of Universitas Negeri Semarang (UNNES) is to become a conservation university with international reputation. Conservation has been the core of the vision itself. Rooted from an education university, we realize that we do not have the same capacity of environmental research if we compare ourselves with other all-round universities which have environment major or green engineering major. At first, it became our main obstacles of achieving our vision. However, we have the commitment to compensate this shortcoming with the adoption of values and cultural conservation like the ones explained by Howland and Rowles (2004). In interpreting the four values of cultural environment we have done the following actions.

In implementing aesthetic value, UNNES has arranged all its gardens and trees in beautiful constellation. We have the belief that tidy arrangement of gardens and trees are the realization of healthy environment.

In implementing recreational values, we have developed eco-education teaching styles around campus. We have conducted classes outdoors. We also invite students from other campuses and even students from high school to visit our university.

In implementing educational and scientific value we have embedded the value of conservations into teaching. Lecturers are obliged to accommodate conservation values to teaching and learning process. Conservation is explicitly mentioned and elaborated in the main curriculum, syllabus, lesson plan and teaching media.

In implementing character-building value, we have assigned each faculty with one main character to develop. In total, there are eight main characters to developed in all faculties: inspiration, humanity, compassion, innovation, creativity, sportiveness, honesty and justice.

## 4 Customized conservation in disruptive era

All those efforts, which have been explained before in cultural values, must be redesigned to fit in the framework of disruptive innovation [7]. Disruptive innovation will not allow the same strategy to be repeated over and over in an ever-changing environment.

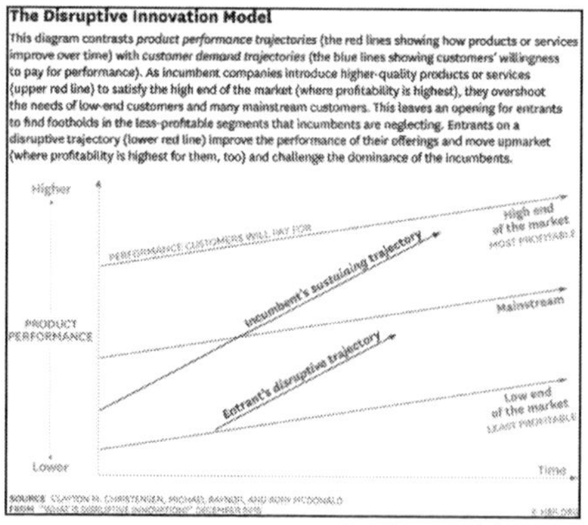


**Fig. 3**. Disruptive theory based on Christensen et al (2015) [2]

According to the disruptive innovation model, new effort of conservation shall be cheaper than the previously available conservation efforts by other universities. While other universities are focusing on green technology and machinery, UNNES tries to disrupt this norm by introducing cultural and characters conservation. The complete model of the disruption is as follows:

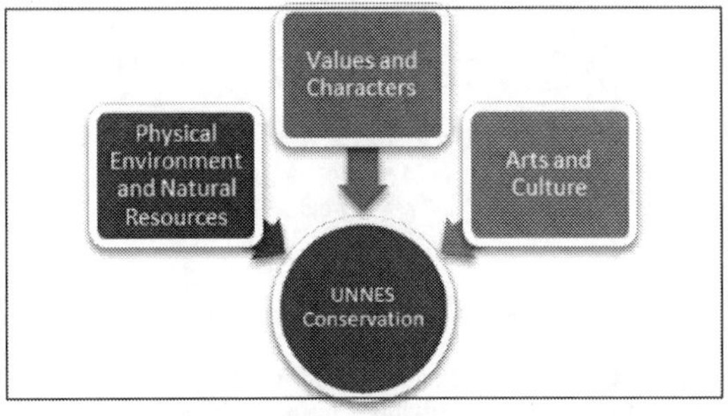


**Fig. 4.** Conservation model at UNNES

This model, we believe have not been adopted by other universities in Indonesia or in the world. Hence, we really hope that the disruptive innovation might work here and other universities will adopt and catch up.

## 5 Conclusions

This paper has a single purpose to deliver and we think that we have delivered it although we have to admit that it is not comprehensive. The theory of disruption by Christensen et al (2015) has led us to believe that UNNES cannot afford the mainstream ways of doing conservation such as providing high-tech waste management machinery or buying expensive solar cells for all the buildings [2]. However, UNNES is able to provide a lower cost conservation i.e. cultural and characters conservation. This type of low point of entrance fits the theory of disruption and we expect that this model will work as well as the more expensive model and even better.

## Reference

1. Akib, N.A.M, *Biodiversity in crisis: conservation challenges in the 21st century*, Available online at https://cgss.usm.my/images/BIODIVERSITY_IN_CRISIS.pdf (2016)
2. Christensen, C. M., Raynor, M. E., & McDonald, R. What is disruptive innovation, *Harvard Business Review*, *93*(12), 44-53 (2015)
3. Geburek, T., & Turok, J., *Conservation and management of forest genetic resources in Europe*. Available online at http://forest-genetic-resources-training-guide.bioversityinternational.org/fileadmin/bioversityDocs/Training/FGR_TG/additional_materials/GeburekTurok/6InsituExsitu496-670.pdf (2005)
4. Haum, R., & Petschow, U., *Lead markets for environmental technologies: The case of the particulate filter for Diesel passenger cars*. IÖW (2003)
5. Holland, A., & Rawles, K., Walking the talk: Philosophy of conservation on the Isle of Rum. *Worldviews: Global Religions, Culture, and Ecology*, *8*(2), 280-297 (2004)
6. Kahneman, D., *Thinking, fast and slow*. Macmillan. (2011)
7. King, A. A., & Baatartogtokh, B., How useful is the theory of disruptive innovation?. *MIT Sloan Management Review*, *57*(1), 77. (2015)
8. Popescu, A. Innovative solutions to prevent and reduce water pollution by application of ecological textile finishing technologies and wastewater treatment. Available online at https://www.interregeurope.eu/fileadmin/user_upload/tx_tevprojects/library/Application%20of%20ecological%20finishing%20and%20wastewater%20treatment.pdf (2017)
9. World Bank, *What a waste: a global review of solid waste management*. Available online at http://siteresources.worldbank.org/INTURBANDEVELOPMENT/Resources/336387-1334852610766/Chap3.pdf (2012)

E3S Web of Conferences **48**, 06004 (2018)
*IWGM 2018*
https://doi.org/10.1051/e3sconf/20184806004

# How the environmental planning of the Universidade Federal de Lavras impacts higher education

*José* Roberto Soares Scolforo[1], *Édila* Vilela de Resende Von Pinho[1], *Antonio* Chalfun-Junior[2], *Adriano* Higino Freire[1], *Leandro* Coelho Naves[3], *Marcio* Machado Ladeira[1,*]

[1] Office of Rector Universidade Federal de Lavras, Campus da UFLA, Lavras, MG, Brazil, 37.200.000
[2] International Office of Universidade Federal de Lavras, Campus da UFLA, Lavras, MG, Brazil, 37.200.000
[3] Department of the Environment, Universidade Federal de Lavras, Campus da UFLA, Lavras, MG, Brazil, 37.200.000

**Abstract.** UFLA was founded in 1908 as an Agriculture School and since that time there was a large concern with environmental issues. During last decade, UFLA has started a huge expansion in different areas of knowledge, becoming a more comprehensive university rather than being focused only in Agriculture Sciences. Due to that, UFLA has faced a need to build many buildings, avenues, improve water and energy supply and deal with different residues produced by their population on campus (including at the laboratories). For this reason, in 2009, UFLA started the idea to implement an Environmental and Structural Planning (ESP), to solve current structural problems and prevent future ones, creating, by that, conditions for keeping expanding and still placing UFLA as an environmentally sustainable University. Moreover, once the Planning was implemented and, even during its implementation, UFLA started to be an excellent example and laboratory for their students. In other words, we could start teaching how to expand keeping been sustainable. For instance, almost 70% of UFLA undergraduate and graduate programs have direct or indirect relation with environment and sustainability, and part of them use facilities and activities created with the ESP implementation. As recognition for the described actions, in 2017, UFLA was ranked in 35th position in UI Green Metric overall world ranking and 1st in Education. In conclusion, the ESP is a great opportunity to improve education and research quality of UFLA.

## 1 Introduction

*Universidade Federal de Lavras*, The Federal University of Lavras (UFLA) was founded in 1908 as the Agricultural School of Lavras. It was renamed as the College of Agriculture of Lavras (ESAL) in 1938, when it started to offer the undergraduate program Agronomy. Then, in 1975 two new undergraduate programs were created and until 1994, when the College of Agriculture became the Federal University of Lavras, it had only 6 undergraduate Programs, ~2,450 students and ~640 employees (faculty, technicians, and administrative staff). In that time, four undergraduate Programs were in Agricultural subject and UFLA had other 10 graduate Programs in this area, which demonstrated the University expertise in Agriculture Science. Nowadays, UFLA has 35 undergraduate Programs and also 35 graduate Programs. Therefore, the University has experienced a huge expansion in the last 2 decades, and the physical and personnel increase experienced by UFLA by its history has required the adoption of new attitudes regarding environmental issues on campus. Therefore, it was necessary to build new buildings, avenues, improve water and energy supply and deal with different residues produced by their daily population and also by the use of laboratories. Currently, our academic community gathers ~13,500 students and ~1,300 employees.

As a recognition of the Environmental Planning implementation, in 2017, UFLA was ranked in 35$^{th}$ position in UI Green Metric overall world ranking, 2$^{nd}$ in Latin America, and 1$^{st}$ in Education [1]. Therefore, this paper has the objective to describe how UFLA Higher Education is linked to environmental and sustainability issues.

## 2 UFLA history and its action on agricultural science

Until 2002 UFLA had 7 undergraduate Programs, being 5 in Agricultural subject (Fig. 1). During the last 15 years with the University expansion, new Programs have been created and it started having careers in different areas of the knowledge, as Life Sciences, Engineering, Medicine, Law, Human Science etc. However, UFLA remains as an important University in Agricultural and Environmental Sciences, having 23 out of the 35 graduate Programs in both areas (Table 1), which helps to explain why UFLA got the first position in Education in UI Green Metric Ranking. It is important to highlight that modern agriculture and livestock production is completely linked with preservation of the environment and sustainability. Some examples are: studies and practices adopted in order to mitigate greenhouse gases emission, improve soil management, decrease

* Corresponding author: mladeira@dzo.ufla.br

https://doi.org/10.1051/e3sconf/20184806004

agricultural defensive use, improve productivity and reduce carbon footprint etc.

**Table 1.** List of graduate programs at UFLA linked to environmental and sustainability.

| **Programs** |
|---|
| Agriculture Engineering |
| Agriculture Microbiology |
| Agro chemistry |
| Agronomy |
| Animal Science |
| Applied Botanic |
| Applied Ecology |
| Biomaterial Engineering |
| Entomology |
| Environmental Technologies and Innovations |
| Feed Science |
| Forestry |
| Genetics and Plant Breeding |
| Phytopathology |
| Plant Physiology |
| Soil Science |
| Sustainable Development |
| Vegetal Biotechnology |
| Veterinary Science |
| Water Resources in Agricultural Systems |
| Wood Science and Technology |

For instance, in 2017, UFLA professors have produced 2,958 academic publications linked to environmental and sustainability and 72.3% of the research funds were used in this area (Table 2). In addition to that, 68.4% of undergraduate and graduate courses offered were related to environmental and sustainability. In addition to traditional careers in Agriculture, new careers created at UFLA have also contributed to undergraduate and graduate programs, such as: Environmental and Sanitary Engineering, Chemical Engineering, Chemistry, Biology, Applied Ecology, Environmental Technologies and Innovations, Sustainable Development, Water Resources in Agricultural Systems.

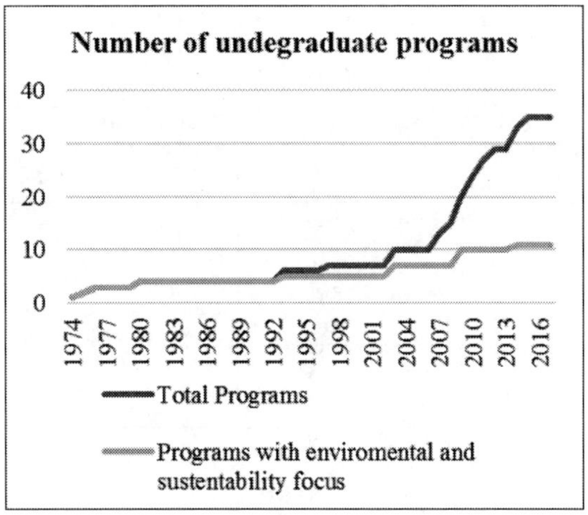


**Fig. 1.** Total number of undergraduate programs and the ones in Agricultural and/or Environmental sciences at UFLA since 1974.

**Table 2.** UFLA numbers in education used by UI GreenMetric ranking.

| **Education item evaluated by UI Green Metric ranking** | |
|---|---|
| Percentage of courses related with enviromental and sustainability | 68.4 |
| Percentage of funds for researches in enviromental and sustainability | 72.3 |
| Number of academic publications in enviromental and sustainability | 2958 |
| Number of events related to enviromental and sustainability | 409 |
| Number of students organizations related to enviromental and sustainability | 151 |

## 3 The environmental planning of UFLA and its impact on higher education

In 2008, UFLA elaborated an Strategic Environmental Planning, in order to create conditions for its well organized expansion. The actions that were part of this plan emerged to solve current problems and to prevent future ones, with the goal of turning UFLA into a benchmark for a Green Campus. In other words, to become UFLA one of the most sustainable and environmentally friendly university in Brazil, and also creating the bases for its community to be among the best ones in the World.

Therefore, during the implementation of the Environmental Planning, UFLA became a laboratory for many students, having direct impact in the academic training. In addition to that, after the total implementation, UFLA reduced costs of water supply, sewage treatment, and with reuse of chemical and organic waste. Total saving was around more than US$ 2.2 million every year and that money goes direct to academic actions.

**Table 3.** Example of actions developed during the implementation of the environmental planning at UFLA and their impacts on higher education.

| **Actions** | **Undergraduate programs directly benefited** |
|---|---|
| UFLA improved its green areas, which included an initial proposal for planting >90,000 native tree species that are essential for protecting water springs and riparian areas and for providing shadow, shelter, and a pleasant place to stay | Forest Engineering, Agronomy, Biology, Animal Science, Environmental and Sanitary Engineering |
| Increase the use of clean energy and its efficient, such as: a) building a new protected electrical network; b) replacing lampposts light bulbs for more-efficient metal halide ones and exchanging mixed bulbs with fluorescent lamps; c) building solar water heating systems in all university-housing buildings etc… | Civil Engineering, Physics |
| Construction of the Chemical Waste Management Laboratory | Environmental and Sanitary Engineering, Chemistry, Chemical Engineering |
| Implementation of a Tissue Digester and a Solid Waste Treatment Experimental Field | Veterinary Medicine, Animal Science, Feed Engineering, Chemistry, Chemical Engineering, Biology, Environmental and Sanitary Engineering, Agronomy |
| Construction of a Sewage Treatment Plant | Chemistry, Chemical Engineering, Biology, Environmental and Sanitary Engineering |
| Improve Water Treatment Plant | Chemistry, Chemical Engineering, Biology, Environmental and Sanitary Engineering |
| Recycling initiatives providing a socially and environmentally friendly final destination for most dry waste generated at UFLA | Every student |
| Construction of 8 kilometers of bike lanes on campus | Every student |

In addition to the examples presented above, UFLA's awareness about the environment is also a major concern in its undergraduate and graduate programs. Undergraduate courses have in their curricula many courses dealing with environmental issues. Such actions are especially relevant taking into account results from studies applied to students in primary and higher education reporting clear evidences that those having deeper awareness (more complex view) on environment and tend to develop more meaningful pro-environmental actions [2].

## 4 Summary/ concluding remarks

The history of UFLA keeps it as one of the most important University in Agricultural Science in Latin America and due to the way modern Agriculture deals with environmental and sustainability issues, it results in a large number of courses offered and manuscripts published. Furthermore, UFLA Environmental Planning has direct and indirect impact in higher education giving to its students a better academic training. Finally, it will contribute to the awareness of society in pursuit of an environmentally sustainable economic development in Brazil and in the world.

## References

1. UI Green Metric, Overall world ranking. Retrieved from: http://greenmetric.ui.ac.id/overall-ranking-2017/ (2017)
2. Freitas, M. R., Macedo, R. L. G., Freitas, M. P. *Environmental perception as a tool to predict pro-environmental actions*. Revista Brasileira de Educação Ambiental (RevBEA), **11**(1), pp. 355-360 (2016)

# Mainstreaming sustainability science to enhance the impacts of "Tridharma" in higher education institution: IPB lessons learned and future direction

*Arif* Satria and *Aceng* Hidayat

Rector, Bogor Agricultural University (IPB), Kampus IPB Darmaga, Bogor 16680, Indonesia

**Abstract.** The adoption of transdisciplinary approaches to address complexity and uncertainty of the problems has been recommended to formulate the effective solution. Although scientists have been working towards the direction, there have been still little integration on policy and concerted implementation to deal particularly with cross-cutting issues on food security, sustainable agriculture, renewable energy and other green development issues. Bogor Agricultural University (IPB) has recently adopted the concept of transdisciplinary and sustainability sciences to better integrate and utilize available resources and management strategies as stated in IPB's Long-term Plan and in recent policy to establish Center for Transdisciplinary and Sustainability Sciences (IPB-CTSS). The drivers for this policy are actually the already established research, academic and community service consortia in which students, scientists, practitioners and policy makers have been working interdisciplinary towards effective solution for achieving better impacts of IPB Tridharma activities. We will discuss the lessons learned, challenges and opportunities identified as solid foundation to implement the concept for enhancing the better impacts of universities extended obligatory tasks on education, research, community services, innovation and business.

## 1 Introduction

In today's world, the development of higher education institution (HEI) is facing more complex problems associated with uncertainties that are requiring new solutions. Based on CCAP (2017), there are practically seven biggest problems the nation faces with respect to its universities' development, including the huge increase in the cost of universities [1]. On the other hand, universities are expected to contribute more actively in addressing nations problems that requires a new approach due to the complexity of the problems. Sustaining our national development requires knowledge that is adaptive to the problems, particularly for IPB Trdidharma activities, such as sustainability issues on agriculture, renewable energy and other land and marine based sector development. Gaps in the implementation are identified due to non-comprehensive approaches leading to the failure of the target achievement. In order to tackle those problems and to enhance the quality of Tridharma and its impacts, IPB has formulated an *IPB Future* vision with the aim to enhance and to support the nation's dignity. The vision of IPB towards 2045 can be stated as "*To be a leading university in strengthening the nation dignity through globally excellent higher education on tropical agriculture, marine and biosciences*".

IPB is coming up with a specialized strategy through the implementation of transdisciplinary and sustainability concepts that has been described many authors among other by Burger and Kamber (2003) where the approaches are consisted of 3 (three) parts, which are problem integration, research integration and solution integration [2]. These parts are made to form a solution to better solve particular problems that come along the way in the middle of the university's development. Problem integration includes the process of describing the problems which may come from the issues on the environment and from the stakeholders. The problems that have been defined in the first phase then will be taken into consideration on the second phase, which is research integration. In research integration, the university with the help of each faculties has to determine research questions, the research itself and research results. In this matter, the university has to build a strong relation and connection between all the disciplines in IPB and to take the multidisciplinary, interdisciplinary and even transdisciplinary approaches into account. The last phase is solution integration in which IPB creates solution based on the research results and to realize the solution using the method of scenario building. The outcome of these phases is to build a certified knowledge and establish creative pool of experts in order to strengthen the university's performance in a changing world. Down-streaming the result results into university business and innovation program has been defined as an additional important pillar to be integrated in the Tridharma business process [3].

 https://doi.org/10.1051/e3sconf/20184806005

Nowadays, IPB comprises 9 (nine) faculties for undergraduate program and 3 (three) schools. The 9 (nine) faculties are Agriculture, Veterinary Medicine, Fishery & Marine Sciences, Animal Sciences, Forestry, Agricultural Engineering & Technology, Mathematics & Natural Sciences, Economics & Management and Human Ecology. The 3 (three) schools are Graduate School, Business School and Vocational School. The total number of students who are pursuing their studies in IPB is around 22 000 students which is consisted of 16 856 undergraduate students and 5 144 postgraduate students. In addition to this number is vocational students with a umber f around 6 000 students. The university's effort in enhancing its education and its employment quality as well as its performance has been paid well and has been acknowledged by World University Rankings and UI Greenmetric. By 2017, IPB is ranked #51 – 100 Top Universities by Subject Agriculture and Forestry by World University Rankings, is also ranked #147 Top Universities by Asia University Rankings and ranked #751 – 1000 Top World Universities by World University Rankings. Moreover, based on UI Greenmetric, IPB is now placed in number 52.

Through utilizing the current organizational capacity, attempts have been carried out to integrate all the resources and strategic management for resulting in better impacts of IPB's Tridharma activities. This paper aims shortly to describe the lessons learned and also challenges that IPB is facing in mainstreaming the sustainability sciences to enhance the impact of HEI's Tridarma activities.

## 2 Global agenda and research priorities

Building on the principle of "leaving no one behind", United Nations has formed the 2030 Agenda for Sustainable Development that includes 17 Sustainable Development Goals (SDGs). As a well-known higher education, IPB is determined to support the SDGs through innovation and research. IPB has put SDGs as its global agenda while developing its university's research agenda in various sectors. IPB has successfully produced Strategic Research Agenda in Food, Energy, Poverty, Ecology and Biomedicine that requires the transdisciplinary approaches.

IPB has recently established a Center for Transdisciplinary & Sustainability Sciences (IPB CTSS) that will in the future be a knowledge hub of existing large research platforms including pool of experts. At the moment, the following examples of research consortia are expected to be integrated for better impacts of the results carried out by interdisciplinary research teams to address SDGs (sustainable development goals) related issues, as follows:

1. **CRC990 – EFFoRTS** with the emphasis on Ecology and Socio-economy and funded by CRC990 and the Ministry of Research, Technology and Higher Education of Indonesia – **SDG No. 15**
2. **REDD+** with the emphasis on Reducing Emission from Deforestation & Forest Degradation funded by DANIDA – **SDG No. 13**
3. **Zoonoses** with the emphasis on Health & Biomedicine funded by JST/SATREPS and JICA – **SDG No. 3**
4. **Australia Indonesia Center (AIC)** with the emphasis on Food/Agriculture & Infrastructure funded by AIC and the Ministry of Research, Technology and Higher Education of Indonesia – **SDG No. 6**
5. **EMBRIO** with the emphasis on Marine Biodiversity funded by IPB and the Ministry of Research, Technology and Higher Education of Indonesia – **SDG No. 14**
6. **LISAT** with the emphasis on Remote Sensing/Satellite for Food Security and Environmental Monitoring funded by LAPAN, IPB and the Ministry of Research, Technology and Higher Education of Indonesia – **SDG No 2**
7. **PETUAH** with the emphasis on Green Knowledge Management, funded by MCA-I – **SDG No. 17**
8. **Peat Fire** with the emphasis on Environment & Climate Change funded by NASA – **SDG No. 13**
9. **Agri_Insurance** with the emphasis on Food Security funded by JST/SATREPS and JICA – **SDG No. 9**
10. **ANBIOCORE** with the emphasis on Food Security & Sovereignty funded by USAID – SHERA – **SDG No. 2**

In addition to research consortia, one of important IPB's infrastructures is IPB Science Park with the core focuses are Technology Transfer and Legal, Business Development, Capacity Building and pool of Expertise, Marketing and Promotion, and ICT. Each focus is dealing with different kind of tasks which will enhance the technology transfer flow with respect to business regulation, innovation support and regulation/incentive support from the government to the process of technology transfer for business (commercial) and for social/non-commercial purposes. IPB Science Park is a subsidiary company of PT BLST (Bogor Life Science and Technology), the Holding Company of IPB woth a main function to manage technology transfer for business purposes (commercial). The technology transfer from various sectors and various levels of company/society delivers its positive impact to social/extension in the form of community services managed by LPPM (Institute of Research and Community Services) and to commercial objectives in the form of business/revenue streams managed by BLST. Moreover, these actions will later on give impacts on economic development and prosperity.

## 3 Lessons learned, challenges & opportunities

Economic development of the country can be enhanced by a more active participation of HEI/Universities by implementing the actionable knowledge proactively through the engagement of wide range of stakeholders. For instance, IPB has developed the new varieties of paddy with higher productivity compared to the previous variety which are called IPB 3S. This new type of paddy was developed by Dr. Hajrial Aswidinnoor initially in order to achieve Indonesia's rice sovereignty in collaboration with MoRTHE (KEMENRISTEKDIKTI), MoA (KEMENTAN), PT. BLST, seed association and farmer groups. Other commodities with positive impacts includes also papaya seed, chili seeds, agricultural machinery, biomedical products (e.g. enzymes, animal pharmaceutical producst etc.).

In addition, IPB is also developing outlets for IPB Products which are located on places around JABODETABEK, such as Botani Square (Bogor), Gandaria Cit (Jakarta), Mal Artha Gading (Jakarta), Kalibata City (Jakarta) and Mal Teras Kota BSD (Tangerang). The outlets are selling IPB's innovative products that are produced from research results (IPR base and non IPR) and from small and medium business enterprises fostered by IPB through community services. The products include food, beverages, fruits, herbal medicine, body care etc.

Due to a rapid change in the world, IPB has to come up with new solutions and innovations to deal with the opportunities, challenges and the future that the changing world brings. There is a growing interest on the nexus of food, health, energy, water and related issues with the SDGs. Therefore, the university has to increase human capacity and competency in conducting research and creating solutions to solve these issues. Moreover, government has also put forward the nation's priority to address issues on rural development, social forestry and renewable energy so that the wealth of the local people can be increased. In order to create a transdisciplinary and sustainability approaches, the important thing to be done by the university is integrating and managing the expertise, resources and the networks. These actions can lead the university to develop broader research possibilities with stakeholders, companies and even with international partner countries. Moreover, the results of the researches can only be real if they are implemented correctly in the society which means the university needs to link/connect scientific discovery to policies and practices through down-streaming processes, e.g. commercial and non-commercial purposes. IPB is on the steering wheel through the established IPB CTSS for future direction on the establishment of community practices (CoP) or forum to address the issues in sustainability.

## 4 Closing remarks

The research and innovation activities have become the hub/center for the start of mainstreaming sustainability in HEI's Tridharma activities. The pools of experts and resources require the suitable mechanism facilitating the flexible flows of skill and knowledge transfer processes. IPB will actively promote transdisciplinary research and innovation for better impacts on nation's economic advancement and people prosperity.

## References

1. CCAP. Seven Challenges Facing Higher Education. https://www.forbes.com/sites/ccap/2017/08/29/seven-challenges-facing-higher-education/#414a24be3180. (2017)
2. Burger, P. and Kamber, R., Cognitive integration in transdisciplinary science: Knowledge as a key notion. *Issues in Interdisciplinary Studies*. **21**, pp. 43-73. (2003)
3. Payumo, J.G., Arasu, P., Fauzi, A.M., Siregar, I.Z. and Noviana, D., An entrepreneurial, research-based university model focused on intellectual property management for economic development in emerging economies: The case of Bogor Agricultural University, Indonesia. *World Patent Information*, ***36***, pp.22-31. (2014).

# NPUST's practice on achieving sustainable development goals

*C.-H.* Tai[1], *C.-L.* Chang[2], and *J.C.* Leong[3*]

[1]President, National Pingtung University of Science and Technology, Taiwan
[2]Dean, Office of General Affairs, National Pingtung University of Science and Technology, Taiwan
[3]Dean, International College, National Pingtung University of Science and Technology, Taiwan

**Abstract.** The National Pingtung University of Science and Technology was established as an agricultural institution in 1924. Since then, this educational institution has been one of the key role players in developing the agricultural advancement in Taiwan. For years, the university has seriously planned and carried out projects that strengthen her identity, "being green". Under the scope of university's uniqueness and responsibility, measures taken in the university directly involves administrative affairs, such as teaching, research, university development, student welfare, etc. As a matter of fact, initiating new policies, measures, and projects is only the beginning of everything. Without proper planning, enforcement, and funding, any one of these policies, measures, and projects may end up going nowhere. In other words, sustainability is absolutely the most critical element that dictates the success of the policies, measures, and projects. This sustainability covers financial allocation, manpower, negotiation, evaluation, and adjustments. The university has invested a lot of funding and effort to achieve the aforementioned goals through various means in the aspects of administration, education, research and development.

## 1 Introduction

The United Nations [1] has compiled 17 Sustainable Development Goals (SDGs) for global welfare which widely cover a range of topics associated to social and economic developments for all countries regardless if the country is a "developed" or "developing" nation. These goals address issues related to poverty, hunger, health, education, climate change, gender equality, water, sanitation, energy, environment and social justice, whose relative importance and exact wording are listed in Table 1. Apparently, poverty, hunger, and health are the three most important ones and they are somewhat very closely connected to each other. Poverty leads to hunger and then undernourishment. This later causes serious health problem. To break this circle of difficulty, one needs to identify its root. As a matter of fact, agriculture itself has provided the largest portion of world employment and has contributed to the largest source of income for poor rural households. It has supported the livelihoods for 40% of the world population. In some parts of Asia and Africa, more than 40% of the agricultural labor force actually come from women. Because of poor agricultural technology and efficiency, crop yield is obviously low, risk for production has become higher lately due to climate change, etc. These lead to poverty, hunger, and poor nutrition. Currently, one out of every two deaths in children under five is owing to poor nutrition. Majority of these cases happen in developing countries. The International Food Policy Research Institute (IFPRI) [2] believes that eliminating hunger and under-nutrition should have higher priority to ending poverty. They suggested that the solutions should be: 1) agriculture-led; 2) social protection- and nutrition intervention-led; or 3) a combination of both of these approaches.

The National Pingtung University of Science and Technology was established in 1924 as an agricultural institution. For almost a century, this educational institution has great contributions in the advancement of Taiwan's agricultural sectors through funded projects from various governmental agencies and private sectors. To be "green", the university undertakes measures involving teaching, research, university development, student welfare, and so on to initiate new policies, to brainstorm the measures, and to carry out the projects. Without proper planning, enforcement, and funding, each of these policies, measures, and projects may eventually go in vain. In the entire process, sustainability is clearly one of the most important questions needs to be raised continuously to ensure the sustainability especially in financial status and workforce. On-going evaluation is then needed for future adjustments.

**Table 1**. List of UN Sustainable Development Goals (SDGs) [1]

| No. | Goal | Exact wording of the goal |
|---|---|---|
| 1 | No Poverty | "End poverty in all its forms everywhere." |
| 2 | Zero Hunger | "End hunger, achieve food security and improved nutrition and promote sustainable |

* Corresponding author: jcleong@mail.npust.edu.tw

https://doi.org/10.1051/e3sconf/20184806006

| | | |
|---|---|---|
| | | agriculture" |
| 3 | Good Health and Well-Being for people | "Ensure healthy lives and promote well-being for all at all ages." |
| 4 | Quality Education | "Ensure inclusive and equitable quality education and promote lifelong learning opportunities for all." |
| 5 | Gender Equality | "Achieve gender equality and empower all women and girls." |
| 6 | Clean Water and Sanitation | "Ensure availability and sustainable management of water and sanitation for all." |
| 7 | Affordable and Clean Energy | "Ensure access to affordable, reliable, sustainable and modern energy for all." |
| 8 | Decent Work and Economic Growth | "Promote sustained, inclusive and sustainable economic growth, full and productive employment and decent work for all." |
| 9 | Industry, Innovation and Infrastructure | "Build resilient infrastructure, promote inclusive and sustainable industrialization and foster innovation." |
| 10 | Reduced Inequalities | "Reduce income inequality within and among countries." |
| 11 | Sustainable Cities and Communities | "Make cities and human settlements inclusive, safe, resilient and sustainable." |
| 12 | Responsible Consumption and Production | "Ensure sustainable consumption and production patterns." |
| 13 | Climate Change | "Take urgent action to combat climate change and its impacts by regulating emissions and promoting developments in renewable energy." |
| 14 | Life Below Water | "Conserve and sustainably use the oceans, seas and marine resources for sustainable development." |
| 15 | Life on Land | "Protect, restore and promote sustainable use of terrestrial ecosystems, sustainably manage forests, combat desertification, and halt and reverse land degradation and halt biodiversity loss." |
| 16 | Peace, Justice and Strong Institutions | "Promote peaceful and inclusive societies for sustainable development, provide access to justice for all and build effective, accountable and inclusive institutions at all levels." |
| 17 | Partnerships for the Goals | "Strengthen the means of implementation and revitalize the global partnership for sustainable development." |

## 2 Approaches

### 2.1 Administration

There are many measures undertaken in the university to achieve sustainable development within the campus. Three of the most successful examples are the implementation of heat pump system in student dormitories, the handling of waste, and the use of electric motorcycles among new intakes. The use of heat pump system in the student dormitories has been a long journey because the university has to go through many processes to get the project going. As a national university, the university budget is not only tight but also under supervision of the government. As a result, the administrators in the university have to come to a consensus recognizing the impact the heat pump system may have on the university's path to being green. Then, the budget required for this system has to be planned, proposed, and approved. Those who mainly involve include the Office of General Affairs, the Office of Student Affairs, and the Accounting Office. The waste handling is critical for any university to protect our environment and to maintain its sustainability. As an agricultural university, NPUST has developed her own strategy to handle different kinds of waste, such as domestic waste, agricultural waste, chemical waste, etc. The promotion of waste handling involves the Office of General Affairs, the Office of Student Affairs, several student bodies, etc. NPUST freshmen used to rely on motorcycles to travel both inside and outside the campus. In recent years, the university has set a goal to reduce the use of gasoline motorcycles by providing subsidy to on-campus/off-campus bus system as well as offering leasing programs for electrical motorcycles to our students. The Dean for the Office of Student Affairs has to ask for financial support from the Department of Transportation, to negotiate leasing programs with electrical motorcycle companies, and to promote the programs among the students and their parents.

## 2.2 Educational and activity

As an educational institution, one of the main goals of NPUST is to educate and to strengthen the understanding of various issues related to environment or sustainability. To raise the awareness, the subjects of target consist of the students enrolled in the university as well as the general public. For the former target, the university puts effort in encouraging the faculty members to relate their course materials to issues about environment or sustainability if the nature of the course permits. The statistical information of these courses is presented in Table 2.

**Table 2**. Statistics of courses related to environment and sustainability

| Year | Number | Percentage† |
|---|---|---|
| 2013 | 1731 | 34.46% |
| 2014 | 1988 | 39.78% |
| 2015 | 2727 | 50.68% |
| 2016 | 2731 | 50.82% |

† number of courses related to environment or sustainability divided by the number of all courses.

In additional to the curriculum, NPUST also promotes the awareness for the importance of environmental and ecology protection through a special unit in the university. The university has established the Center for Environment Protection, Safety and Health (CEPSH) whose main responsibility is to treat and recycle the wastewater generated in the 8 student dormitories and 2 student restaurants. After an upgrade in 2008, this center is capable of treating up to 800 $m^3$ wastewater on a daily basis. Through CEPSH, NPUST organizes activities to educate both NPUST students and the public communities how natural environments function, and particularly, how human beings can manage behavior and ecosystems to attain sustainable mankind. This helps us transitioning to a society that is knowledgeable of the environment and its associated problems, aware of the solutions to these problems, and motivated enough to solve them. It is no doubt a complicated field integrating multiple disciplines ranging from science to engineering and management to humanity. Only through simple yet interesting activities, as listed in Table 3, can CEPSH help inject momentum into the public awareness on ecological, environmental, and social sustainability. Figures 1 are the pictures of the activity.

**Table 3.** CEPSH educational activities in 2015

| Activities | Number | Total Number of Attendees |
|---|---|---|
| Tours/Visits | 53 | 1,538 |
| Workshops/Seminars | 21 | 990 |

**Fig. 1.** CEPSH engagement in public education: community visit (top), kindergartners visit (middle), off-campus education (bottom)

## 2.3 Research and development

The third approach NPUST undertakes is related to research and development. Unlike the previous two approaches, this approach may seem to have very little effect on the university itself. However, when successful, it may have a great contribution over the world when the technology developed is widely used in the relevant industry. The followings are some examples:

- Circular economy in forest areas
- Electric car for sloping area maneuver
- High-yield rice cultivation
- Hydroponics
- Power generation using manure
- Recycle of aquaculture water
- Water-saving irrigation technology

# 3 Summary/ concluding remarks

It is important that the university should have a vision to help solve the problems currently faced in the world. Also, the university should have the right attitude to lead the society in solving the problems. Since UN has identified the sustainable development goals, NPUST recognizes the importance of these goals and their impacts on our future generation. NPUST vows to keep on the pace to initiate more projects that supports sustainable development goals. Although those identified by UN are too much for a university to tackle, the university can help preparing the essential elements.

## References

1. *Sustainable Development Knowledge Platform*. United Nations Department of Public Information. Available online at https://sustainabledevelopment.un.org/sdgs, accessed on 23 Februari, (2018)
2. Fan, Shenggen. *Ending world hunger and undernutrition by 2025*. IFPRI BLOG. Available online at http://www.ifpri.org/blog/ending-world-hunger-and-undernutrition-2025, accessed on 25 Feb., 2018. (2014)

# Bringing sustainability to the heart of a university through teaching, research and service

*Esra* Gençtürk[1] and *M. Pinar* Mengüç[2,*]

[1]Rector, Ozyegin University, Istanbul, Turkey
[2]Director, Center for Energy, Environment and Economy, Ozyegin University, Istanbul, Turkey

**Abstract.** In this short paper, we summarize our targeted efforts at Ozyegin University in Istanbul, Turkey for establishing a sustainable research, teaching and learning environment. The University is striving to have highest level impact on sustainable education, energy, architecture, built environment, business and life-long learning practices. The strategic plan of the University puts the principles of sustainability at the cornerstone of its efforts, with the aim to aspire not only our students and staff, but also the community at large.

## 1 Introduction

Ozyegin University (OzU) was established in 2007 with the goal of becoming one of the leading research universities in Turkey [1]. The University had its first student intake in 2008. In 2012, it moved to its modern campus at Çekmekoy; one of the new suburbs of Istanbul. OzU have always had the vision to impact the society with innovation and entrepreneurship. The university-wide efforts on sustainability, both in scientific and practical sense, have made the university well known among the higher education institutes in Turkey as a visionary place to work and learn.

These efforts have paid off as OzU has consistently been ranked among the top ten list of the Entrepreneurial and Innovative University Index of Turkish Science and Technology Foundation (TUBITAK) since 2011 (out of more than 170 Universities in Turkey). This ranking is simply a reflection of faculty productivity in research as well as in patents and companies they have established. Among the grants received, those related to sustainable infrastructure and sustainable energy have considerably contributed to OzU's sustainability goals. To this end, six EU-FP7 and Horizon 2020 grants received and completed by Center for Energy, Environment and Economy (CEEE) is worth mentioning [2].

The teaching environment at OzU is also designed differently than many other universities as it starts with a coherent sustainability emphasis in the freshman curriculum, allowing all students to have a grounded awareness for the global and local sustainability goals. In addition, the Faculty of Social Sciences in general and its Psychology Department in particular are known for outstanding programs geared towards gender-equality at the University and in the Society. Several innovative energy courses were also introduced at the University over the years, including 'Zero Istanbul 2050' (chosen as one of the Global Initiatives by IBM) and the dual undergraduate and graduate level 'Sustainable Materials, Energy, and Systems' course co-organized by CEEE along with IMSAD, the largest umbrella NGO on construction and building materials in Turkey [2].

The University is also the lead contributor to the business world on establishing deep-rooted sustainable development goals for establishments of various sizes and at different sectors [3]. The launch of the Sustainable Development Forum at OzU in partnership with TUSIAD, the largest NGO for the Turkish Business World, is unique among the Turkish universities. More details are given below.

## 2 Pursuing sustainability

Ozyegin University has been emphasizing the principles of sustainable architecture, engineering, education, living and practices since it was established in 2007. Over the last ten years, several strong initiatives have been started by the University administration and by the faculty members which were aligned with the UN Sustainable Development Goals (SDG) even before the subsequent announcement of the SDGs [3].

The following seven SDG's have actively been pursued by OzU: *SDG4: Quality Education; SDG 5: Gender Equality; SDG7: Affordable and Sustainable Energy; SDG 9: Industry, Innovation and Infrastructure; SDG 11: Sustainable Cities and Communities; SDG 13: Climate Action; and SDG 17: Partnerships to Achieve these Goals.* The spirit of these goals has explicitly and implicitly influenced all the decisions made and the strategies formulated by the University so far. Below, we discuss specific OzU efforts geared towards *Constructing a Green Campus*, on *Research and Teaching*, and on *Partnerships*. We also list the specific *Awards* given to some of the University members for their contributions to the sustainability science and applications.

* Corresponding author: pinar.menguc@ozyegin.edu.tr

## 2.1 Constructing a green campus

Since its first days, Ozyegin University has aimed to build a sustainable and exemplary education environment. The University built a modern campus in Cekmekoy, at the suburbs of Istanbul, in 2012. The first three buildings erected at the campus were all awarded with LEED certificates, showing the dedication of OzU to environmental sustainability. All other buildings were also built as sustainable and energy efficient buildings. In addition to LEED certificates, OzU also has ISO 14001 and OHSAS 18001 certificates.

The fourth academic building of OzU was constructed with more stringent rules than a LEED certificate would dictate. SCOLA building was designed and built as part of an European Union Framework Programme 7 (EU-FP7) project called NEED4B (New Energy Efficient Demonstration for Buildings) [4]. This major effort was conducted by Center for Energy, Environment and Economy (CEEE), which allowed the detailed and integrated engineering and architecture principles to be applied to the construction as well as to the operations of SCOLA building. These efforts are in line with the Sustainable Development Goals 7, 9 and 11 mentioned above. Currently, the SCOLA building is considered as one of the most energy efficient and sustainable buildings in Turkey [5].

Interestingly, the building constructed by the University after SCOLA, The Law School building, has become the second most efficient building on the campus although CEEE is not involved in the process. All engineers and architects involved in NEED4B project have applied their experiences effectively to the new construction. Through teaching by practice and example, OzU had another sustainable building, with very low carbon footprint. This action is indeed in line with the Sustainable Development Goals of 13 and 17.

The sustainable engineering efforts undertaken by the University are not limited to the academic buildings. Before the campus was built, the University Board of Trustees approved the construction of the Energy Distribution Center (EDS) in 2010 paving the way to centralized heating and cooling modalities. EDS is in the core of regional heating/cooling concept empahasized by the European Union FP7 and Horizon 2020 programs to achieve high energy performance on more than a single building. EDS at OzU makes the campus very efficient in energy use with very low carbon footptint.

Currently, the University is working on the construction of a signature School of Architecture building. This challenging, complicated, multi-purpose structure is being designed with explicit sustainability goals and expected to be an example educational building with high energy performance.

The sustainable energy concepts can be seen everywhere at the campus. There are solar panels to provide about 7% of the energy from the sun, and there are green roof systems to help the energy efficiency of all buildings. Several other new concepts developed by the CEEE researchers are also used in several buildings, as outlined in the next section. At Ozu, there is extensive water and waste management systems in place, which are inherent in the sustainable eco-system of the campus. With all these special buildings and initiatives, OzU is considered as the pioneer of innovative and sustainable higher-education institutions in Turkey.

## 2.2 Education and research

Ozyegin University has been prioritizing the depth and the breadth of education it provides in line with the principles of sustainability principles since it was established in 2007. The University has introduced various topics, engineering practices, campus operation modalities as well as integrated research projects which incorporated sustainable development goals in almost all fields. Both in regular class-room courses or as part of the research related courses, the University has a great interest in placing multi-, inter-and trans-disciplinary studies in all of its academic units. This strategy also creates and fosters an interdisciplinary atmosphere at the University, allowing faculty as well as students to think beyond their own specific fields of study.

Almost all departments at OzU include the sustainability vision into their curriculum, with the help of umbrella courses offered by the University. Probably the most unifying theme is due to the teaching of English in the University's School of Languages (SCOLA) [5]. The language of instruction at OzU is primarily English. SCOLA is responsible for preparing incoming students to have strong English communication skills. To this end, sustainability is designated as the substantive topic of one of the English courses which is mandatory for all OzU students. The instructors responsible with this course use academic as well as popular press articles addressing different aspects of sustainability. Later, students use this know-how in their advanced studies.

Sustainability is also covered at several other courses taught at OzU beyond the SCOLA curriculum. Some examples are listed below :

- Energy, Environment and Economy for Sustainable Istanbul 2050
- Sustainability in Higher Education
- Environmental Control
- Sustainable Design and Environment through Biomimicry
- Green Buildings and Sustainable Design
- Energy Efficient Building Design and Renovation
- Certificate Sustainability in Business
- Social Entrepreneurship, Social Innovation and Innovating Social Change
- Environmental Law
- Power management for Advanced Energy Storage Systems
- Ecology and Production of Food
- Sustainable Development for Hotels
- Ecotourism
- Energy Licenses
- Partnership in Energy Markets
- Renewable Energy & Renewable Energy Law
- Sustainable Energy Materials and Systems: An Integrated Approach for Buildings.

Beyond undergraduate courses and student projects, several graduate level research projects are also deep-rooted in the principles of sustainability. These projects have clearly created a knowledge and awareness platform, improved the experience of the University members and revealed the needs of the research environment in especially at the local scale. Around 98 specific projects, which have been carried out so far at OzU, are related to the 7 SDGs mentioned above. The University is continuously working on new initiatives and new courses to involve more students and faculty in the practicing towards these specific SDGs.

OzU students have to take part in social responsibility projects to meet the civic engagement requirement of their degrees. These activities are performed under a platform named *Duyarlı* OZU (*Sensitive* OzU) which was founded by students in 2011 as a social initiative for developing volunteering opportunities. Several projects in different locations in and outside of Istanbul, Turkey has given students the chance of gaining awareness in social sustainability and helping others, such as children, elders, animals and environment.

The University has a focused effort on sustainable energy and infrastructure for building and cities. The research and development projects carried out by the Center for Energy, Environment and Economy (CEEE) have been supported through six major European Union grants received so far. One of them, as mentioned above, is the EU-FP7 Project called NEED4B [4]. This allowed, the University to have one of the most energy efficient buildings in Turkey. Another project conducted by CEEE is to renovate a major hospital building in Aydin Province of Turkey, at Adnan Menderes University. The third major EU Grant to CEEE is TRIBE project funded by the Horizon 2020 program, which allowed the researchers to study human-building interactions in depth. This research brings the sustainable energy concepts to the user level and aims to have significant reduction in carbon footprint by increasing the awareness of building occupants. The concepts developed from this TRIBE project is currently applied to the Engineering Building at OzU [2].

All these concepts can be most useful if the know-how developed can be transferred to technicians who actually build the required infrastructure to make buildings more energy efficient and comfortable. CEEE is working with three local vocational schools to improve their curriculum to achieve these. Note that all these efforts by CEEE can be listed under SDGs 4, 7, 9, 11, 13 and 17.

Due to these coordinated and structured emphases on sustainability at OzU as related to sciences, engineering, architecture, law, and societal issues, the students, faculty and staff have developed quite high level of awareness and knowledge on the topic. As the members apply and share this knowledge within their daily routines, a chain of additional sustainable practices naturally follow. This has resulted in an unifying grass roots culture around the campus on what sustainability is and how it can be internalized in everything we do.

## 2.3 Partnership

Ozyegin University believes in the mission of serving the public by generating knowledge and sharing it. In this direction, OzU cares about the integration of partnerships into education and research. The University has developed several initiatives with professionals and stakeholders. Among them, SKF (Sustainable Development Forum) founded in 2015 with the collaboration of TUSIAD (Turkish Industry and Business Association) focuses mainly on sustainable development in the business world within the framework of SDGs. By being a part of this global agenda, OzU's objective is to build a strong dialogue, support the business world with its unique research competences while nourishing and directing its research areas through the developments and needs of the business world.

OzU has a broader participation on societal problems related to Istanbul and Turkey. In 2010, OzU CEEE has organized the first *Workshop on Istanbul'2050*, which brought together 80 people from all walks of society [6]. It was also attended by the EU Energy Ministry, and at the end the Workshop highlighted the challenges Istanbul has been facing. Interestingly, 'the protection of the soul of Istanbul' was found to be the most pressing issue; in essence this was the prelude to 'sustainable cities' emphasis of the United Nations, before the UN SDGs were declared. Later in 2011, CEEE has organized a Conference on Integrated Engineering and Architecture for Sustainable Buildings. It was co-funded by the US National Science Foundation (NSF) and the Turkish Science and Technology Council (TUBITAK), and had more than 30 lectures from both countries. Since then CEEE is routinely organizing conferences and workshops on sustainable energy and human-building interactions for sustainable buildings. Since 2017, CEEE is also involved with the major effort called Sustainable Anatolia, which is organized in the city of Ilgaz, in Central Turkey. This year, OzU CEEE will be one of the main sponsors of the event.

OzU Technology Transfer Office (TTO) was established in 2013 to provide support to all projects and the intellectual properties of OzU faculty [7]. When established, OzU TTO was one of the ten initial such offices supported by TUBITAK. Although their effort is not related to SDGs directly, the presence of a well funded TTO at OzU is important for the establishment and continuity of a sustainable research environment at the University.

Sustainability requires innovation and determined leadership. With this, the University has recently opened a new building, fondly named OzU-X, where innovation and entrepreneurship are pursued by the students and the faculty. The University has established a partnership with General Electric (GE), to develop further initiatives, which will enhance the sustainability and societal impact of research conducted by OzU members. In fact these effort are paying off, as OzU received the first Industry 4.0 funding from the Ministry of Science, Industry and Technology [8]. The project "Digital Transformation of SMEs in Turkey through Establishment of the Industry 4.0. Competence Center" presented under the

"Manufacturing Industry" category was selected among the priority projects to receive funding for the second phase of the Competitive Sectors Programme run jointly by the Turkish Government, the General Directorate of European Union and Foreign Affairs, and the EU Financial Programs Department

### 2.4 Awards to show the efforts

Detailed and dedicated studies of several OzU members who work on the general framework of sustainability have been rewarded with several awards both in local and global scale. The University was ranked quite high in Greenmetric rankings even though it is a very Below are some of the most significant awards received by the OzU researchers and the students:

- 2016: Özyeğin University was included in the annual report of international sustainable campus best practices as one of the 19 universities with selected cases on "Capacity Building" and "Innovation for Best Practices" reflecting how universities are developing their sustainability skills and building capacities in collaboration with corporate partners.
- 2017: Özyeğin University was featured as one of the best practice examples by the International Sustainable Campus Network (ISCN) and Global University Leader Forum (GULF), which is the academic platform of World Economic Forum, and included in the annual report of 2017 "Educating for Sustainability" in recognition of its current practice "Innovative Three-Layered Sustainability Education for the Business World".
- 2017: A team by Elif Gizem Tunçel and Ebru Tatar, two women graduate students at the Mechanical Engineering Department, who are also among the CEEE researchers, have become one of the twelve teams to be invited to the finals of "Go Green in the City" competition organized by Schneider. They were anong 24 individuals selected out of 19,722 competing worldwide. Working with Professor M. P. Mengüç, Ebru and Elif Gizem have also received the **"Women in Business Award"** at the ceremony in Paris, France [9].
- 2017: Azadeh Didari, a post-doctoral fellow at CEEE was chosen to receive the Elsevier/JQSRT Raymond Viskanta Young Scientist Award, which is presented to only one researcher each year world-wide [10]. The award ceremony was held at KAIST, Daejeon, Korea. Her work is on sustainable energy at nanoscales. She received her PhD from OzU under M.P. Mengüç.

## 3 Future plans

The University has been preparing its new 5-year strategic plan, which will be in effect by the 2018-2019 academic year. The University Administration and the Board of Trustees have agreed unanimously on making sustainability one of the pillars of University. To this end, the University plans to develop more inter-, multi- and trans-disciplinary studies that combine the efforts of different academic and administrative units. These new initiatives will not only cover the academic curriculum but also include specific action plans to elevate the research output and impact to a higher level.

Two OzU buildings are planned to have further contributions to the University sustainability efforts. As mentioned above, OzU-X, a new open innovation co-working hub opened at the heart of the campus Also, a new Architecture Building is being constructed. This building is designed by combining all sustainable engineering-architecture principles with the sustainable energy emphasis.

The University is continually reaching out to the world outside of the campus to transfer the new knowledge developed by its faculty and students to different sectors and businesses. Collaborations with TUSIAD, the plans of the new Center of Excellence for Industry 4.0 Competence [8], and the research and the teaching efforts of the Center for Energy, Environment and Economy are all helping OzU to be one of the most impactful sustainable universities in Turkey and beyond.

**ACKNOWLEDGEMENTS.** We thank to a number of people at the University who helped us in preparing this short document. Prof. Dr. Hulya Turgut has been leading the work of the Sustainable Development Forum since 2017. Ms. Nergis Uyan is the Director of SCOLA, who has been responsible for implementing and tracking the sustainability initiatives at ENG courses. Ms. Gokce Ersel, who is studying Sustainable City concept at the College of Architecture under M.P. Mengüç is to be credited with streamlining the details provided in this document.

## References

1. Ozyegin University Foundation History, Retrieved March 30, 2017, from https://www.ozyegin.edu.tr/en/about-us/foundation-history (2017).
2. Center for Energy, Environment and Economy (CEEE) at Ozyegin University, Retrieved March 30, 2017 https://ecem.ozyegin.edu.tr/en (2017).
3. UN Sustainable Development Goals, retrieved from http://unsdsn.org/what-we-do/sustainable-development-goals/about-the-sdgs/ (2017).
4. New Energy Efficiency Demonstration For Buildings, NEED4B, EU-FP-7 Project, Retrieved March 30, 2017, from http://www.need4b.eu/ (2017).
5. School of Languages, SCOLA, Ozyegin University, Retrieved March 30, 2017, https://www.ozyegin.edu.tr/en/school-languages-scola, (2017).
6. Zero-Istanbul 2050 Initiative, CEEE, Ozyegin University, Retrieved March 30, 2017, https://www.ozyegin.edu.tr/en/etkinlikler/50015023 (2010).
7. Ozyegin University, Technology Transfer Ofisi, OzU-TTO, Retrieved March 30, 2017, http://tto.ozyegin.edu.tr/en (2017).

8. Ozyegin University "Digital Transformation of SMEs in Turkey through Establishment of the Industry 4.0. Competence Center", Retrieved March 30, 2017, https://www.ozyegin.edu.tr/en/announcements/2358/our-industry-40-competence-center-project-admitted-into-the-priority-group-to-receive-funding (2017).
9. Schneider Go Green in the City, Women in Business Award. Retrieved from https://www.schneider-electric.com/ww/en/documents/Press/2017/08/09-release-go-green-in-the-city-tcm50-323141.pdf (2017).
10. Elsevier JQSRT Viskanta Young Scientist Awards, Retrieved March 30, 2017, https://www.journals.elsevier.com/journal-of-quantitative-spectroscopy-and-radiative-transfer/awards-winners/winners-of-the-2017-elsevierjqsrt-raymond-viskanta-awards-el (2017).

E3S Web of Conferences **48**, 06008 (2018) https://doi.org/10.1051/e3sconf/20184806008
*IWGM 2018*

# RTU approach to sustainable development: governance and measuring impact

*Juris* Iljins[1]* *and Leonids* Ribickis[2]

[1] Juris Iljins*, Riga Technical University, Faculty of Engineering Economics and Management
[2] Prof. Leonids Ribickis, Rector of Riga Technical University

**Abstract.** The aim of this article is to explore Riga Technical University (RTU) approach to Sustainable Development, sustainability governance practice and its impact. In order to successfully implement sustainability as an essential part of organizations daily operations, processes should be properly governed, communicated and adequate resources should be allocated. Implementation of sustainable thinking has helped RTU to re-evaluate processes, goals and outcomes. The concept of RTU in long-term is to move towards building holistic sustainability concept thus impacting social, economic and environmental issues. For the achievement of these goals, governance of sustainability is a crucial factor.

## 1 Introduction

In order to govern and implement sustainability concept, RTU has adopted two Strategic planning documents. One of these documents is RTU Strategy, where Sustainable development is seen as an important cornerstone, and it is integrated in main strategic objectives of university. Additionally, RTU Senate has adopted RTU Excellence approach to achieve organizational excellence and increase sustainability effort. The Approach defines process steps to set required results, plan approach, implement and assess outcomes. This approach defines RTU sustainability targets and priorities. RTU has set Sustainable Development Goals (SDG) where university tends to make an impact. RTU as its priorities has defined seven SDG. Within these actives RTU has prioritized its actions to make impact not only by contribution to research, projects and activities, but also by implementing these goals in the work culture of RTU. Impact of each SDG is measured. Additionally, in order to facilitate green thinking and reduce environmental impact, RTU has adopted its Green Policy, that focuses on three core criteria: Green Energy, Eco thinking and Sustainability – Competiveness.

## 2 RTU approach to sustainable development

RTU acts to promote sustainable development of Latvia, it facilitates achieving the goals of the European Higher Education Area, society, employers, students, alumni, and other stakeholders. The University has integrated the concept of sustainable development within its Strategy, value chain, and processes; it provides the necessary resources to achieve these goals. Sustainable development is inextricably connected with concerns about sustaining the environment and mitigation of negative environmental impacts. Within the framework of RTU Environmental Policy, measurements are taken to optimize the consequences of activities, product life cycle and services, and to mitigate their impact on public health, security and environment.

Introducing its approach to pursuing excellence, Riga Technical University has defined the stages of the process leading to excellence. The stages of RTU Excellence Approach (Fig. 1) reflect organizational culture of RTU and serve as a common language that ensures common understanding of the University quality issues, which, in its turn, promotes sustainable development and achievement of the University goals.

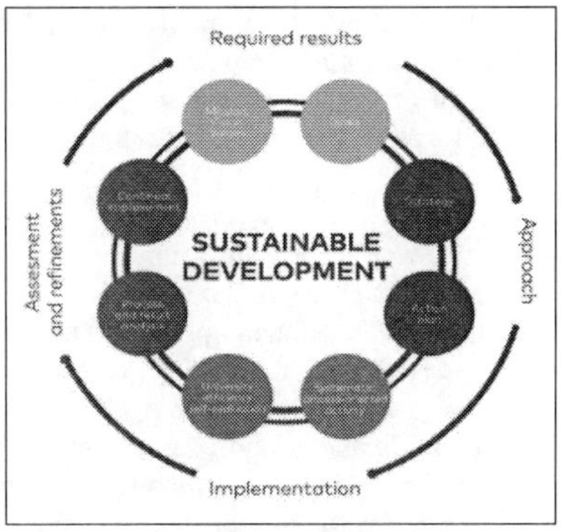


**Fig. 1.** Stages of RTU approach to pursuing excellence.
**Source**: RTU approach to pursuing excellence (2017) [2].

Fig. 1 presents the process cycle of ensuring excellence. Initially, RTU Mission, Vision and Values are set by RTU Senate and administration. They are mainly focused on setting achievable goals, developing and

* Corresponding author: juris.iljins@rtu.lv

reinforcing more competitive national economy through creation of additional value for perspective graduates and existing students. Keeping this in mind, RTU has set its goals that are implemented within RTU Strategy, which focuses on high quality study process, excellence in research and sustainable valorization [1]. Once all strategic objectives are set, action plan is drawn up by listing concrete activities on how to improve university performance and to help achieve the targets set previously. Implementation phase begins with systematic process-oriented review of the real time data. Analysed data is used to further enhance the process or make conclusions on the set targets. This is conducted by RTU Quality Management Unit. Once a year, the university prepares a self-evaluation report to analyse the processes and consider possible improvements in the future in terms of Action Plan and Strategy implementation process. After self-evaluation and analysis of performance data, proposals for administration to introduce further changes for continuous improvement are made [4]. Sustainability is core component in different strategic documents of RTU. For example: RTU Constitution, Strategy, Quality Policy and external quality concepts or standards like EFQM excellence model, UN SDG and European Standards and Guidelines.

## 2.1 Priority SDG

RTU Deans Council has defined the most important UN SDG goals that university focuses attention on. Within these sustainable development goals RTU plans strategic development objectives [3].

**Goal 4. Ensure inclusive and equitable quality education and promote lifelong learning opportunities for all.**
One of the main objectives of RTU Strategy is to plan and ensure high quality study process, provide students and other stakeholders with knowledge, which will help them meet future goals and promote their talent. RTU offers life-long learning opportunities to everyone. Ensuring high quality education, RTU provides opportunities to students from all over the world.

**Goal 6. Ensure availability and sustainable management of water and sanitation for all.**
Scientific activities of Riga Technical University and research on water treatment provide the basis for sustainable development. RTU researchers are developing new water treatment equipment, which will purify water and make it more accessible in various conditions. Water treatment equipment developed at RTU will be available to individual consumers.

**Goal 7. Ensure access to affordable, reliable, sustainable and modern energy for all.**
Ii is difficult to imagine contemporary life and business without electricity, heat energy and fuel, in other words, without power supply. Economic development, world population growth and urbanization impose ever growing requirements on power supply systems. Recognizing that the volume of non-renewable energy sources is gradually decreasing (oil, coal, natural gas, etc.), sustainable electric power supply has become one of the main challenges. RTU researchers working within the research platform "Energy and Environment" develop technologies and solutions for cheap and accessible energy.

**Goal 9. Build a sustainable infrastructure, promote inclusive and sustainable industrialization and foster innovation.**
Infrastructure and industrialization are a research priority of Riga Technical University. The scope of research conducted at the University in the field materials is very wide, ranging from nanoparticle synthesis to large, practically applicable construction calculations, structural strength testing and long-term forecasting.

**Goal 11. Make cities and populated areas inclusive, safe, resilient and sustainable.**
Contemporary cities are a vivid proof of developments in science and technology, that is why urbanization issues are topical on the global scale, and researchers all around the world address these issues. Most of the world resources are consumed in cities, where the largest part of world population resides. Cities as the drives of economic development, culture and innovation centres attract people to live, study and work there. At the same time, concentration of population poses challenges for specialists in urban development. Sustaining of cultural and historical heritage, dwelling, mobility and infrastructure provision, employment and social issues – this wide range of issues illustrates a complex cross disciplinary character of urban development.

**Goal 12. Ensure sustainable consumption and production patterns.**
Research in the field of security and defence is inextricably connected with the state interests; therefore, active collaboration with state governance structures is vital. RTU research platform "Security and Defence" has been established to promote cross-disciplinary research in the fields connected with individual and state security. Research areas in the field of defence cover a wide range of problems, as contemporary world threats are complex. Their elimination is a great challenge for researchers and various industries.

**Goal 17. Revitalize the global partnership for sustainable development.**
In order to reach the above-mentioned goals, Riga Technical University is actively looking for cooperation partners in Latvia and abroad. RTU partner institutions can be found on all continents and are united by common interests and sustainable solutions.

### 2.2 Governance

RTU as Quality Model system has European Foundation for Quality Management excellence model. Principles defined in the EFQM Excellence Model and various higher education and quality standards are horizontally integrated with the stages of RTU Excellence Approach. These principles characterize the approach to pursuing excellence and sustainable development that RTU has adopted Fig. 2.

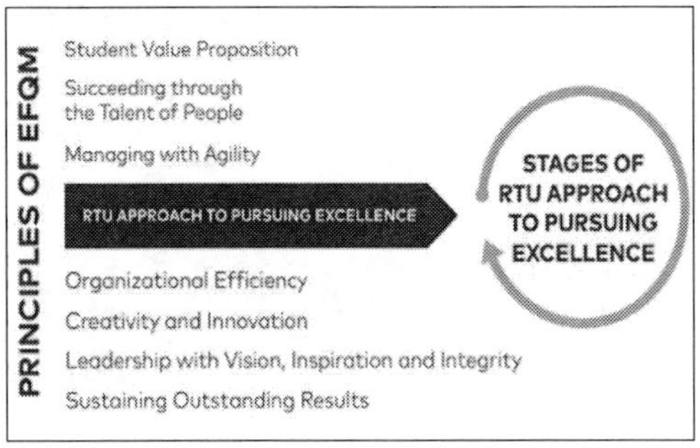


**Fig. 2.** Integration of the principles of EFQM excellence model in RTU approach to pursuing excellence
**Source**: RTU approach to pursuing excellence (2017) [2].

The structure of RTU Excellence Approach provides the basis for maintaining high level of University performance, a prerequisite for continual development of RTU to achieve sustainable performance results and excellence.

RTU has adopted its Green Policy which focuses on three main corner stones with sub-targets:

1. Climate and Green Energy
   a. Energy resource optimization
   b. Building energy efficiency
2. Eco - Thinking
   a. Recycling
   b. Toxic waste utilization
   c. Social contribution in environmental clean-up and thinking
3. Sustainability and Competiveness
   a. Green Thinking in all areas of University operation
   b. Green Campus concept
   c. Participation in international sustainability networks

Within GreenMetrics Rankings RTU has managed to achieve the best indication in area of Energy and Climate Change.

### 2.3 Measuring impact

To achieve the sustainable development KPI that are set in RTU Strategy, different environmental protection processes are adopted at RTU. Achievable targets are monitored, and corresponding policy impact is evaluated each year. The most important issue is to reduce universities environmental impact in energy consumption rate. RTU as a public entity has some limitations in this aspect.

In waste management RTU is monitoring waste output and recycling targets. In transportation RTU is moving towards using only electrical vehicles and bicycles within RTU campus. In table 1 and table 2 some targets with impact can be seen.

**Table 1.** Energy consumption targets at RTU, year 2016 - 2018

| Resource | Target in 2016 against (year 2015), % | Target in 2017 against (year 2015), % | Target in 2018 against (year 2015), % |
|---|---|---|---|
| Power | +5% | +2% | -1% |
| Heat Energy | +15% | +10% | +5% |
| Water | -3% | -7% | -11% |

As it can be seen all energy consumption rates are planned against benchmarking year. Next benchmarking year will be 2018 and future targets will be planned against the indicator.

**Table 2.** Transport efficiency targets at RTU, year 2016 – 2018

| Activity | Target in year 2016 | Target in year 2017 | Target in year 2018 |
|---|---|---|---|
| Usage of electric cars at campus | 80 000 km | 90 000 km | 100 000 km |
| Bicycle parking spaces | 60 % buildings accessible | 70 % buildings accessible | 80 % buildings accessible |

Targets in table 1 and table 2 are fallowed by action plan on how to achieve these targets. Additionally, there are key targets that are not included in this article for waste management and green thinking information campaigns. In similar fashion each SDG targets sub-targets planned indicators are prepared and measured.

## 3 Conclusion

Recent development of RTU has been mainly achieved due to the adopted RTU Development Strategy 2014 – 2020 and introduction of the European Foundation for Quality Management (EFQM) management model. RTU Strategy focuses on three pillars – high quality study process, research excellence and sustainable innovation and commercialization. RTU has introduced the EFQM management system as the basis for operations to support implementation of the strategy and to ensure process efficiency and quality. The proposed model and actions to introduce sustainable integrated concept are based on the Planning Stage, Approach, Implementation, Assessment and Refinement. The method on how to define action plan priorities has been adopted to suit the needs of RTU. Further RTU plans to develop its Green Campus concept to create university community that focuses on sustainability. This includes not only mentioned green thinking concept, driving electro vehicles and bicycles, green projects, but also to develop green entrepreneurial thinking. This will include new green technology innovation, support for green technology start-up businesses and environmentally friendly solutions within university itself for daily operations.

## References

1. Riga Technical University, *Strategy of Riga Technical University 2014-2020*, Approved by the Decision of the Senate of RTU, (2013)
2. Riga Technical University, *RTU Excellence approach*, Approved by the Decision of the Senate of RTU, (2017)
3. Riga Technical University, Available online at https://www.rtu.lv/en/university/strategy/sustainable-development/rtu-goals
4. Iljins, J., Zeps, A., Ribickis, L., *RTU Approach to Pursuing Excellence: Sustainable Integration of Internal Quality System in the Strategy Development. Pilot Project Review*, Proceedings of the 45th SEFI Annual Conference, Portugal, Angra do Heroismo, pp.661-668. ISBN 978-989-98875-7-2. (2017).

# Green transportation system to promote sustainable lifestyle in Chulalongkorn University

*Jirarat* Pinthong[1], *Korb* Limsuwan[1], and *Boonchai* Stitmannaithum[2,*]

[1]Energy and Environment Unit, Office of Physical Resources Management, Chulalongkorn University, Thailand
[2]Office of the University, Chulalongkorn University, Thailand

**Abstract.** Chulalongkorn University (CU) is located at the heart of Bangkok, which is one of the most traffic congested cities in the world. It is very crucial for the university to develop a green and clean transportation system that is good for both the CU community and the whole society. To reduce on-campus traffic, the university provides four parking buildings on the edge of four corners of the campus to serve visitors, students, faculties and staffs who travel by private cars. While providing added convenience, these parking garages reduce traffic congestion on campus and, thus, pollutions from harmful emissions and traffic noises. To promote eco-friendly transportation in the campus, the university provides "CU Shuttle Bus" - an electric shuttle bus service that cover not only campus area, but also reach out to public sky train and subway stations around the campus. The CU Shuttle Bus's mobile application, developed by engineering students, helps improve user experience by showing all useful information including campus map, bus routing, and real-time locations of all buses. To encourage walking and cycling within the campus and to promote good health and fitness, the university has been constructing covered walkways and bike lanes throughout the campus. In addition, "CU Bike" - a bike sharing program, was first introduced in 2014 and has quickly grown in popularity among CU students since. A new "CU Toyota Hamo", an electric vehicle rental program, is another great option of green transportations for those who cannot ride a bicycle and for older people of the aging society. All these projects help promote the development of innovations and practices that are both sustainable and protective of the environment of Chulalongkorn University, as well as the surrounding community, the country and planet as a whole.

## 1 Introduction

Transportation is a big part of everyday life. Commute eco-friendly is the best way to show our responsibility to the environment and society. Bangkok city suffers tremendously from the traffic congestion and air pollution [1-2]. Located at the center of Bangkok, Chulalongkorn University (CU) is right at the heart of the problem area. Therefore, a green transportation system is a very important part of the CU 100$^{th}$ Anniversary University Development Plan [3]. The goals of the university master plan are to reduce on-campus traffic and air pollution together with promoting eco-friendly transportation among CU community.

CU "Green University" policy was implemented in 2004 to improve physical environment and facility of the university to be more eco-friendly. In 2008, CU had set a new goal to become a "Sustainable University". The policy focuses mainly on cultivating mind set and concern about environment and sustainability. This way CU students can become role models and drive a whole society toward sustainability. It is very important that the university takes action to set an example and to inspire both students and the rest of the society.

## 2 Parking system

Chulalongkorn University has implemented "Green University" policies to limit the number of vehicles that enter and park on campus as well as to provide shuttle transport within the grounds and promote walking and riding bicycles. To encourage green transportation, the university started the university parking building project (Fig. 1), because the number of faculties, staffs and students, who drive to the university have significantly increased in recent years. This has led to problems on campus, such as noise and air pollution, traffic congestion and a lack of parking spaces. To solve those problems following the university development plan, the parking buildings have been constructed at the four corners of the campus to achieve the following: (i) reduce vehicle traffic within the campus, (ii) reduce $CO_2$ emission and air pollution inside the campus area, (iii) reduce noise pollution from on-campus traffic, and (iv) provide sufficient parking spaces for those who still need to travel using private cars. People can commute from each parking building to various campus locations as well as popular attractions around the university by walking, cycling or the free shuttle bus service. The car park building has solved the lack of parking spaces and has also reduced traffic on campus. Not only do they enhance the

* Corresponding author: boonchai.s@chula.ac.th

campus atmosphere, they make the university a safer place for people living inside and around the campus.

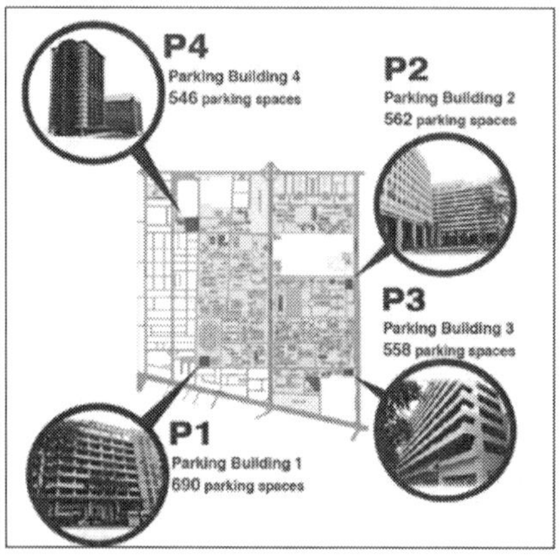


**Fig. 1.** Parking buildings on four corners of the campus

## 3 Mass transportation

As located at a city center, the university encourages CU students and staffs to use public transportation to help reduce the traffic problem of the city as well as the campus. To serve the community, there is "CU Shuttle Bus" that provides a free shuttle service for the CU community and visitors (Fig. 2). The 26 fully- electric buses are operated on 6 routes that not only cover all campus area but also reach out to connect with the public transportation systems of the city. People using the Bangkok Mass Transit System (BTS) skytrain can take the shuttles on route no.1, no. 2 and no.4 getting into the campus. While people using the Metropolitan Rapid Transit (MRT) subway can get to a hub of the CU Shuttle Bus within a several minute's walk.

**Fig. 2.** CU Shuttle Bus

Integrating Information Technology to improve a service, the CU Shuttle Bus mobile application has been developed by a group of engineering students and available for free-download online for both Android and iOS mobile devices. The application provides all useful information to assist all shuttle users including university map, shuttle bus station and routing, real-time location of buses. For young generation like CU students, a social media such as Facebook is used to promote mass transportation on campus and receive feedback from all passengers to keep improving the service that both serves CU community and is safe for our environment.

## 4 Walking and biking in campus

To cultivate sustainable lifestyle among CU community, the university has promoted walking and cycling within the campus as the greenest mode of transportation. Both walking and biking are not only super clean for the environment, but also beneficial for a good health and wellbeing of individuals. To encourage cycling, the university provides bicycle lanes and parking racks throughout the campus ground. Since 2010, over a thousand of bicycles have been distributed to each and every units of the university for staffs to commute within the campus. Most of those bicycles were sponsored by Metropolitan Electricity Authority as a corporate social responsibility project with the following objectives: (i) reduce carbon monoxide emission from vehicles which is a major cause to global warming and air pollution, (ii) promote exercise via cycling, (iii) reduce vehicle traffic within and around campus, and (iv) promote efficient use of energy. For campaigning, there is an annual event led by the president of the university to promote cycling in campus (Fig. 3).

**Fig. 3.** The annual event to promote cycling in campus

In 2014, "CU Bike", a bike sharing program was implemented on campus with 5 stations and 120 bicycles made available (Fig. 4). The CU Bike is very eco-friendly. Paddling a bike simultaneously generate electricity and store within a battery to supply a lighting system, a GPS system and a control panel of the bike. The solar panels also a renewable-energy source for a communication system at the bike stations. In addition, the bike has a chain guard and front storage which make it very practical and user-friendly. As a result, this sharing bike program has grown in popularity among CU students with average of over 77,000 rental each year. In 2016, three additional CU Bike stations has been constructed to expand the service.

Fig. 4. CU Bike

## 5 Electric vehicle (EV)

"CU Toyota Hamo" is a collaborative project between Chulalongkorn University and Toyota Motor Thailand with an objective to initiate the first EV-sharing system in Thailand. "Hamo" stands for "Harmonious Mobility Network" which is an innovation for compact EV-sharing system connecting and integrating personal mobility with public mass transit networks (Fig. 5).

**Fig. 5.** CU Toyota Hamo

Thailand is currently facing various mobility issues. Quality of life is also negatively affected by the mobility issues. EV-sharing is a new option which connects public transportation system to destinations. Moreover, it increases economic value (sharing economy), reduces pollution, and helps people to drive when necessary.

CU Toyota Hamo is another great option of green transportations for those who cannot ride a bicycle and for older people of the aging society. At the same time, it supports a sustainable society and promotes a sustainable lifestyle in the university.

## 6 Concluding remarks

In conclusion, CU promotes green transportation and sustainable lifestyle on campus via CU Bike, CU Shuttle Bus, CU Toyota Hamo service, along with parking facilities and covered walkways. Implementing clean and green transportation system on campus encourages both CU community and visitors to cultivate their personal habits of commuting with responsibility. As a result, nowadays, the on-campus traffic has been significantly reduced and the air quality has become much cleaner and healthier.

Chulalongkorn University has initiated many projects and activities to achieve its goal of becoming a Green and Sustainable University that not only provides a very best environment for university life, but also inspires students to take a responsibility in driving our society together toward sustainability.

## References

1. INRIX. INRIX Global Traffic Scorecard. Available online at http://inrix.com/scorecard/, accessed on 2018, February 2. (2017)
2. The Nation. Bangkok air pollution remains at harmful level. Available online at http://www.nationmultimedia.com/detail/national/30338660, accessed on 2018, February 2. (2018)
3. Chulalongkorn University, Faculty of Architecture and Faculty of Engineering, *Report of Chulalongkorn University's 100-year Anniversary Development Plan*. Chulalongkorn University. (2000)

https://doi.org/10.1051/e3sconf/20184807002

# Good mobility practices at Poltificia Universidad Javeriana (transportation)

*Alexandra* Leal [1,*], and *Sandra Catalina* Acosta[2]

1 Pontificia Universidad Javeriana, Carrera 7 # 40-62, Bogotá – Colombia
2 Pontificia Universidad Javeriana, Carrera 7 # 40-62, Bogotá – Colombia

**Abstract.** This paper presents the strategies that the Pontificia Universidad Javeriana has been implementing since 2012, aimed to contributing from the academy to the mobility of the city of Bogotá, in order to improve the quality of life of our university community and to reduce the impact on the environment.

## 1 Introduction

The Pontificia Universidad Javeriana is located in the city of Bogotá, capital of Colombia. It is a city with about 10 million inhabitants counting the annexed municipalities. It occupies an area of 1,587 square kilometres. The city along with its rapid growth has experienced traffic problems that became critical in the last decade of the 20th century; problems that were worsen by not having a real urban public transport system that would serve as an alternative to the use of the private vehicle. The city had low levels of competitiveness and an unsatisfactory quality of life for the vast majority of its inhabitants. In 2000, the city completed the TRANSMILENIO mass transport system project, which included the construction of a special infrastructure specifically and exclusively for its operation. This infrastructure was based on specialized trunk corridors and was equipped with single-use lanes, stations, bridges, bike paths and special pedestrian access platforms; all of them designed to facilitate the use of the system by users [1]. TRANSMILENIO started its operation in the year 2000. Currently, the company TRANSMILENIO, owned by the city, has 113 kilometres of trunk/main roads in operation and is the manager of the integrated public transport system IPTS (in Spanish SITP; here after used as such) of the city, which in turn integrates different types of transport (urban, special, complementary, trunk, feeder) seeking the effective coverage of transportation in Bogotá. The SITP has a unique means of payment through smart cards [1].

Despite the implementation of the integrated system and the great efforts of local governments, these have not been sufficient to meet efficiently the transportation problems of the city. According to studies carried out by the Observatory of Mobility of Bogota, the city has about 2,200,000 vehicles, of which 94% correspond to private vehicles, which only mobilize 24% of the population. 69% is mobilized in the integrated system [2]. The average speed of vehicles in the city was 31.1 kilometres per hour in 2008. This rate has been decreasing due to the growth in the number of vehicles, and in 2016, it was 24.5 kilometres per hour. According to the last global report of the firm INRIX that analyses 1360 cities of 38 countries, Bogota ranks sixth among the most traffic-densed cities. The same study concludes that the solution to the problem should be viewed collectively through alternative solutions to the vehicle. This solution would aim at strengthening the public transport system by improving the conditions for users to have more ways to use alternative means such as cycling and walking, and modifying the habits regarding the use of personal vehicle.

## 2 The problem of mobility

The Pontificia Universidad Javeriana is not oblivious to this problem. The university community is made up of 30,000 people according to what is indicated in table 1:

**Table 1.** Javeriana university community

| Description | Quantity |
|---|---|
| Undergraduate and graduate students | 24,203 |
| Teachers | 3,458 |
| Administrative employees | 1,637 |
| Total community | 29,298 |

Source: Planning Secretariat, Pontificia Universidad Javeriana.

The size of the University and its location in the city of Bogotá cause that the members of the university community invest very long times in their commuting to and from the University. Moreover, productivity and quality of life within the University are effected generating as well traffic jams in all its area of influence. Being aware of this, and in the framework of environmental sustainability and social responsibility programs, in 2012 the University formally gave life to the "Javeriana Mobility Program". This program led and managed by the University Services Directorate embraces the following objectives as guiding principles: a. Contribute to the improvement of the mobility of the city of Bogotá, b. Encourage the use of alternative and sustainable means of transport. c. Sensitize the university community about good practices in mobility, d. Improve the quality of life of the members of the University community. Since then, the "Javeriana Mobility Program" is part of university planning, particularly by the University Services Department.

## 3 Strategies for mobility

In 2012, the University was linked to the PEMS Network (in Spanish, Sustainable Mobility Enterprise Plans, and hereafter used as such), which promotes the implementation of corporate policies and practical actions aimed at reducing the impact of employees' commuting/mobility, to and from work; in their quality of life and in the mobility of the city [3]. With this initiative, the private sector has the opportunity to be an active agent that contributes to a more sustainable mobility. The PEMS network was born in 2010 as a collaborative project between the private sector and the academy and managed to link 40 companies from different sectors. A Sustainable Mobility Business Plan brings together strategies that aimed at rationalizing journeys to the workplace and mitigating the negative impacts that these generate in the three areas of sustainability: economic, social and environmental, in this case benefiting the university community and actively contributing to the city.

The companies of the PEMS Network have worked together for the mobility of the city, consolidating a culture of sustainable development by exchanging experiences, and best practices. Through workshops with experts, awareness campaigns, training sessions, good practices have been shared and strategies have been designed to provide solutions to the mobility problem.

In 2015, the Javeriana University, along with Network experts, made its mobility diagnosis based on a survey and interviews that were carried out with some areas of the University. The survey gathered 3,091 people from the university community, (approximately 10% of the population). It ensured the statistical representation of the results of the same. In 2013, a first diagnosis was presented, yet the sample was not representative; as such, the results did not generate a possibility of comparison. The survey asked members of the university community aspects such as: a. Time distribution on trips/commuting. b. Modes of mobilization to and from the University. c. Distances, times, costs of communting. d. Infrastructure for mobility.

The results of this diagnosis allowed us to obtain relevant information about the mobility of the university community and, along with the methodology applied by network experts, four synthesis indicators were obtained that allowed quantifying the social and environmental impact of employee and students' mobility of the Javeriana University [4]. These four indicators are carbon footprints, energy, quality of life and equity. The measurement of these four footprints generated the following results:

- The carbon footprint indicates the tons of CO2 equivalent emitted per capita and the amount of land with vegetation cover necessary to absorb these emissions. Each year, the University's employee and students emit 4,848 tons of CO2 equivalent, for which 3,068 hectares of forest are needed for their absorption.
- The energy footprint corresponds to fuel consumption for trips to and from the University. For these trips, 713,925 gallons of fuel are consumed per year, of which 67% is due to the mobility of people in personal vehicles (63% vehicle and 4% motorcycle).
- The quality of life footprint refers to the average time a person spends on commuting for work and study per year. On average, an employee or a student spends 12 days a year moving to and from the University.
- The equity footprint refers to the percentage of income that a person must allocate to pay for transportation to and from the University. On average, an employee or student at the University spends 15% of their monthly income on transportation.
- Other findings that were evidenced in the diagnosis made in 2015 were the following:
- 37.6% of trips are made in the integrated transport system of the city, 20.3% in other types of buses that are not part of the integrated system, 17.3% in automobiles, 7.3 % on bicycle and 8.9% on foot. Regarding the vehicles that arrive at the University, the average of passengers is 2; in one day, employees and students travel an average of 7.8 kilometers per trip to get to the University from home; each trip has an average duration of 44 minutes.
- The aforementioned indicators or traces, as well as the findings found in the diagnosis made, were the basis to define the strategies to continue as well as the new ones to be implemented. The following typology of strategies was defined: a. Efficient and responsible use of the car. b. Use of alternative means of transport. c. Change of mobility habits and d. Infrastructure improvements.

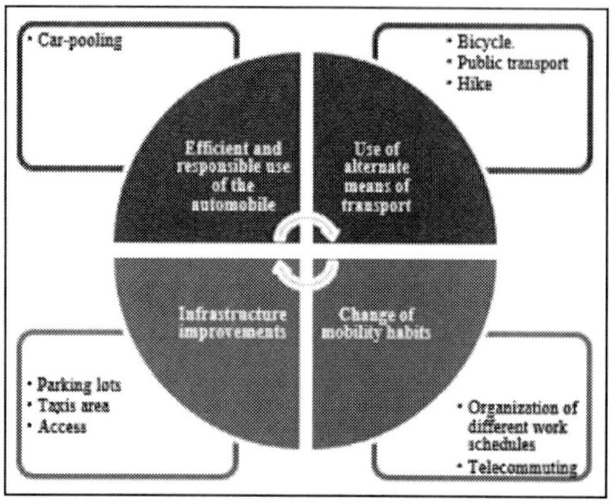


**Fig. 1.** Typology of strategies – PEMS

From these four strategies, between 2013 and 2018, the following activities have been generated and implemented:

a. Efficient and responsible care use
   - Encourage the use of the shared vehicle known as car-pooling
   - Implementation of the app (Javeriana Mobility Application) for car-pooling, taxi, motorcycles, bicycles and walks.
   - Promote plan for the users of the application.
   - Active participation in the vehicle-restriction day planned by the Bogota City Hall.

b. Use of alternative means of transport
   - Encourage the use of bicycles and other means of transport.
   - Incentive plan for bicycle users.
   - Organization of caravans for bicycles and walks.
   - Bicycle loan.
   - Bicycle repair site at the University.
   - Encourage the use of shared public transport and procurement of a provider to offer different routes to and from the university.
   - Encourage the use of the integrated transport system by the location of a selling and reloading smart-cards site.

c. Changing mobility habits
   - Pilot tests are being carried out so that some job-positions can telecommute some days of the week.
   - Organization of different working schedules to reduce the mobility during the peak hours.

d. Infrastructure improvements
   - Increase in bicycle parking spaces.
   - Improvement of bicycle parking infrastructure (lockers, bicycle repair site, water fountain).
   - Definition of preferential parking spaces for users who share their vehicle.

The implementation of the four mentioned strategies is supported in a transversal pillar conceive as "Culture and education". This pillar seeks to sensitize the university community, generate awareness, and modify habits that produce changes and improve the social and environmental impacts that the problem of mobility generates for the city of Bogotá and its inhabitants. This has been done with the following actions: awareness and information campaigns to the university community on mobility issues, conferences and discussions with the Mobility Secretariat and experts on good practices, Javeriana mobility weeks, competitions, creation of a digital community on Mobility (social networks), information to users through the different communication channels of the University and socialization of the results of fingerprint measurements and mobility diagnosis.

## 4 Discussion

During these years, we have reached certain goals that we consider important to be shared, both at the university level and at the city level:

- The number of spaces available for bicycle parking was significantly increased. Today we have three bike parking areas, increasing from 48 places in 2012 to 382 places in 2018.
- The number of registered bicycle users increased significantly, from 180 users in 2012 to 917 in 2017, thanks to awareness campaigns and activities such as "Semana mejor en Bici" "a week, better by biking" and "Javeriana en Bici" "Javeriana by Bike".
- The use of the bicycle loan service has increased. Thanks to this service, several opportunities are offered to the members of the university community to carry out their bike trips to know their advantages and eventually to convert them into permanent users. In 2015, 252 users made 910 round-trips through this service.
- In the days of the "car-restriction day" planned by the Mayor of Bogotá, the number of users transported to the University by bicycle increased by 55%.
- There have been 2.830 downloads of the Javeriana Mobility application, a tool that allows members of the Javeriana university community to share their vehicle, and to organize 1) routes for walks and 2) mobilization by bicycle or motorcycle.
- The use of the shared vehicle or car-pooling has increased by 27%, between 2016 and 2017.
- The "Golded seal" was received from the Bogota Mobility Secretariat for having a bicycle-parking infrastructure with standards of excellence in quality, safety and service.
- Recognition was received by the Mobility Secretariat of the city for having mobilized the largest number of managers on bicycles in 2017.
- Due to the achievements, the Javeriana University received the recognition of the District Mobility Secretary and the District Department of the

https://doi.org/10.1051/e3sconf/20184807002

Environment in 2016 for its active contributions that have actively contributed to improving the mobility of the city through the different and permanent strategies implemented.

## 5 Conclusions

The problem of mobility in the city of Bogotá is of great impact for the lives of those who live there; it is everyone's problem. That is why all support has been given from the management levels of the institution to the "Javeriana Mobility Program", which supported in generating culture, education and awareness, has designed and implemented strategies that mitigate the effects of mobility in the university community. In the year 2018, another measurement will be carried out in order to have a comparison parameter with the diagnosis made in 2015 and thus determine whether the strategies and actions implemented have had the expected effect or make the corresponding adjustments. The University has a high commitment to 5.

the solution of this problem and it hopes to contribute the best way, also being frame under the Encyclical of Pope Francis "Laudato si" on care for our common home.

## References

1. Transmilenio, mass transport system. Available online at: http://www.transmilenio.gov.co/
2. Observatory of Mobility of Bogotá. Integrated Information System on Regional Urban Mobility SIMUR. Available online at http://www.simur.gov.co/SimurVisorBoletinWA/2016.html#PDF/1
3. Universidad de Los Andes, Fundación Chévrolet, Andi Seccional Cundinamarca-Boyacá, PEMS. Diagnóstico de Movilidad, Resultados Julio 2016.
4. Planes Empresariales de Movilidad Sostenible – PEMS. Diagnóstico de Movilidad Pontificia Universidad Javeriana (2016).

# Reengineering sustainability at Universiti Tun Hussein Onn Malaysia (UTHM): travel & transportation

*Wahid* Razally[1], *Noralfishah* Sulaiman[2*], *Munzilah* Md. Rohani[3], *Basil* David Daniel[4], *Indera Syahrul* Mat Radzuan[5] and *Thuwaibah* Mohd Junaid[6]

[1]Vice Chancellor, Vice Chancellor's Office, Universiti Tun Hussein Onn Malaysia, 86400 Parit Raja, Batu Pahat, Johor, Malaysia.
[2]Head of Unit, Sustainable Campus Unit, Universiti Tun Hussein Onn Malaysia, 86400 Parit Raja, Batu Pahat, Johor, Malaysia.
[3]Associate Professor, Dept. of Infrastructure & Geomatic Engineering, Faculty of Civil & Environmental Engineering, Universiti Tun Hussein Onn Malaysia, 86400 Parit Raja, Batu Pahat, Johor, Malaysia.
[4]Senior Lecturer, Dept. of Infrastructure & Geomatic Engineering, Faculty of Civil & Environmental Engineering, Universiti Tun Hussein Onn Malaysia, 86400 Parit Raja, Batu Pahat, Johor, Malaysia.
[5]Head of Department, Sustainable Campus Unit, Universiti Tun Hussein Onn Malaysia, 86400 Parit Raja, Batu Pahat, Johor, Malaysia.
[6]Language Teacher, Dept. of English Language and Linguistics, Center for Language Studies, Universiti Tun Hussein Onn Malaysia, 86400 Parit Raja, Batu Pahat, Johor, Malaysia.

**Abstract.** Conveniently situated in Parit Raja town which is also known as the "Education Town" in Johor, Universiti Tun Hussien Onn Malaysia (UTHM) is a member of Malaysian Technical University Network (MTUN). In pursuit of turning UTHM as a Green Campus, UTHM has its own ambitious targets for sustainability. It is committed to provide sustainable, efficient, convenient, and safe travel and transport to all students, staffs and visitors. The inception of Sustainable Campus Unit (SCU) in May 2010, showed that UTHM has doubled its green efforts and initiatives to provide sustainable travel and transport in the campus when and where possible. The formation of Strategy in Operation (SIO5): Travel & Transport in the SCU has encouraged active transportation and healthy commuting options in the campus by providing a wealth of innovative green transit options including bus, bicycle, pedestrian, and car. These options indeed will improve the quality of life on campus, strengthens UTHM ties to the neighbouring community, and reduces impacts on the environment as travel and transport system plays an important role on the carbon emission and pollutant level in university. This paper is therefore, trying to discuss some issues, measures and innovative strategies employed at UTHM as an integral part to achieve success in climbing the UI GreenMetric World University Ranking ladder.

## 1 Introduction

Sustainable Campus Unit (SCU) is a unit responsible for planning, developing and implementing smart and innovative sustainable solutions in the pursuit of an environmentally sustainable campus in Universiti Tun Hussein Onn Malaysia (UTHM), Johor, Malaysia. With the aphorism "Reengineering Sustainability", in February 2017, SCU has re-engineered its structure by re-branding it to become more resilient strategic entity. Ever since then, SCU has strived to continuously rethink and redesign its unique sustainable green business process in order to achieve fundamental improvements in essential measures of performance such as cost, quality of service delivery and speed of changing. Through SCU's Sustainable Champion Team (SCT), an established team consists of professional/expert individuals who seek to lead green efforts in UTHM, SCU serves as a green hub to catalyse and coordinate a large-scale university green effort towards resilient and sustainable campus ecosystem.

In SCU, there are 8 Strategy in Operation (SIO) employed to frame the process of in-campus sustainability reengineering known as SIO 1 (Corporate Governance), SIO 2 (Waste & Recycling), SIO 3 (Energy), SIO 4 (Water), SIO 5 (Travel & Transport), SIO 6 (Education & Research), SIO 7 (Development, Climate & Resiliency) and SIO 8 (Awareness, Training & Communication).

## 2 Strategy in Operation (SIO) 5: travel & transport

### 2.1 In-campus travel & transport activities

With more than 16,000 students and staffs, the UTHM campus is a hive of activity during the weekdays. Private vehicles (i.e. cars and motorcycles) are the primary choice of travel within the campus. The large number of vehicles in the campus is attributed to the ever-increasing student intake and the high dependency on private vehicles. As more vehicles enter the UTHM campus, the impact on environmental quality of the

* Corresponding author: nora@uthm.edu.my

campus has become one of the most critical issues that calls for a comprehensive sustainable transportation system. Hence, a sustainable Travel & Transport strategic aims are established as below:

- To promote sustainability in travel and transport options across the university;
- To encourage the use of means of transport other than private vehicles; and
- To facilitate active participation from the campus community in reducing the environmental impacts of transportation

Along with the strategic aims, the Strategy in Operation (SIO) 5: Travel & Transport is also designed to accomplish the following objectives:

- To reduce emission of greenhouse gases from public and private vehicles in the university;
- To encourage students and staff to use public transportation as a preferred mode of travel within the campus;
- To promote non-motorised travel in the campus as a healthy and sustainable travel option among students and staffs; and
- To minimise single occupancy vehicle use and encourage ride-sharing programs in the campus.

## 2.2 Promoting active transportation

Active transportation is a term used to describe the type of travel that uses human power, i.e. walking and cycling. Also known as non-motorised transport, active transportation has many health, social and environmental benefits, such as providing basic mobility, access to public transit, affordable and sustainable transportation, and physical fitness [1].

UTHM has seen an increase in students' population over the years, resulting in great dependency on private vehicles that has caused traffic congestions, parking problems and environmental issues. Efforts are being stepped up to reduce the high numbers of private vehicles on campus, through the promotion of active transportation as a preferred choice of travel. The newly opened UTHM campus in Pagoh advocates active transportation by offering facilities for walking and cycling, while banning the entry of student-owned private vehicles.

## 2.3 Creating and improving pedestrian spaces to encourage walking on campus

Walking is undeniably the most affordable mode of transport, and the construction of walking facilities provides an elementary approach to support travel by foot [2]. Ensuring acceptable levels of walkability on campus is vital to the efficient and sustainable transportation of students and staff within the UTHM campus.

UTHM is continuously striving to create pedestrian spaces that provide clean, safe and comfortable environments suitable for walking. Efforts undertaken by the university to encourage walking include the construction of sheltered walkways, connecting walkways to bus stops and increasing the appeal of walkways.

### *2.3.1 Construction of sheltered walkways*

Walkways are perceived by pedestrians as comfortable if the walkways are adequately wide, unobstructed and well-maintained. Pedestrians find walkways safe if they are segregated from roads and motorised traffic [3].

The tropical climate makes walking on campus a big challenge for students and staffs, with sweltering heat and torrential rain occurring every now and then. Sheltered walkways not only remedy this problem, but also have been found to encourage walking on campuses [4]. For better utilisation and convenience of use, UTHM has constructed sheltered walkways that form linkages between buildings on campus (Fig. 1 - 3).

Students and staffs have found it more convenient to travel by foot since the introduction of sheltered walkways on campus, seeing as they are protected from harsh weather.

**Fig. 1.** Newly built sheltered walkway linking Tun Dr. Ismail Residential College to the Faculty of Mechanical and Manufacturing Engineering in the UTHM main campus.

**Fig. 2.** A sheltered walkway built in the newly opened UTHM campus in Pagoh.

**Fig. 3.** One of several sheltered walkways connecting to the Lecture Complex in the UTHM main campus.

### *2.3.2 Connecting walkways to bus stops*

Pedestrian facilities are vital for the support of public transportation. Facilities that are of great quality help foster the use of public transit services since transit riders are also pedestrians at each end of a trip [5].

As part of its commitment to sustainable travel, UTHM provides bus shuttle services for students within the main campus area, and to and from off-campus residential colleges. In order to encourage more students to ride the bus on campus, the university has constructed pedestrian walkways that directly link with the bus stops (Fig. 4 and 5).

**Fig. 4.** Walkway linking to bus stops along Persiaran Tun Ghafar Baba in the UTHM main campus.

**Fig. 5.** Walkway linking to a bus stop along Persiaran Tun Tan Siew Sin in the UTHM main campus.

This is a key approach taken by the university to enhance pedestrian connectivity with buildings, public spaces and transit services on campus.

### *2.3.3 Increasing the appeal of walkways*

Creating a walkable environment is not just about providing safe and comfortable infrastructure to support walking. The walkable environment should also be interesting, in a sense that it integrates walking routes and destinations that are lively and attractive. These added attributes result in enjoyable walking experience and help promote the culture of walking [6].

UTHM through its facilities management constantly upkeeps pedestrian walkways to ensure that they are clean and conducive for walking. Trees and plants that add aesthetic value and provide shade to walkways are regularly pruned to prevent overgrowth of foliage that may obstruct pedestrians.

Walkways have been built by lakesides to create scenic routes that allow pedestrians to enjoy nature while walking to class or taking brisk walks (Fig. 6).

**Fig. 6.** Walkway by the lakeside.

The installation of outdoor exercise equipment along walkways invites more people to walk, not just as a means of travel, but for keeping fit and healthy (Fig. 7). Aesthetic enhancement of walkways has also been undertaken by UTHM. One such example is the central walkway along Persiaran Tun Ghazali Shafie, with the fitting of in-ground water fountains and attractive landscaping (Fig. 8).

**Fig. 7.** Walkway with outdoor exercise equipment.

**Fig. 8.** Aesthetic enhancement of the walkway along Persiaran Tun Ghazali Shafie.

An ongoing park beautification project adjacent to the Sultan Ibrahim Mosque has been designed to include walkways, gazebos and benches that complement each other (Fig. 9). Studies have shown that walking in the midst of nature can relief stress [7], overcome mental fatigue [8], improve memory, and sharpen thinking and creativity [9].

**Fig. 9.** Ongoing park beautification project that includes walkways, gazebos and benches.

## 2.4 Encouraging active green cycling facilities

UTHM is also implementing initiatives to encourage active green cycling transportation as a viable alternative to private vehicle and as a method of promoting a more active and healthy lifestyle. Reducing the number of motor vehicles on the campus decreases the quantity of pollutants released into the atmosphere by motor vehicles. The effects of climate change can also be reduced by encouraging drivers to use other modes of transport such as bicycle [10]. The following section discusses some of the key initiatives taken by UTHM to its newly opened branch campus, the Pagoh Edu Hub located in Muar, Johor.

### *2.4.1 Providing OBike facilities*

In December 2017, UTHM completed the state-of-the-art the '*oBike*' green cycling facilities with a vision to move the campus community forward into an active transportation strategies and initiatives (Fig. 10). '*oBike*' is a bicycle sharing services platform, allowing everyone to change how transportation is viewed locally. '*oBike*' helps to decrease the traffic congestion, reduce environmental pollution, and make UTHM campus greener and a better place to study and live in. As of 15 January 2018, the total signed-up users for the '*oBike*' App are 4,294 with the total bikes provided are 600 units and total trip of 16,059 (Table 1).

**Fig. 10.** The state-of-the-art the '*oBike*' cycling facilities at UTHM Pagoh Edu Hub.

**Table 1.** OBike rides at UTHM Pagoh Edu Hub.

| | |
|---|---|
| Total Signed-up User | 4,294 |
| Total Bikes Deploy | 600 bikes |
| Total Trips | 16,059 |

### *2.4.2 Reducing total carbon emission*

Aside from being popular recreational activity, cycling is also efficient, affordable, environmentally-friendly and accessible means of transportation. The wider benefits of cycling include: reduce road congestion and greenhouse gas emissions; cheaper infrastructure, lower maintenance costs; road safety improvements; and lower user costs as compared to motorised vehicles [11].

UTHM Pagoh Edu Hub is currently working towards a sustainable green campus environment and this is closely monitored by the SCU. As of 15 January 2018, the distance ride of '*oBike*' with 42,128 km indicates that good cycling infrastructure is integral to overall traffic and transport experience in the campus (Table 2). The most visible effect of the '*oBike*' facilities at UTHM Pagoh Edu Hub can be seen with the reduction of the total carbon emission to 11,315 kg. These facts clearly demonstrate the potential to increase the number of bicycle usage within the campus compound.

**Table 2.** OBike and carbon emission reduction.

| | |
|---|---|
| Total Distance Ride | 42,128 km |
| Total Carbon Emission Saved | 11,315 kg |

### 2.5 Efficient UTHM internal bus services

UTHM provides a unique on-campus bus ride for internal use of its students (Fig. 11). It is mainly provided to transport students from university residential areas to the main campus of UTHM. The operation of the bus is managed by Student Housing and Transportation Centre (P2P UTHM). The services are free and it is available from 0700 hours to 2300 hours daily [12]. Besides, the services also intended to minimise single occupancy vehicle use among the students and encourage ride-sharing programs in the campus.

**Fig. 11.** UTHM on-campus bus service.

UTHM has located many bus stops adjacent to the location of students' activity centre, such as at the library, faculties and lecture hall buildings. Providing bus stops within walking distance of lecture halls or main areas where students congregate are of paramount importance for UTHM. This is to increase students' accessibility [13] and encourage more students to use the bus services as their main sources of in campus mobility. In addition, UTHM bus stop facilities are designed to suit with environment characteristics (Fig. 12) and passenger needs. This includes bus passenger shelter and seating, security (including lighting), information (timetables and maps), and pedestrian footways.

**Fig. 12.** UTHM bus stops are designed to suit environment characteristic and accessibility.

UTHM has taken some measures to improve internal bus services by introducing several schemes, including the introduction of UTHM-KATSANA (Fig. 13). UTHM-KATSANA is an industry-proven GPS bus tracking system & fleet management solutions that tracks or secure vehicles, make fleet operations easier and more efficient. It is an android app that helps students to track the real time of the bus position and movement. The system is operated 24 hours and can be accessed with smart phone and internet. The system allows UTHM to monitor bus movements and makes the bus schedule accessible through UTHM website. It is proven that this service has increased students' satisfaction as the bus passengers and allows better time management among students.

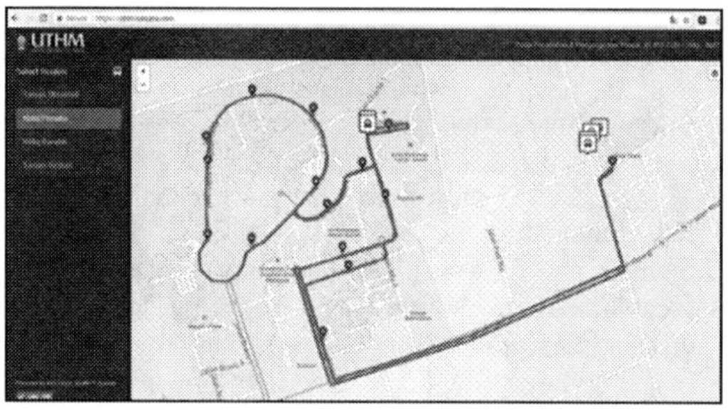


**Fig. 13.** UTHM KATSANA system.

Students' satisfaction is UTHM's priority. UTHM has ensured that the bus service operations have to efficiently meet daily demands of the students. Although it is reported on several cases that number of buses are not sufficient, trip frequency and bus schedule are sometimes not matched with students' time tabling and requirements, ratio of students to bus services is high and inefficient and unsafe behaviour of bus drivers, improving efficiency of the bus services operation and increasing students' satisfaction, comfort and safety are among the greatest challenges faced by Strategy in Operation (SIO) 5: Travel & Transport team at UTHM.

### 2.6 Research & development

To deal with the problems related to drivers' behaviour and passengers' safety, UTHM Smart Driving Research Center (SDRC) has recently conducted an innovative research study named Bus Smart DRV (Bus smart Driving Variation Indicator) (Fig. 14). Smart DRV is an in-vehicle device designed to function as an indicator for bus smart driving. The functions of the device are to display real-time driver feedback instantaneously and to warn drivers if they are any infringing limits occur while driving. It is designed specifically for Malaysian bus drivers, primarily to encourage good driving behaviour as well as for the attainment of economical and safer driving. This project is still at its initial stages and if the project succeeds, UTHM will definitely enhance the current internal bus service quality and efficiency.

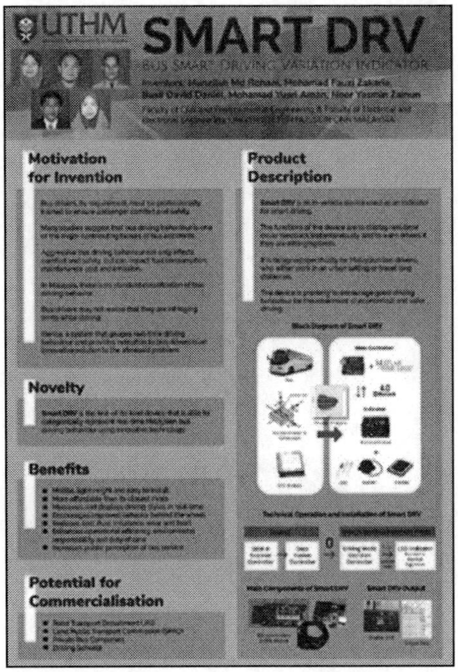


**Fig. 14.** UTHM Smart DRV research projects

## 3 Conclusion

The "Reengineering Sustainability" aphorism has made UTHM to focus not only on the achievement of SIO 5: Travel and Transport in UI GreenMetric World University Ranking but also for the other 7 SIOs namely SIO 1 (Corporate Governance), SIO 2 (Waste & Recycling), SIO 3 (Energy), SIO 4 (Water), SIO 5 (Travel & Transport), SIO 6 (Education & Research), SIO 7 (Development, Climate & Resiliency) and SIO 8 (Awareness, Training & Communication). Practical, efficient and innovative green strategies are made by UTHM in order to improve the provision of SIO5 such as promoting active transportation, creating and improving pedestrian spaces, encouraging active green cycling facilities, efficient internal bus services and tackling other travel and transportation issues by conducting research and development. Many other methods are underway on campus such as improvising current travel and transport policies and guidelines, disseminating awareness through SCU's website, Facebook and Instagram, collaborating with private sectors and NGOs in promoting and encouraging sustainable travel options and activities as well as trying to reduce cost of on-campus travel and transportation innovatively.

## References

1. T.Litman, *Evaluating Active Transport Benefits and Costs: Guide to Valuing Walking and Cycling Improvements and Encouragement Programs*, Victoria Transport Policy Institute (2017).
2. A. Frsyth, Urban Design International, *What is a Walkable Place? The Walkability Debate in Urban Design*, **20** (4), 274-292 (2015).
3. B.D. Daniel, S.N.M. Nor, M. Md Rohani, J. Prasetijo, M.Y. Aman, K. Ambak, MATEC Web of Conferences, *Pedestrian Footpath Level of Service (FOOT-LOS) Model for Johor Bahru*, **47**, 03006 (2016).
4. M.Y. Merrill, P. Burkhardt-Holm, C.H. Chang, M.S. Islam, Y. Chang, *Education and Sustainability: Paradigms, Policies and Practices in Asia*, Routledge (2018).
5. L. Bain, B. Gray, D. Rodgers, *Living Streets: Strategies for Crafting Public Space*, John Wiley & Sons (2012).
6. Victoria Walks, *Developing a Walking Strategy: A Guide for Councils*, Victoria Walks Inc. (2013).
7. Q. Li, Environmental Health & Preventive Medicine, *Effect of Forest Bathing Trips on Human Immune Function*, **15** (1), 9-17 (2010).
8. R. Berto, Environmental Psychology, *Exposure to Restorative Environments Helps Restore Attentional Capacity*, **25** (3), 249-259 (2005).
9. M.G. Berman, J. Jonides, S. Kaplan, Psychological Science, *The Cognitive Benefits of Interacting with Nature*, **19** (12), 1207-1212 (2008).
10. Shaheen, Susan A., & Lipman, Timothy E. *Reducing Greenhouse Emissions and Fuel Consumption: Sustainable Approaches for Surface Transportation*. IATSS Research, **31**(1), 6-20. (2007).
11. Des McKibbin. *The Wider Contribution of Cycling and Its Potential to Replace Car Journeys*. Research and Information Service Research Paper: Nothern Ireland Assembly (2011).
12. Rohani, Munzilah *Bus driving behaviour and fuel consumption* University of Southampton, Faculty of Engineering and the Environment, Doctoral Thesis, 405pp. (2012).
13. Public Transport Team *Accessible Bus Stop Design Guidance,* Transport for London, Report, 51pp (2017)

# AUTHOR INDEX

# AUTHOR INDEX